Coal, crisis and conflict

Coal, crisis and conflict

The 1984–85 miners' strike in Yorkshire

Jonathan and Ruth Winterton

Manchester University Press
Manchester and New York
Distributed exclusively in the USA and Canada by St. Martin's Press

Published by Manchester University Press
Oxford Road, Manchester M13 9PL, UK
and Room 400, 175 Fifth Avenue,
New York, NY 10010, USA

Distributed exclusively in the USA and Canada
by St. Martin's Press, Inc.,
175 Fifth Avenue, New York, NY 10010, USA

British Library cataloguing in publication data
Coal, crisis and conflict : the 1984–85
miners' strike in Yorkshire
1. West Yorkshire. Coal industries. Miners.
Industrial relations. Strikes. 1984–5
I. Title II. Winterton, Ruth
331.89'282233'094281

Library of Congress cataloging in publication data
Winterton, Jonathan.
Coal, crisis and conflict : the 1984–85 miners' strike in
Yorkshire / Jonathan and Ruth Winterton.
p. cm.
Bibliography: p.324.
Includes index.
ISBN 0-7190-2548-6 : $40.00 (est.)
1. Coal strike, Great Britain, 1984–1985. 2. Strikes and
lockouts—Coal mining—England—Yorkshire—Case studies. 3. Coal
mines and mining—Great Britain. 4. Coal trade—Great Britain.
5. Industrial relations—Great Britain. I. Winterton, Ruth.
II. Title.
HD5365.M6152 1984
[.W56 1988]
331.89'2822334'0—dc19 88-9659

ISBN 0 7190 2548 6 *hardback*
0 7190 2830 2 *paperback*

Phototypeset in Great Britain
by Megaron, Cardiff, Wales

Printed in Great Britain
by Biddles Ltd, Guildford and King's Lynn

Contents

Tables

Figures

Foreword

For over a decade we have been investigating strike activity and were researching the coal crisis for almost two years before the strike began. It was natural for this research to carry over into the strike and we were enthusiastically supported by miners we had been privileged to teach and others we came to know during the strike. We are indebted to all the miners and MPs who granted us extensive interviews, to the people who helped in the distribution of questionnaires and to those who have commented on successive drafts. The study could not have been completed without the individuals mentioned in the list of interviews. We hope that they recognise their strike in what we have written.

Several academic colleagues have been a source of inspiration as well as encouragement; the final manuscript has benefited enormously from the advice of Professors Vic Allen (Leeds) and Huw Beynon (Manchester). Colleagues teaching the miners' day-release courses, particularly Keith Forrester (Leeds) and Phil Wright (Sheffield), provided moral support and further coalfield contacts. Max Madden MP generously found time between his extensive constituency and parliamentary commitments to facilitate interviews at Westminster. Dave Feickert from the NUM national office and Eric Illsley MP, formerly at Yorkshire Area headquarters, were able to provide information and advice. Liz and Steve Vickers, from Featherstone, shared their strike with us and gave comradely criticism of successive drafts. The advice and support of Alec McAulay, editor at MUP, ensured that the manuscript met the stringent standards of the Press Committee. For a variety of reasons we are unable to record our debt to some people; they know who they are and we are nevertheless grateful to them.

In contrast to the support of individuals, we failed to obtain any research awards to free us from teaching commitments, provide clerical or computing support, or contribute to the cost of travelling some 6,000 miles within the Yorkshire coalfield. We are grateful for the support which our respective institutions were able to grant: Northern College for six weeks' sabbatical leave and Bradford University for printing questionnaires and providing computer time.

Like those about whom and for whom this book has been written, the miners' strike profoundly affected our lives. Our children, Benjamin and Rachel, were deprived of time we should have liked to have spent with them; we hope they will forgive the sacrifice and agree it was worthwhile. Inevitably the strike altered our perspectives, providing simultaneously an inspiration of industrial action and many painful lessons of defeat. We hope we have captured the ambivalence and contradictions of the strike in a sensitive and meaningful way, partial to the miners' case while remaining objective in our methods.

Jonathan Winterton
Ruth Winterton

Bradford, August 1987

Abbreviations

ACAS	Advisory, Conciliation and Arbitration Service
ACPO	Association of Chief Police Officers
AES	Alternative Economic Strategy
AGR	Advanced Gas-cooled Reactor (nuclear power station)
AIS	Area Incentive Scheme
ASL	Anderton Shearer Loader (coal-face machine)
ASLEF	Associated Society of Locomotive Engineers and Firemen
AUEW	Amalgamated Union of Engineering Workers
BACM	British Association of Colliery Management
BC	British Coal (formerly NCB)
BFAWU	Bakers, Food and Allied Workers' Union
BL	British Leyland
BR	British Rail
BS	British Steel
BWAPC	Barnsley Women Against Pit Closures
CAB	Citizens' Advice Bureau
CBI	Confederation of British Industry
CCC	Coalfield Communities' Campaign
CCU	Civil Contingencies Unit
CEGB	Central Electricity Generating Board
CGT	Confédération Générale du Travail (France)
CIBTO	Coal Industry Benevolent Trust Organisation
CINCC	Coal Industry National Consultative Committee
CISWO	Coal Industry Social Welfare Organisation
CLP	Constituency Labour Party
COBRA	Cabinet Office Briefing Room
COSA	Colliery Officials and Staffs Association
CRP	Colliery Review Procedure
CTAWA	Colliery Tradesmen and Allied Workers' Association
DEA	Department of Economic Affairs
DEn	Department of Energy
DHSS	Department of Health and Social Security
EAC	Extraordinary Annual Conference (NUM)
ECC	Emergencies Co-ordinating Committee

EEC	European Economic Community
EETPU	Electrical, Electronic, Telecommunication and Plumbing Union
ELSIE	Electronic Signalling and Indicating Equipment
EMA	Engineers and Managers Association
FBU	Fire Brigades Union
FIDO	Face Information Digested On-line (see MINOS)
GDP	Gross Domestic Product
GJ	Giga Joules (measure of energy content)
GMBATU	General, Municipal, Boilermakers' and Allied Trades Union
H C Deb	House of Commons Debate (*Hansard*)
HSE	Health and Safety Executive
ICTU	International Confederation of Trade-Unions
IDMC	Indigenous Deep-Mined Coal
IGBE	West German Metal Workers' Union
IRB	Independent Review Body
ISTC	Iron and Steel Trades Confederation
ITF	International Transport Workers' Federation
JNNC	Joint National Negotiating Committee
LP	Labour Party
LRD	Labour Research Department
MBC	Metropolitan Borough Council
MCC	Metropolitan County Council
MCG	Miners' Campaign Group
MCRP	Modified Colliery Review Procedure
MEP	Member of European Parliament
MINOS	Mine Operating System
MMC	Monopolies and Mergers Commission
MRDE	Mining Research and Development Establishment (NCB)
MSF	Miners' Solidarity Fund
NAC	National Amnesty Campaign
NACAB	National Association of Citizens' Advice Bureaux
NACODS	National Association of Colliery Overmen, Deputies and Shotfirers
Nat WMC	National Working Miners' Committee
NBPI	National Board for Prices and Incomes
NCB	National Coal Board (now BC)
NCB(E)	National Coal Board (Enterprises) Ltd
NCC	National Co-ordinating Committee (NUM)
NCCL	National Council for Civil Liberties
NEC	National Executive Committee (NUM)
NGA	National Graphical Association
Notts WMC	Nottinghamshire Working Miners' Committee
NPLA	National Power-Loading Agreement
NRC	National Reporting Centre
NUM	National Union of Mineworkers

NUR	National Union of Railwaymen
NUS	National Union of Seamen
NYWAPC	North Yorkshire Women Against Pit Closures
OPEC	Organisation of Petroleum Exporting Countries
ORE	Operational Research Executive (NCB)
PEP	Political and Economic Planning
PLP	Parliamentary Labour Party
PSU	Police Support Unit
PWR	Pressurised Water Reactor (nuclear power station)
RCP	Revolutionary Communist Party
RMPS	Redundant Mineworkers Payment Scheme
ROLF	Remotely-Operated Longwall Face
SCC	Strike Co-ordinating Committee (Yorkshire Area NUM)
SCEBTA	Scottish Colliery Enginemen, Boiler-Firemen and Tradesmen's Association
SDC	Special Delegate Conference
SDP	Social Democratic Party
SOGAT82	Society of Graphical and Allied Trades
SSEB	South of Scotland Electricity Board
SWP	Socialist Workers' Party
SYDC	South Yorkshire Defence Committee
TCPA	Town and Country Planning Association
TGWU	Transport and General Workers' Union
TNC	Trans-National Corporation
TUC	Trades Union Congress
TUCRIC	Trade Union and Community Resource and Information Centre (Leeds)
UDM	Union of Democratic Mineworkers
UMW	United Mineworkers of America
UPW	Union of Postal Workers
WAG	Women's Action Group
WAPC	Women Against Pit Closures
WERG	Working Environment Research Group (University of Bradford)
WMMCC	West Midlands Metropolitan County Council
WPIS	Weekly-paid Industrial Staff
WRP	Workers' Revolutionary Party
WYMCC	West Yorkshire Metropolitan County Council

Introduction

The 1984–85 coal strike represented the most important industrial stoppage since the General Strike and coal lock-out of 1926. The 1984 strike entailed a loss of about 38m working days in the coal industry alone, compared with a previous post-war record for *all* industries of 29.5m working days in 1979. In the 1926 stoppage over 150m days were lost, but there were almost one million miners compared with 181,000 in March 1984, so the 1984 strike entailed substantially greater endurance. Like the 1926 dispute, the political significance of the 1984 strike far outweighs these statistics, for both stoppages assumed the character of historic battles between labour and capital. The 1984 conflict could have been fought out in almost any industry because the same conditions were afflicting the whole economy. The Union was attempting to resist unemployment, preserve communities and protect long-term national policy. The Government was aiming to subordinate labour, restructure industry and impose the imperatives of the market. The coal industry became the terrain for the confrontation because of the symbolic importance of the miners in the labour movement and because the effect of Government policies was becoming acutely apparent in coal.

Objectives

More has been written about the 1984 miners' strike than any conflict since 1926, and the existing literature offers diverse perspectives and concerns.[1] A definitive study of a social phenomenon of such complexity is impossible, so this book seeks to contribute to our understanding in three ways. Firstly, although the strike is our central concern, we attempt to analyse the coal crisis, in which the strike was merely one episode. Secondly, recognising that no single book could capture both the diversity and detail of experience in all areas, we concentrate upon Yorkshire and its interaction with the National Union. Thirdly, we are concerned to identify the lessons of the strike for the labour movement as a whole and particularly for trade-union activists facing the challenge of strike organisation.

In addressing the coal crisis, it is necessary to trace its origins through developments in the industry since 1973 and their effect upon relations between the NUM and NCB. The nature and extent of the current crisis is examined in some detail because it conditioned the character of the 1984 strike. Moreover, many aspects of the crisis remain relevant and unresolved in the struggle which continues as this book goes to press thirty months after the strike ended.

The Yorkshire focus was dictated by our own contacts within the Union and the geographical advantages afforded by our bases in Bradford and Barnsley. However, as our analysis makes clear, the Yorkshire Area was the key to mobilisation and had such an influence upon strike organisation that events in Yorkshire must be examined to understand the dynamics of the strike. Throughout the book we consider the national context and major national events but focus upon the Yorkshire Area for specific examples. As the largest coal-mining area, Yorkshire also offers such a diversity of situations that it can be viewed in many ways as a miniature of the national situation. While it is precarious to generalise between the coal areas, hypotheses tested within Yorkshire are liable to prove more transferable because of this diversity (see below).

In attempting to draw lessons from the strike for trade-union activists, we are concerned to identify the strategies and tactics used by the various parties at different stages, and examine the strengths and weaknesses of the strike. The miners' strike may be viewed as a watershed in industrial relations, so it is essential for the future for the labour movement to debate the strategies and tactics of the strike. The analytical framework for this action research incorporates the work of Hiller,[2] who identified phases of the strike process, Steuben,[3] who assessed the strike strategies of American employers, and McIlroy,[4] who has provided a modern hand-book of strike organisation for trade-unionists.

Methodology

The Yorkshire focus and the objective of identifying lessons for future strike organisation have both necessitated and facilitated our methodological approach. While secondary sources such as media accounts and other published material have been consulted, we consider our primary sources to be most important. Different qualities of information were obtained from each of the four primary sources: documentation, observation, interviews and questionnaires.

We had to rely upon documentary material, in the form of minutes, for accounts of meetings of the National and Yorkshire Area delegates, Executives and strike co-ordinating bodies, to which we could not be admitted. Such documentation is authoritative, but limited to the extent that the Union bureaucracy mediates and sterilises debate and dissent. Participant observation was important in shaping our views, making us more aware of the miners' problems and allowing us to experience police aggression and media distortion for ourselves. We gained valuable insights from picket lines, strike centres, kitchens, informal activist meetings, clubs, pubs and individuals' houses. Because information from such sources could not be gathered systematically, it remained little more than anecdotal evidence, for all its richness. The framework for the systematic collection and analysis of information was, nevertheless, affected by our observations of the strike as much as by our own preconceptions.

The interviews and questionnaires employed for systematic data collection had distinct objectives. The interview aimed to elicit high-quality, detailed information about the strike at individual collieries from key activists. One individual was interviewed from each Yorkshire pit in accordance with the schedule reproduced in Appendix A. Each interview took a minimum of three hours, and interviewees were encouraged to elaborate on their responses to questions. A number of individuals volunteered to collect further information on our behalf to supplement what they had told us in the interview. From the interviews we established comparable information for each colliery about how the strike began, the main features and deficiencies of strike organisation and maintenance, and industrial relations at colliery level in the immediate aftermath of the strike. The most important information concerned the number and background (family, occupation, etc.) of those participating in various strike and strike-breaking activities.

The questionnaire aimed to gather a high volume of lower level data from individual strikers. The questionnaire was designed to allow responses to be coded for computer processing using the Statistical Package for the Social Sciences.[5] Approximately 6,000 questionnaires were distributed through colliery contacts and women's action groups, and 1,830 were completed and returned, an overall response rate of thirty per cent. A questionnaire was produced for distribution during the third month of the strike but, at the request of the Yorkshire Area Executive, was not distributed until the strike was ended, otherwise a higher response rate would almost certainly have been achieved. Nevertheless, we are satisfied that the sample is sufficiently large and reasonably representative of the coalfield. The questionnaire, which is reproduced as Appendix B, sought to establish the respondents' involvement and attitudes towards the strike, sources of information and finance, politics and personal background. Part of the questionnaire, for completion by the miner's domestic partner, was concerned with women's involvement and perceptions, and with relations in the household. From the questionnaire data it is possible to investigate whether, for example, perceptions of the strike or strike activism varied with occupation or age. It was necessary to establish the respondents' strike involvement for three reasons. Firstly, this information made it possible to investigate differences between the perceptions of active and passive strikers. Secondly, it was inevitable that activists would be over-represented in our sample and we wished to avoid the problems which Hartley and colleagues encountered in their study of the 1980 steel strike, where their questionnaire analysis suggested that ninety-four per cent of manual steel workers were involved in picketing.[6] By analysing active and passive strikers' responses separately, a weighted average could be calculated using the participation rates established by the interviews. Thirdly, since each woman's questionnaire was paired with a miner's, it was possible to examine the relationship between men's and women's involvement in the strike.

Outline

The structure of the book has been dictated by its objectives and the analytical framework. Chapter one traces the roots of the coal crisis to 1973, and relates the basis of the 1984 conflict, uneconomic pits, to the factors creating surplus capacity. Demand and supply projections are combined to estimate the scale of surplus capacity, and the future size of the coal industry under different scenarios. Chapter two considers the succession of false starts from January 1980 to the crucial mobilisation of Yorkshire in March 1984 and the domino strategy which was expected to bring about a complete national stoppage. Chapter three describes the evolution of strike organisation and identifies deficiencies in the relationship between the national and area unions. Variations across the Yorkshire coalfield in branch strike organisation are examined, and patterns of activism analysed. Chapter four is concerned with the parallel problems of strike maintenance in the longest ever strike of such proportions. The lack of support from many leaders of the labour movement is contrasted with the support organised by rank-and-file trade-unionists and political activists. The duration of the strike is explained largely by the community cohesion fostered by women's action groups, which occupied the vacuum left by the Union. Chapter five traces the separate strands of the Government's strategy against organised labour and their synthesis during the miners' strike. In contrast with NUM strategy, Government contingency planning is shown to have been modified as a result of earlier miners' strikes. Changes in labour law are compared with more antagonistic policing of industrial disputes. Media hostility to the miners' strike is related to other strike-breaking initiatives, while the back-to-work movement is traced to the tactics of American steel employers in the 1930s. We analyse the patterns of strike-breaking across the Yorkshire coalfield and suggest ways in which unions could more effectively combat such an offensive. Chapter six examines factors influencing the progress towards settlement of the strike. The costs of the strike and the levels of coal stocks are analysed to assess the conditions necessary for the strike to have been won. We describe the progress, and failure, of negotiations, and examine the role of third parties. Strike-breaking in Yorkshire after Christmas is analysed, along with other pressures which led to the organised return. The return without agreement left the position of sacked miners unresolved and necessitated the amnesty campaign which preoccupied activists long after the strike ended. Chapter seven is concerned with the aftermath of the strike. Key managerial objectives are shown to have been at the root of the new industrial relations and the breakaway Union of Democratic Mineworkers. The post-strike NUM, with a deterioration of the activist and women's networks in Yorkshire, has been unable to resist pit closures through the review procedure. Finally, we consider how the local authorities' concern with pit closures was channelled into a national campaign, and conclude with our prognosis for the coal industry. The thematic treatment of the strike necessitates some departure from the order in which events occurred, so for the reader's guidance a Strike chronology is provided after the Notes.

The Yorkshire coalfield

The reader unfamiliar with the Yorkshire coalfield may find the following brief description of the collieries at the time the strike began a helpful preliminary to the book. The Yorkshire coalfield represents the northern portion of the central coal deposits that continue through Derbyshire and the East Midlands. The coal measures outcrop along the eastern edge of the Pennines and dip towards the east. Over the years there has been a continual eastward movement of coal-mining as the outcrops and shallower seams were gradually worked out and technological developments enabled the mining of deeper seams. Of the Yorkshire collieries working on nationalisation, the proportion that still exists increases to the east, while the remaining westerly pits are currently the least secure.

NCB operations in Yorkshire were formerly divided into four Areas: North Yorkshire, Doncaster, Barnsley and South Yorkshire. The size of the coalfield and the tendency of collieries to act autonomously led to the development of a tier of Union government unique to Yorkshire. In addition to sending delegates to Area Council, as in other areas of the NUM, branch delegates attend panel meetings, where matters likely to affect neighbouring collieries are discussed. The four panels were established to coincide with the NCB areas and each had an appointed agent, responsible to the Area general secretary.[7] In October 1985 a new North Yorkshire Area was created from the merger of the old North Yorkshire with Barnsley, and the other two areas constituted the new South Yorkshire Area. We retain the distinction between the four former NCB areas which operated throughout the strike. The boundaries of the areas and the location of individual collieries are shown in Fig. 1.

The former North Yorkshire NCB Area extends from Leeds to north-east of Selby and south of Wakefield. Earliest workings were along the outcrop of coal seams west of Castleford, while later mines were sunk in the east to reach the coal concealed beneath the rocks deposited after the Carboniferous period. The Area, therefore, contained a mixture of Victorian collieries like Glasshoughton and Fryston, the modern Kellingley colliery, and the Selby 'super-pit' complex. In the extreme north-east, the five Selby mines represent the most advanced mining complex in the world. The thick, high-quality *Barnsley* seam, which was the basis of the prosperity of earlier Yorkshire coal-mining, deteriorates in the northern part of the old coalfield but recovers north of Selby. Kellingley colliery is, in several respects, a forerunner of Selby; the largest pit in the country has substantial reserves, in good geological conditions, which are not complicated by older workings. Of the remaining collieries, five were established in the 1870s, two in the 1880s and two in the first decade of the twentieth century.

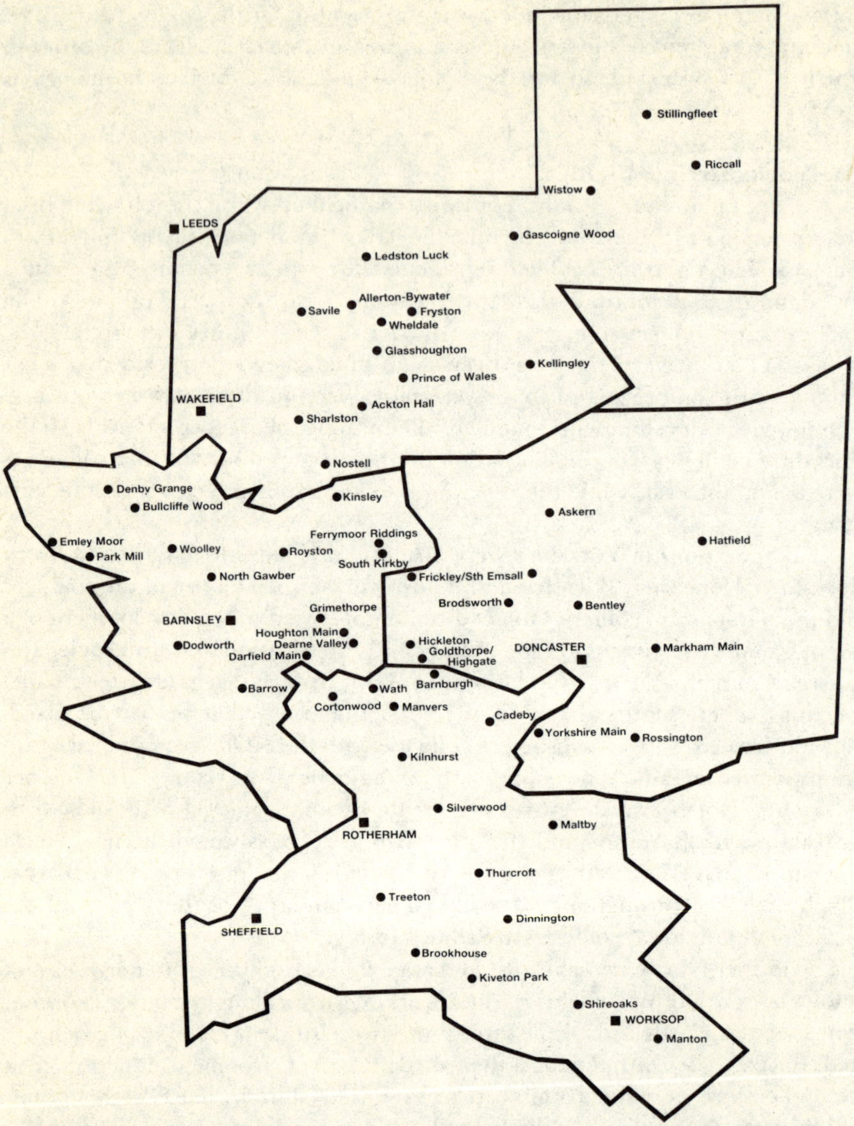

Fig 1 The Yorkshire coalfield

Several are nearing exhaustion, facing geological problems or have shafts or surface facilities which have deteriorated.

North Yorkshire as a whole has a reputation for moderation, and the panel has traditionally been controlled by the right. The collieries around Castleford, especially Glasshoughton and Savile, along with Nostell to the south, represent the political centre where most of the old right leadership was spawned. The

Featherstone pits, Ackton Hall and Ackworth, had a more militant reputation, as did Kellingley, whose original workforce largely comprised miners who had transferred from Scotland during the contraction of the 1960s. It is too early to assess the politics of the Selby NUM branches, but in addition to good mining conditions and high bonus earnings, two other factors may operate as a brake upon militancy. Firstly, the NCB has allowed moderate collieries (including Glasshoughton and Savile) to close before exhaustion so that their workforce can transfer to Selby, whereas Ackton Hall miners have been dispersed to Allerton Bywater and Prince of Wales. Secondly, there is no traditional mining community around Selby; miners either commute from elsewhere in the coalfield or live in private housing integrated into the wider community.

The former NCB Barnsley Area adjoined the three other Yorkshire Areas: its boundary with North Yorkshire was south of Wakefield; its boundary with Doncaster passes through Ackworth, South Elmsall and Dearne, while Wombwell and parts of Rotherham fall on the Barnsley side of its boundary with South Yorkshire. The northern and western parts of the area, where many seams outcrop, have been heavily worked in the past by shallow pits and surface drifts. Thicker, high-quality seams were next worked extensively in the south end of the area. As a result, three groups of collieries may be distinguished. In the north-west are the oldest, shallow collieries, sunk from the 1850s, like Emley Moor which was the last Yorkshire colliery to be mechanised. To the south-east of these are the deeper late-Victorian pits around Barnsley town and along the Dearne Valley. In the north-east of the area are three newer drift mines, such as Royston, opened, or reopened, in the 1970s to gain access to further reserves. The current reserves of the area are mainly in a large number of thin seams.

In recent years Barnsley has been regarded as the second most militant area, but political variations within the area reflect these geographical and historical groupings. At the old West Yorkshire collieries, like Emley Moor, miners were often also small-holding farmers who followed the political direction of the North Yorkshire right. Most support for the left was found in the Barnsley and Dearne pits like North Gawber, Grimethorpe and Darfield Main. The drift mines within the Barnsley belt, Kinsley and Ferrymoor/Riddings, tended to be part of the left, while Royston was often subject to the moderating influence of Nostell through its proximity to Wakefield.

The former NCB Doncaster Area was bounded to the north by the North Yorkshire Area, roughly parallel to the M62; to the west by Barnsley Area, taking in South Elmsall and Dearne; and to the south by South Yorkshire and North Nottinghamshire Areas. In the Doncaster Area the coalfield becomes hidden beneath the later rocks, so all but one of the collieries, Goldthorpe/Highgate on the western edge of the area, are over 500m deep and three are at depths of around 800m. All but one other colliery, Hickleton which is also on the western edge, were sunk in the first decade of this century. With the exception of these two collieries and the third on the western boundary, Frickley/South Elmsall, all the area's collieries were sunk through the Permo-Trias strata. The late development of Doncaster Area resulted in the 'model village'. The South Yorkshire collieries of Denaby Main (1863) and Cadeby (1889), which were the most easterly pits until the Doncaster coalfield was opened up, were sunk at some distance from existing settlements, and remote company villages were established nearby. When the Doncaster Area was developed, the model village concept resulted in the creation of Edlington (Yorkshire Main), New Rossington (Rossington) and Woodlands (Brodsworth).

Doncaster was regarded as the most militant area in Yorkshire and the Doncaster Panel had a reputation for powerful collective action. The left leadership developed here, particularly at Hatfield Main, Brodsworth and Markham Main, where the Communist Party had most following. The most westerly of the Doncaster collieries were equally militant, but less remote and more akin to the Dearne communities. Yorkshire Main and Rossington, moreover, are equally typical Doncaster settlements, yet have more moderate traditions. While the remoteness of Doncaster mining communities is probably an important influence, it would be an over-simplification to view the isolated community as a structural determinant of social behaviour.[8]

The former NCB South Yorkshire Area extended from the southern boundaries of Barnsley and Doncaster Areas, including Wombwell and Bolton-Upon-Dearne, to North Derbyshire, taking in Sheffield, and North Nottinghamshire, including Worksop. Seven of the fifteen collieries in the area were sunk before 1880; these included all the most westerly pits. The familiar eastward development and move to deeper reserves occurred with Cadeby (1889), Manton (1898) and a cluster of twentieth-century collieries, from Silverwood (1900), to Thurcroft (1909). The earlier pits south of this cluster, Shireoaks (1856) and Kiveton Park (1866), work the Clowne seam and are geologically closer to North Derbyshire.

The South Yorkshire Area was generally as moderate as North Yorkshire, with little panel cohesion and a tendency for branches to act independently. Any concerted action usually involved Wath, Manvers, Barnburgh and Maltby, which are part of the Dearne family. Cadeby, Silverwood, Thurcroft and Dinnington, while resembling Doncaster collieries, seldom took collective action. The pits to the south of the Area, from Brookhouse to Manton, which resemble Midlands collieries, were generally to the right politically, as was Cortonwood, although in other respects it was a typical Barnsley pit. The events of the 1984–85 coal strike, however, were to defy many characterisations of the coalfield.

Chapter one

The coal crisis

The 1984 coal strike can only be understood in the context of the crisis to which it was a reaction. In this chapter the roots of the coal crisis are traced to NCB strategy since the 1973 oil crisis. Both the nature and the extent of the crisis are considered. The central question of 'uneconomic pits' (more accurately 'surplus capacity') is analysed to establish the theoretical basis of the conflict. The scale of surplus capacity can only be assessed by comparing projections of demand and supply. On the demand side, market projections for indigenous deep-mined coal are shown to be determined by the overall level of economic activity, nuclear power, oil prices, imported coal, and indigenous opencast production. On the supply side, the restructuring of British coal production is shown to entail the introduction of new capacity, the elimination of jobs through automation, and the closure of surplus capacity. The size of the future coal industry under different scenarios is estimated by combining the demand- and supply-side projections.

1.1 The roots of the crisis

It may appear a paradox that the roots of the coal crisis can be traced to the first oil shock in 1973, which ended fifteen years of declining coal demand and enhanced the miners' bargaining power. As part of the 1973 pay claim, conference committed the NEC to seek a return to the 1 November settlement date and Yorkshire Area Council called for a national overtime ban when the NCB refused. On 26 October an SDC endorsed the NEC's rejection of the seven per cent offer, and agreed an overtime ban and a strike ballot. Electricity supply workers were also in dispute over their pay claim, and the Government proclaimed a state of emergency on 13 November, introducing restrictions on the use of electricity. On 18 December, after the power-workers' dispute was settled, further emergency measures were introduced to conserve fuel, including a three-day working week to be imposed from 1 January 1974.

The Miron Report

The NCB re-evaluated their strategy towards the NUM during the 1973 dispute. Wilfred Miron, NCB member and former chairman of the East Midlands Division, prepared a report which he sent to NCB Chairman, Sir Derek Ezra with a covering letter on 6 December 1973: 'For reasons which will become obvious to you when you read the enclosed, I am sending this to your private address . . . The enclosed paper is anonymous . . . How far, if at all, any of the observations should be exposed to Peter Walker . . . is debatable.' Ezra arranged to meet Miron on 19 December over lunch; he had read the report 'with great interest and would very much like to discuss the implications'. The central argument of Miron's report was as follows:

INDUSTRIAL RELATIONS

1. We must keep in mind that the strategy of the NUM's Executive will become increasingly politically orientated and that its Left Wing (Communists, Marxists and their ilk, however organisationally fragmented) will maintain a unified strategy towards the ideological end – the overthrow of the present 'System' . . .
4. One has only to examine the present composition of the NEC to see the number of younger men – Communists, Marxists or the like – who will exercise authority for longer than some of the so-called 'moderates' will remain on the NEC. Scargill, in his mid-thirties, is a key figure – he aims to succeed Gormley as President: Whelan (Notts), Tait (Yorks) and McGahey (Scotland) are known Communist Party members, the first two are young men: Emlyn Williams (South Wales) and Heathfield (North Derbyshire) are younger Marxists or members of 'fringe' Marxist groups: Dai Francis (South Wales) is a Communist, much older than the others but likely to be succeeded in a few years by another younger Left-Winger as Area Secretary: McClean (Scotland) and Collins (Kent) are Communists and Clarke (Scotland) is Marxist: McKie (Staffs) is Left Wing who votes counter to the Midland Area traditions: Daly's aims are well known.

   In the field, at the pits, there are other young men, Marxist indoctrinated, who within the next 10–15 years will emerge as a 'new generation' augmented by student (University and Polytechnic) tuition to take the places of older men who will be retiring or moving upwards . . .

   Moreover, over the next 10–15 years or so . . . the NUM's own organisation . . . will change so as to diminish the numbers and strengths of the so-called 'moderates' or moderate Areas . . .
5. This quasi-political conspectus serves to demonstrate that the traditionally 'moderate' elements in the NUM . . . should be reinforced and restored to their basic anti-Left position. The Left-Wing authority of the NEC should be more and more emasculated . . .
9. The large majority of membership of the NUM are Labour Party members who eschew Marxism in any form. It is suggested that the aim should be to ensure that such membership has satisfactory, secure and progressive levels of wages and conditions . . . so as to constrain or neutralise the political ammunition of the Left and the influence of the Marxist.
10. One step in this direction would be to revert to some form of local pit or district incentives based upon Agreements negotiable locally which would restore to some of the traditionally moderate leaders the authority and influence they have increasingly lacked since the NPLA and the Third Wage Structure; to strengthen them thereby and cause them not to need to ally themselves with the 'Leftists' over so great a field as at present . . .

11. The position of NACODS vis-à-vis BACM must be considered in the total context. The more supervisory staff who can be channelled into BACM the better. NACODS is a member of the TUC: BACM is not. NACODS have demonstrated recently that they are less and less a real part of management –

OPERATIONS

13. So, it should be the endeavour to get coal, or energy from coal, either without or with a minimum of mineworkers, or with skilled engineers or supervisory staff on the one hand, or by processing or conversion without extraction on the other.
14. . . . Is it possible to set out the Selby mine(s) on ROLF or its up-to-date like? . . .
    ROLF or like operations, manless at the coalface, would or could involve only supervisors, engineers, craftsmen, WPIS [weekly paid industrial staff], so as to reinforce non-coalmining trade union membership but membership of BACM or at least NUM Power Group (Ottey) or COSA (Storey).
17. . . . The aims should be to:
    (a) limit the future manning of the industry to restrict, to neutralize, alien or subversive political influences;
    (b) ensure that of those employed in the mining industry the maximum number should be outwith the NUM.[1]

While there is no evidence that the Miron Report ever became official NCB policy, one former NCB official has admitted 'it reflects what was being said, and much of it has come to pass'.[2] There is a remarkable congruence between the content of the Miron Report and key features of NCB strategy after 1973, particularly in relation to NCB industrial relations and wages policy. Miron outlined key NCB objectives in the introduction of new technology and the restructuring of the industry (see below). Moreover, in the Miron Report can be found the origins of the NCB's union-busting (Chapter five) and management's influence upon the struggle between opposing factions for control of the NUM (Chapter seven).

Although the Miron Report was not widely publicised until 1986, the media had been hostile to the miners throughout the 1973–74 dispute, accusing the leadership, especially McGahey, of pursuing political ends. Right-wing union leaders were conscripted to vilify left-wingers on the NEC. Tom Jackson, UPW General Secretary, claimed 'the fight is not about wages but about revolution'.[3] Frank Smith, Leicestershire Area Secretary, claimed the 'dispute was communist-inspired . . . if there was a free vote tomorrow the majority of men would support me and call this dispute off'.[4] Leicestershire miners responded with unofficial strikes and Smith was obliged to apologise.

By 23 January 1974, Yorkshire Area Executive concluded that, given the emergency arrangements, the level of fuel stocks could only be reduced by full-scale strike action.[5] The following day the NEC authorised a strike ballot. Before the ballot, but after the Miron Report, the Conservative Party Research Department had published a briefing note attacking left-wingers on the NEC:

THE EXTREMIST THREAT

It is worth noting that Rule Three of the rule-book of the National Union of Mineworkers lists, among the Union's objectives 'to join with other organizations for the purpose of, and with a view to, the complete abolition of capitalism'. These words may appear archaic to many coalminers, yet the fact is that there are six card-carrying members of the Communist Party on the national executive of the NUM.

The key figure is Mr Mick McGahey, vice-president of the NUM and president of the Scottish branch of the Union, who is also a member of the Communist Party National Executive and, still more importantly, of the Party's 'inner circle' political committee. He makes no secret of his determination to smash Phase Three, to bring down the Government and to abolish capitalism . . .

There are in addition to the Communists on the NUM Executive four or five extreme left wingers, the most notable of whom is Mr Arthur Scargill. Although he left the Communist Party in 1962, 'he still maintains a Marxist view of life, and his early training was unmistakably Communist' (*Times*, 4 January 1974). He organized the violent picketing of the coke depot at Saltley, Birmingham, during the 1972 coal strike, and forced the authorities to close it.[6]

Once the ballot was announced, the media attack intensified with the participation of Prime Minister Edward Heath.[7] The Government's strategy attempted to 'transform the dispute into a constitutional issue' and isolate the left on the NEC.[8]

Management provocation during the overtime ban and media hostility only increased the miners' determination, and a majority of eighty-one per cent nationally, ninety per cent in Yorkshire, supported strike action. On 7 February 1974 the Government announced a general election for 28 February, but the NEC meeting the following day called a strike from midnight 9 February. The NEC decided to maintain a low-profile strike to prevent the Conservatives gaining support in an election which Heath had launched on the theme 'Who governs Britain?'. The strike was continually overshadowed by the general election, which gave Labour a majority of only nineteen seats *with* Liberal support, and this could be lost by the votes of twenty-three others at any time. The Wilson Government quickly sought a solution to the dispute which entailed meeting the miners' claim and work was resumed on 11 March 1974.

Plan for coal

After the 1974 strike was settled, the NCB submitted to the Labour Government a *Plan for Coal*, which required a further £600m investment in addition to the £800m (January 1974 prices) already committed to maintain capacity.[9] Following tripartite discussion of the NCB *Plan for Coal*, the Final Report on the *Coal Industry Examination* reaffirmed Government support, emphasising the need for an incentive scheme and endorsing the NCB's research and development programme.[10] Feasibility studies were begun for major new developments like the Vale of Belvoir, and a planning application was made for the Selby complex. In implementing the 1974 *Plan for Coal*, the NCB developed strategies which would reduce the influence of the NUM membership and designed technology to reduce the miners' control over the labour process.

In 1976 the NCB developed its strategy for beyond 1985 in *Plan 2000* which estimated a range of coal demand between 135m tons (the 1981 level) and 200m tons; a mid-point of 170m tons was taken as a target, 20m tons of this to be open-cast.[11] When the progress of the 1974 *Plan* was reviewed by the tripartite group, investment costs were found to have escalated from £1,400m to £3,200m and a further £200m was needed for the Selby project which was running behind schedule. *Coal for the Future* approved the new investment,[12] and the Coal Industry Act 1977 increased the borrowing powers of the NCB and provided financial support to maintain the CEGB coal burn.

Plan for Coal had recognised that 'the need to close pits on economic grounds should be much reduced', but the 1977 Act also provided finance to 'assist in the redeployment of the manpower resources of the Board and the elimination of uneconomic colliery capacity'. After five years of Labour administration there were thirty-six fewer collieries than the 259 inherited from the Heath era, representing a contraction of fourteen per cent in the number of pits and under seven per cent in manpower. Tony Benn recalls that when he was Secretary of State for Energy, he 'offered the NUM a veto on pit closures, in order to deal with the very dishonest way in which the matter had been handled earlier'.[13] Although *Plan 2000* was not formally adopted by the Government, the Green Paper *Energy Policy* reaffirmed commitment to the programmes outlined in *Plan for Coal* and *Coal for the Future* but warned of the need 'to keep coal competitive'.[14] The new policy for coal was vindicated by the 1978 Iranian oil crisis which was followed by the first increase in coal consumption since 1957 (apart from the temporary increase brought about by the reflationary policies before the 1964 election). In 1979 coal supplied thirty-six per cent of the total UK energy requirements, sixty-nine per cent of which was for the power-station market.

Following the introduction of power-loading, productivity settled down to about 2·25 tonnes per man shift overall and to about 7·75 tonnes per man shift at the face. In the 1960s, attempts to introduce stand-alone automation such as ROLF (Remotely Operated Longwall Face) and ELSIE (Electronic Signalling and Indicating Equipment) had met with little success because of local NUM resistance and the limitations of the hardware then available. Miron recommended the establishment of a 'high-powered confidential Working Group at Headquarters' to develop and implement the new automation strategy. Shortly afterwards the Central Planning Unit was established to co-ordinate investment projects under *Plan for Coal*.[15] The Operational Research Executive (ORE) reviewed the NCB's activities to identify how the productivity increases of the 1960s might be continued. The adoption of systems engineering directed the attention of designers not just towards items of machinery but towards all colliery operations. This change of emphasis is demonstrated by the Mining Research and Development Establishment (MRDE) expenditure pattern between 1975 and 1982; while research on coal face projects fell from thirty per cent to fifteen per cent of total expenditure, research on computer-based comprehensive monitoring increased from ten per cent to twenty-five per cent of total expenditure.[16] Systems analysts broke down colliery activities into distinct but related logical subsystems, which were further broken down where necessary into functional subsystems with objectives that computer systems designers and engineers could develop operational systems to meet.[17]

In the late 1970s, the systems engineering programme led the MRDE to design MINOS (Mine Operating System), a computer system for centralised monitoring and control of colliery activities.[18] The application of control engineering and the computerised information systems which comprise MINOS have been facilitated by developments in microelectronics; unlike ROLF and ELSIE, MINOS can make use of microprocessors which are cheap, reliable, robust and versatile. MINOS comprises various subsystems devoted to particular functions such as coal clearance, face delays and coal preparation, all of which are overseen by minicomputers in a surface control room. The screens of the control console present mimic displays of the coal-face, conveyor systems and other operations, indicating the status of items of plant. When the conveyors are not running, when the shearer is stopped or when ventilation fans have failed, the system alerts control room operators. Much of the information that is passed to the control room is analysed by a secondary computer accessible only to management and, at the top of the system hierarchy, Compower, a wholly-owned NCB subsidiary, collates the information obtained from individual pits to compare performance; Fig. 2 gives a schematic view of the system.

The design philosophy of MINOS embodies three principles which have serious implications for the NUM. Firstly, the system is hierarchical and centralised; it mirrors the existing management structure but is designed to bring accurate, detailed information from all areas of the pit to the highest levels of management on a continuous basis.[19] This design feature reflects the objective of increasing managerial control. Secondly as many functions as possible are subject to closed-loop automatic control, with operator intervention limited to simple keyboard procedures.[20] This feature reflects the objectives of increasing labour productivity[21] and return on capital,[22] and of deskilling which facilitates increased management control. The third significant principle of the MINOS design is its modularity. This systems characteristic enabled the separate subsystems to be evaluated in a variety of situations, but also concealed the overall extent of automation from the NUM.[23] Before the strike a total of 125 MINOS applications were identified; about eighty pits had at least one MINOS facility, but none had the entire system. The first complete MINOS installation was to have been the Woolley complex in Barnsley, planned to be operational in May 1984 but for the overtime ban and

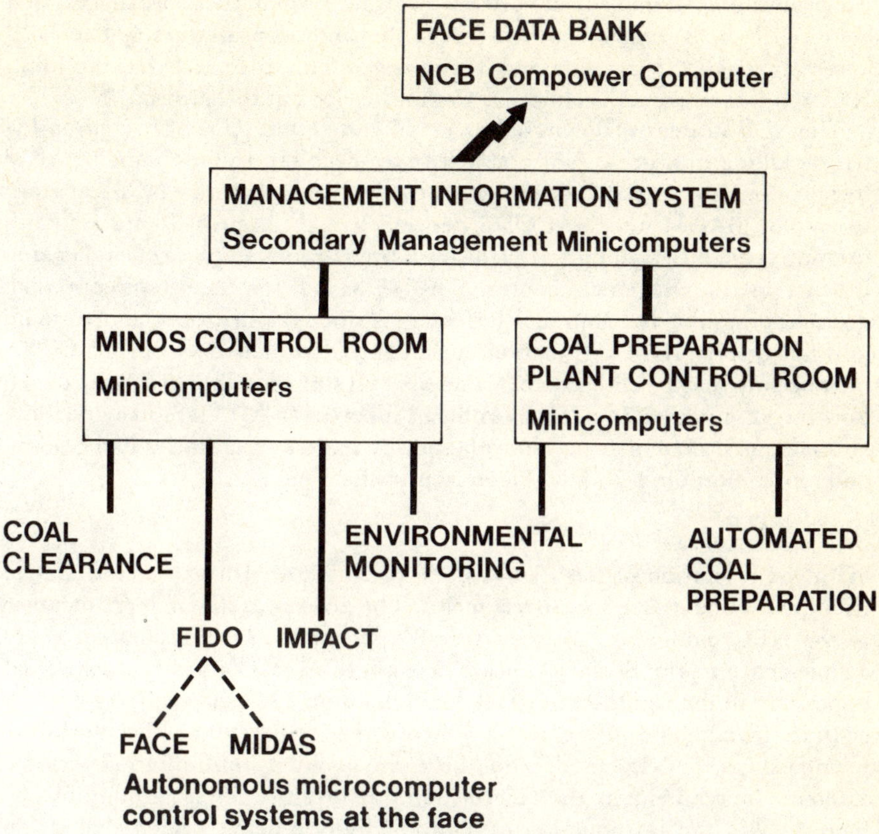

Fig 2 Schematic representation of MINOS
Source: A. Burns *et al.*, 'The miners and new technology', *Industrial Relations Journal*, Vol. 14, 1983, No. 4, p. 13.

strike. The Selby complex has been designed around MINOS entirely.

Strategies and tactics of resistance to new technology were slow to develop within the NUM compared with the rate at which the NCB was able to install MINOS.[24] National and area union leaders were unaware of MINOS until a dispute over a new coal preparation plant at South Kirkby led to the establishment of a national Joint Committee on Grading and Technology to resolve future problems over the grading of jobs affected by new technology in *specific applications*. The NUM side of the committee was given no overview of the NCB's automation programme and only became aware of the introduction of new technology in cases where a dispute arose over grading. In 1981 Yorkshire Area NUM Secretary, Owen Briscoe wrote to National Union Secretary, Lawrence Daly following a review meeting where the NCB revealed that a 'massive Remote Control Centre' was to be installed at West Side: 'I understand that not a word of communication regarding this innovation has been communicated to you at national level and I would therefore suggest that a strong letter is sent to the Board regarding non-consultation as had been agreed through the Grading and Technology Committee and that the Joint, National Grading and Technology Committee look at this project.'[25]

The NCB strategy of downgrading jobs diverted the NUM from confronting the deskilling and drew the committee into negotiating the maintenance of socially-constructed skill labels for work that was deskilled. The NCB's attempt to employ BACM and NACODS personnel in the control rooms similarly prompted the NUM to pursue exclusive representation in place of challenging the increased management control. The NUM had some success in opposing downgrading and the National Reference Tribunal ruled that control room jobs fall under NUM jurisdiction, although individuals may opt for COSA membership and several control rooms are still staffed with BACM personnel. Such negotiation left intact the problems inherent in MINOS, like deskilling and increased management control, and, by removing obstacles like grading and representation, may have eased its introduction.

Area incentive schemes

With the resolution of the 1974 dispute there followed five years of relative tranquillity under Labour Governments. The peace was due in large measure to the NEC moderates' manoeuvres around conference decisions to secure acquiescence to the Social Contract. Yorkshire Area Council had expressed opposition to the Contract but a national ballot in 1975 showed sixty-one per cent in favour and there was a narrow majority within the Yorkshire membership. For NEC moderates, the key to avoiding confrontation with the Labour Government in the face of militant pressure for a substantial pay increase was the introduction of a productivity scheme, which the Miron Report had identified as a key management objective. In 1974 a membership ballot had rejected an incentive scheme by sixty-two per cent, a view reaffirmed at the 1977 Conference. Another ballot in October rejected incentive payments by fifty-six per cent nationally and seventy-seven per cent in Yorkshire. In November 1977 the NEC ruled that there could be no national scheme but invited areas where a majority had voted for this to negotiate their own schemes. Such a contradiction of the membership ballot was only possible because moderates were overrepresented on the NEC. In 1979 the fifteen moderates (62·5% of the NEC votes) represented only forty-six per cent of the membership. Kent, Yorkshire and South Wales failed to obtain an injunction to prevent local schemes being negotiated in the moderate areas and once the schemes were operating, other areas followed.

Apart from fragmenting pay bargaining, the Area Incentive Scheme (AIS) was designed to increase productivity and act as a catalyst for the introduction of MINOS. From work measurement during the 1970s, the NCB calculated that the available shift time divided into three approximately equal portions: machine running time, delays caused by operational difficulties such as conveyor or shearer failure, and 'lost time', or delays arising from natural breaks taken by face teams. The NCB Method Study Branch confirmed that 'lost time' accounted for between thirty-seven per cent and thirty-two per cent of available shift time in the years 1977 to 1982.[26] FIDO (Face Information Digested On-line) was developed to monitor the man-made delays: 'The first, we knew about FIDO was when this voice from the gods asked us, from the pit top, why the machine was stood.'[27] General purpose monitoring outstations at the coal face send digital information to the control room computer which indicates on the console VDU the position of the shearer along the face, as well as the cause and duration of any delays: 'The controller knows immediately the face is stopped; he does not have to rely on a message from underground . . . People underground are aware of the resources that are available to the controller. They realise that he will not be fobbed off with imprecise information.'[28] Once identified by FIDO, delays of less than twenty minutes are attributed to the men and adversely affect their bonus pay, whereas delays of twenty minutes or more are allowed as contingencies under the scheme (provided they are found to have arisen out of operational difficulties beyond the control of the face team). Operating like a continuous tacograph, FIDO, the 'watchdog' as it has been dubbed by face-workers, has been designed to distinguish between two successive delays of ten minutes (a 'split delay') and one continuous delay of twenty minutes.

As delays are reduced and machine running time is increased, fewer faces will be required to produce a given volume of coal. Manpower, particularly face-workers and craftsmen, will be reduced and, for those who remain, work will entail less skill and autonomy than before.[29] Moreover, the reduction of delay time may weaken the basis of the solidiarity of the miners. While comradeship is usually attributed to the mutual dependence of workers in a dangerous environment, it is also fostered by a situation where workers have opportunities for lengthy discussion. One reason why face-workers and craftsmen are so prominent in the NUM is because they have more opportunities to develop agitational skills as a result of their autonomy and the inevitability of delays. Goodrich quoted leaders of the American mineworkers' union who recognised the importance of the *gob-pile oration*: '"It's the delays that make the miners debaters." It is certainly the delays that bring them together in the small groups on the gob piles, and very likely it is true that the "orations" that are delivered there are of more importance in setting the manner of life in the mines than all the speeches of the official union leaders outside.'[30]

Resistance to the introduction of MINOS was blunted by the AIS which guaranteed that miners would receive a proportion of the benefits of increased productivity. Incentive earnings constituted an increasingly important component of pay and were susceptible to local pressure. As collective bargaining was fragmented, national wage unity was destroyed, so a strike in support of a national pay claim became improbable. Differences in bonus earnings between the central and peripheral coalfields and within Yorkshire between super-pits and those with geological difficulties added to the inherent divisiveness of pit closures, so that after 1974 there was no major coal conflict until 1980 (see Chapter two).

The left in power

The Miners' Forum, formed in 1967 to promote the left within the NUM, provided the catalyst to translate into action grievances over pit closures and declining relative earnings. After the Yorkshire Miners' Forum had assumed responsibility for the local organisation of the 1972 strike, Arthur Scargill was elected compensation agent and, in May 1973, Area President. The ascendancy of the Yorkshire Left was confirmed in September 1973 when Owen Briscoe from Armthorpe, centre of Doncaster militancy, was elected Secretary against Bill O'Brien from Glasshoughton, a moderate pit even by North Yorkshire standards.

Moderate control of the NEC was destined to end once the left had come to power in Yorkshire, but the effect of the moderates' era, and particularly the AIS, was to impose a lasting constraint on the new leadership. The political transformation of the NUM was completed with the election of Scargill to the national presidency in December 1981. Scargill was the obvious candidate for the Forum because Gormley had delayed retirement to ensure that McGahey was too old to stand. Scargill also had the resources and support of Yorkshire, whereas Peter Heathfield, the other possible contender, was from the smaller North Derbyshire Area. Scargill was already well known as the media had built his image, but he campaigned systematically well before the election and secured the nomination of twelve NUM areas representing over eighty per cent of the membership. His performance at rallies around the coalfields was brilliant, and especially successful in captivating young miners who had entered the industry since 1974. Scargill fought the election on substantial left-wing policies which he had elaborated in *Miners in the Eighties*, published three months before the NUM election.[31] The pamphlet established the nature of the economic crisis which would cause further contraction, and argued the need to oppose pit closures other than on grounds of exhaustion and to pursue improvements in miners' terms and conditions. Scargill argued for changes in the government of the NUM and, most significantly, declared that miners would inevitably have to resort to industrial action to pursue these policies.

In contrast to the unity of the left, the right-wing caucus was in disarray when Gormley announced his retirement. Eventually the right fielded three candidates: Trevor Bell, of the white-collar section COSA; Ray Chadburn from Nottinghamshire, and Bernard Donaghy from Lancashire. Donaghy was unknown outside Lancashire and failed to campaign beyond the area. Chadburn was potentially the strongest of the three but his campaign was equally ineffective, probably because he could not penetrate Yorkshire, whereas Scargill enjoyed the support of the majority of activists in Nottinghamshire. Bell was unlikely to succeed as a white-collar representative; he was also from Yorkshire so lacked a geographic base from which to challenge Scargill. Bell's campaign was professional and anti-political, calling for a continuation of Gormley's policies which Scargill was showing to be bankrupt in the crisis of the 1980s.

The election result was a foregone conclusion but the size of the margin surprised even the activists who had worked so hard for Scargill's victory. In a characteristically high poll, Scargill obtained 70·3% of the votes cast, against Bell's 17·3%, a majority unprecedented in elections for NUM officials. At the 1982 conference in his first presidential address, Scargill interpreted his landslide election as evidence of widespread commitment to the policies on which he had campaigned. Nevertheless, there was an obvious need to formulate strategies and tactics to pursue these policies and insufficient time for the quality of campaign required. Scargill's reforms of the NUM were necessary to create the machinery which could develop an effective strategy, but these ran into problems on two fronts. Firstly, national office staff rebelled against the imposition of the new régime which was designed to bring appointed officers under the direction of elected officials. Few of the London staff moved with the national office from Euston Road to Sheffield. The inexperience of new headquarters staff hampered the campaign and limited the effectiveness of central support available during the 1984 strike. Secondly, there was resistance from the area leaders. The government of the NUM had been established as a loose federation of area unions, each having considerable autonomy and representation on the NEC. Scargill argued that the NEC should be made more representative and more accountable to the supreme policy-making body, the delegate conference; the administration of decisions could then be more centralised without undermining democracy. However, NUM democracy was was founded upon federal autonomy as left and right factions used the areas as their power bases, which none of the area leaders was prepared to relinquish. The struggle between area and central interests was to have disastrous consequences for the organisation of the 1984 strike (see Chapter three). Although the National Union now had left leadership, this did not alter the fact that the industry was moving rapidly towards crisis and that serious divisions still existed within the NUM.

Scargill's election to the national presidency ensured that research underway in Yorkshire into new technology in coal-mining would be considered at national level. In September 1982 the Working Environment Research Group (WERG) was invited to present its findings to the NUM side of the Grading and Technology Committee, which recommended that a report be prepared and presented to the full National Executive with the utmost urgency. At the Executive meeting of December 1982 an *Interim Assessment of MINOS*[32] was presented and supported by the new National President, who moved that the report should be widely circulated within the union and that educational initiatives on new technology should be organised. Disseminating the overall assessment of MINOS assisted activists to raise membership consciousness over new technology and established a dialogue between workers at pits experiencing *different* MINOS facilities.

At national level, the concern over the *Interim Assessment* focused upon the job loss projections which confirmed the existence of a major pit closure programme. Rank-and-file miners who were experiencing the impact of MINOS upon the labour process, however, were equally concerned with the issues of deskilling and control. It was from the Yorkshire Area weekend school in March 1983 that pressure developed for a new technology agreement; a motion from South Kirkby branch was passed at Yorkshire Area Council and accepted at the 1983 NUM conference. The draft agreement sought to establish a procedure for negotiating technological change with the *status quo* prevailing until agreement is reached. The agreement would have preserved jobs through reductions in working time: a four-day, twenty-eight-hour working week; early retirement at fifty-five, and longer holidays. Moreover, it would have eliminated computer-based work-monitoring systems like FIDO which would be unlawful under the Swedish and Norwegian Work Environment Acts.

Coal under Thatcher

The election of the first Thatcher Government in 1979 effectively ended coal's revival. Further pit closures became inevitable but the origins of the conflict were as much a part of *Plan for Coal* as of Thatcherism. The control of public expenditure created new financial objectives for the NCB: to remain within the External Financing Limit and to break even after interest and receipt of government grants. The Coal Industry Act 1980 increased the borrowing powers of the NCB and provided for grants to be made to reduce deficit for a financial year. In *The Miners and the Battle for Britain*, the NUM denounced the Act:

> Although appearing to support the *Plan for Coal* by an extension of NCB borrowing powers and provision for deferred interest loans on major capital projects, the real effect of the Act was to replace the production targets, towards which the industry has been working since 1974 with new financial targets. Neither the unions in the industry nor the NCB itself believe that these targets could be met at a time of industrial recession without huge price increases, unprecedented productivity increases or the closure of collieries which are presently making a financial loss.[33]

In the recession brought about by monetarist policies, the NCB was unable to meet its financial obligations and attempted to solve the problem by accelerating pit closures in 1981 (see Chapter two). Following tripartite discussions, the Government increased the money available to the industry in the Coal Industry Act 1982. As a result, output from NCB deep mines was maintained into the 1980s, and CEGB coal burning was maintained until power stations represented seventy-four per cent of total coal demand. The 1980 Act extended grants in connection with pit closures to the end of March 1984 and increased the aggregate amount available to £170m, which was raised to £200m in the 1982 Act. Payments to redundant miners were similarly extended to the end of March 1984 in the 1980 Act and the aggregate amount increased to £220m, which was raised to £300m in 1982.

The Government's intention of closing collieries on financial grounds as a solution to the coal crisis it had helped to create became ever more apparent in 1982. On 3 March 1982 the Department of Trade and Industry referred to the Monopolies and Mergers Commission (MMC) questions relating to the efficiency and costs of the NCB. Paul Routledge speculated over the future of forty-five pits 'in jeopardy' on the basis of NCB colliery operating results.[34] A document containing evidence submitted to the MMC and other details of colliery costs was leaked from the NCB to the NUM on 1 November. An Annex to the MMC submission entitled 'Reduction of capacity', referred to 'premature closure (i.e. earlier than natural exhaustion) . . . of uneconomic collieries' and contained a table illustrating the effect of the 'progressive closure of those collieries by the year 1990/91', described as 'a very sensitive area . . . not for publication'.[35] A paper on 'The variability of colliery costs' showed forty-four of ninety-three peripheral collieries and thirty-one of one hundred and fourteen central block collieries to be 'short-life' pits. The third paper of the leaked documents, 'Unprofitable collieries', defined one hundred and thirty-eight of the two hundred operating collieries in 1981 as loss-makers, and attributed £228m of the NCB's total losses of £253m to the thirty pits with the greatest losses.

Following discussion at the NEC meeting of 2 November 1982, Daly wrote to NCB Chairman, James Cowan, asking for an explanation of the documents, the names of the collieries concerned and a meeting with the NCB to discuss these. The leaked information appeared in media reports with Union estimates that seventy-five pits were to be closed with a loss of 70,000 jobs in the next eight years. The controversy caused the House of Commons Select Committee on Energy to conduct an inquiry into these allegations of pit closures. After detailing the basis of the Union's estimate of the scale of closures planned, the NUM President accused the NCB of duplicity and called for the dismissal of Board members.[36] In its evidence the NCB denied there was a hit list: 'The obsession of one man with the production of a list is getting us all into these difficulties. We have a substantial number of pits under discussion, not necessarily for closure but for improvement or co-operation with the unions at local level.'[37] The Energy Committee concluded that the NCB did not maintain 'a "hit list" in the sense implied in the NUM's evidence' but warned that the Union's attitude towards closures could necessitate national Board involvement 'in decisions about the future of individual collieries'.[38] Arguing that coal stocks approaching 50m tonnes and high-cost surplus capacity were imposing a drain upon public funds, the Energy Committee recommended 'that the Board must take steps to bring its capacity more into line with existing and expected demands for coal'.[39]

The Energy Committee's recommendations were reinforced six months later in the MMC Report on NCB efficiency and costs. The MMC Report examined the means available to the NCB of meeting the financial objectives established in the 1983 Act, and concluded:

> We consider that the NCB should continue to impose the strict controls that have been applied in recent years so that the deep mine budget is set with the aim of reducing unit operating costs in real terms in line with the increasing capital intensity of the industry. This will involve a maximization of output from low-cost capacity with relatively low marginal costs and the reduction of high-cost capacity.[40]

The Coal Industry Act 1983, which received the Royal Assent on 21 December, further increased NCB borrowing powers and deficit grants but required the NCB to break even by the end of March 1986. Grants in connection with pit closures were extended to the end of March 1986 and the aggregate amount available was increased to £400m, while payments to redundant mineworkers were increased to £1,200m. By the end of March 1984 there were fifty-three fewer collieries than the 223 in operation when the Conservatives came to power in 1979, a colliery contraction rate of twenty-four per cent (compared with the fourteen per cent contraction over the previous five years of Labour) and a manpower contraction of twenty-three per cent (compared with seven per cent under Labour).

Conservative strategy was aimed not at avoiding the crisis but rather at delaying its manifestation until the state was prepared (see Chapter five). The deadline for the conflict was progressively extended by statute while the industry was being steadily run down. The dimensions of the crisis had such a determining influence upon the character of the 1984–85 strike and its aftermath that they are analysed in detail below.

1.2 Uneconomic pits or strategic resources?

At the end of March 1984 the NCB controlled 170 operating collieries, only eighteen per cent of the 958 at nationalisation. The number of miners had dwindled to 181,000 from a peak of 716,500 in 1948, a loss of seventy-five per cent of the former manpower. The contraction had shown little party allegiance as successive governments accepted the 'logic' of the market. It is difficult to comprehend the miners' resistance to further closures when so little of the industry remained, except in so far as this was *all they had left to defend*. The refusal to accept further closures embodied more than defensive action, although this was the major characteristic of the strike; it also entailed an implicit rejection of the market ideology which had ruled the coal industry through private and public ownership alike. In this sense the strike was indeed 'political', and the conflicting ideologies found expression in the debates surrounding uneconomic pits and the sufficiency of natural resources, more commonly the former, reflecting the considerable success which the Government had in determining the agenda. Although the provenance of both arguments can be traced to the industry's earlier history, during the strike the formal challenge to the uneconomic pits question seldom extricated itself from the bounds of bourgeois market economics to promote the alternative resource argument.

Economics of overcapacity

The amount of overcapacity and its relationship with high-cost collieries had been demonstrated in the MMC Report graph of operating costs against cumulative output, reproduced as Fig. 3.[41] The report had suggested the scale of closures required to meet the Government's financial objectives in noting that seventy collieries producing only eighteeen per cent of the output had operating losses in excess of £10 per tonne.[42] Although the NCB never publicly declared the overall scale of the closure programme, numerous economists offered their interpretations of how many pits *should* be closed.

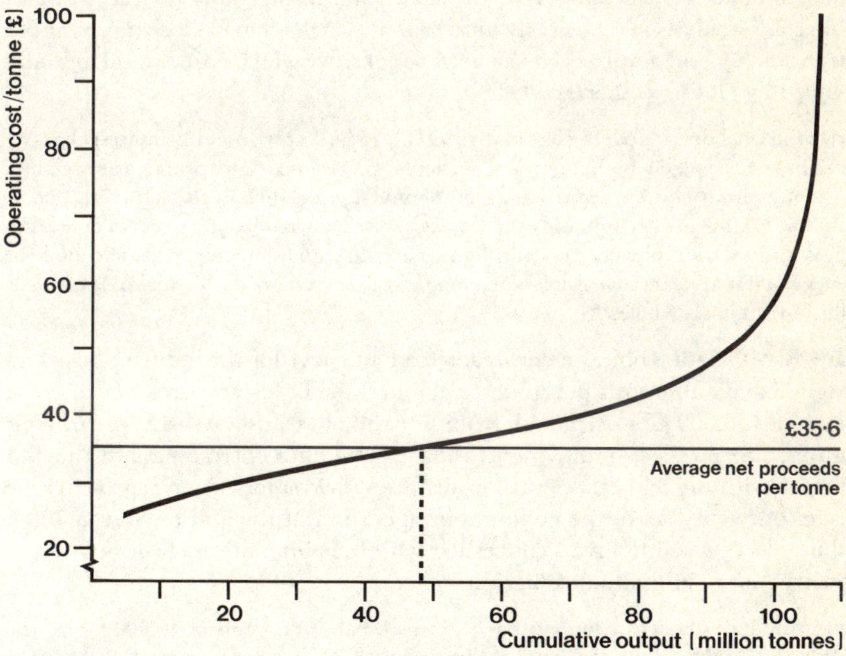

Fig 3 Unit operating costs against cumulative output

Source: Monopolies and Mergers Commission, *A Report on the efficiency and costs in the development, production and supply of coal by the NCB*, Vol. 1, Cmnd. 8920, London, HMSO, 1983, p.367.

The most aggressive Monetarist stance was adopted by Professor Minford, Liverpool University economist and adviser to the Thatcher Government. In the eighth month of the strike (when NACODS appeared poised to join the strike – see Chapter six) Minford and Kung assessed the costs of the dispute and concluded that these would be:

lower national unemployment from closing high cost pits; the rise in productive efficiency and the improvement in tax structure resulting from the closure of only 4 million tonnes of capacity would alone create 18,000 more jobs in total at national level (ie. over and above the job gains that offset the job losses in mining). This figure would be roughly trebled if all uneconomic pits were to be closed.[43]

The resources (capital and wages) employed in high-cost pits exceeded the value of their product by £180m per year. According to Minford's theory of labour market equilibration,[44] diverting this £180m to more efficient industries would create an additonal 3,000 jobs. A further 15,000 jobs would also be created by reduced energy costs and low income tax, according to the Liverpool macroeconomic model of the UK.[45]

It was also in October 1984 that Dr Bill Robinson of the London Business School offered his assessment of the economic background to the dispute. Robinson's analysis was basically akin to that which a chain store might apply to its branches and emphasised the need to consider which costs would actually be saved by closing collieries:

Part of the Coal Board's overall loss is attributable to sunk costs, notably interest charges, which have to be paid whether or not a particular pit is closed. Any proper assessment of the savings from closing a particular pit should ignore such costs. Thus if a pit is profitable taking into account only the *avoidable* costs incurred by keeping it in operation (wages and salaries, power, heat and light, necessary maintenance) it should be kept open, even if it appears *unprofitable* when made to bear its share of the unavoidable costs of the industry as a whole.[46]

Using MMC data Robinson constructed a cost curve for the industry based on marginal avoidable costs per miner and found the industry's break-even point to exist at 165,000 jobs. Although Robinson did not translate this figure into pit closures, this level of employment using MMC data corresponds to the 144 collieries with the highest operating surpluses (lowest losses), so represents the closure of twenty-six of the collieries in operation at the end of March 1984. Robinson's first solution was concerned with the elimination of subsidy to the industry, not to individual collieries:

even if the industry had been cut back to this break even point of 165,000 jobs, the subsidy to workers in the marginal colliery was £4,745 per man in 1981–2. In other words, even if the mining industry as a whole is not receiving any subsidy, the cross-subsidy within the industry is very large – some very efficient pits are subsidizing some very inefficient ones. This point is presumably not lost on the miners of Nottinghamshire, where a high proportion of the profitable pits are located.[47]

In December Robinson refined the analysis and concluded that when unavoidable costs and fluctuations were taken into account, 160,000 miners could remain employed with all collieries showing a surplus in most years.[48] However, Robinson also noted that the industry's surpluses could be maximised by cutting back to 60,000 miners: 'any expansion beyond this point reduces the cash available to the Coal Board for investment in efficient pits and eventually plunges the industry as a whole into deficit'.[49] Examination of the MMC data upon which Robinson's analysis was based reveals that 160,000 miners equates with 140 collieries, while 60,000 miners represents fifty-nine collieries. In other words, Robinson's calculations set the parameters of a closure programme at anything between thirty and 111 collieries.

Commentators seeking to provide economic legitimisation of pit closures gained momentum as the strike progressed: hence Robinson's December analysis was more severe than his October one and worse than Minford's. One analysis which preceded these claimed to take into account the interests of the NUM as well as those of the NCB and the Government. In September 1984 Davies and Metcalf prepared a report for LWT's *Weekend World* which proposed that collieries should close 'when their *real value to the whole economy* in net output and social terms is no longer positive'.[50] While the authors rejected the principle that pits should be worked to exhaustion, they also argued for the inclusion of the costs of unemployment, the resource cost of not producing (uneconomic) coal, and the responsibility of the NCB towards coalfield communities. On this basis they calculated that twelve per cent of NCB capacity, fifty to sixty pits, should be closed over the next five years. Presented as a compromise between the NUM exhaustion principle and what would be necessary for the NCB to break even, John Lloyd supported the solution as 'one which could commend itself to future Labour governments'.[51] Lloyd's observation about Labour governments may regrettably be true, but the Davies and Metcalf formula is deficient from a socialist perspective. The central assumption is that redundant miners find alternative employment and within ten years the labour market 'equilibrates', i.e. an equivalent number of jobs is created to replace those taken out of the economy: 'redundant mine-workers would *eventually* create extra net jobs in the economy, either through entrepreneurial skills and/or through depressing the general level of the real wage'.[52] Although their conclusions are entirely dependent upon the assumption of equilibration, no evidence is provided to support the assertion that within ten years 'some of the redundant miners are presumably going to find alternative jobs without displacing other workers'. Exactly the same criticism may be made of Minford and Kung who disguise their assumptions in the magic 'black box' of the Liverpool macroeconomic model.

Costs of pit closures

Before any of these papers appeared, an opposite view as forcefully advanced by Andrew Glyn in a paper which was widely circulated in its various drafts and published by the NUM in October 1984 as the *Economic Case Against Pit Closures*.[53] In fact the case had been put first at the Energy Committee on 25 November 1982 by Scargill, who said that the NUM research department had calculated that:

> if the Board keep all these pits open between now and 1990–91 and assume that the losses continued, it could cost approximately £2,000 million. On the other hand, my research

department calculate that the closure of these pits with the loss of approximately 70,000 jobs will cost our nation, the British taxpayer, £4,488 million, if we take into account the recent government figures detailing the cost of making the mineworkers redundant over a 10 year period.[54]

The NUM case against colliery closures presented to the Energy Committee centred upon the costs of increasing unemployment and the fact that many of the costs assumed to make pits unprofitable would not be saved by their closure. Many of the arguments focused upon NCB accounting practices. The NUM claimed that the peculiar method of financing the NCB, through borrowing rather than Government Equity, imposed a heavy burden of interest charges, amounting at that time to some £2.78 per tonne of coal produced. The scale of depreciation was also held to be excessive, adding a further £2.13 per tonne of coal. Stocks held, however, were undervalued in the accounts in comparison with the prices at which they were subsequently sold. Between the years ended March 1981 and 1982, increases in the costs of subsidence, early retirement, interest and depreciation and other expenses were 'sufficient to account for the entire loss of deep-mined coal production'.[55] The NUM also argued that the way in which headquarters overheads were allocated to collieries affected their operating results. Most of the costs considered would not be substantially reduced by pit closures, while a 'different approach to accounting practices might render invalid the already dubious notion of uneconomic collieries'.[56]

Glyn's report began from the similar premise that many components of the NCB's operating losses were not costs of production and would, therefore, still have to be met if pits were closed. Glyn challenged the inclusion of subsidence, pensions and interest payments in charges on the NCB and argued that redundancy payments and unemployment benefits should be considered costs of closure, or costs of *not mining* coal.[57] Adding the lost revenue from taxation to the costs of redundancy pay and costs of others made unemployed as a result of pit closures, Glyn calculated that the costs of closure far outweighed the subsidy required to maintain production.[58] Even after five years when RMPS payments to an older miner were assumed to cease and be replaced with costs of keeping a school leaver on the dole, Glyn shows a saving of almost £4,000 per person per year from *maintaining* mining employment (the savings in the first year are about £10,000 higher). The calculation was performed for every colliery to demonstrate that 'there are no pits whatsoever where closures would benefit Government revenue'; even the twenty collieries with the highest operating losses per tonne amounting to £153m per year would cost £267m to close, a net loss of £114m to the Government.[59]

Following the publication of Andrew Glyn's report, many academics joined in the criticism of NCB accounts. A paper by Tony Berry and colleagues (including Professors of Accounting at UMIST and Sheffield) was reported in the media during November 1984, but was not published until January 1985. Berry and colleagues concluded that the NCB accounts were inadequate for making informed decisions about colliery profitability:

The standard accounting statement for pits – the 'F23' is, as far as can be ascertained, used for making budgets, projections of pit profit and loss in the future and, more importantly, to justify decision-making, at least in public debate. Its decision-making rationale cannot therefore be ignored. Yet for this purpose . . . the F23 is fundamentally flawed . . . The F23 may be used for control purposes but does not provide a sensible basis for pit closure decisions or public debate on them.[60]

Brewer, a Coal Board accountant, confirmed the inadequacy of the F23 five months after the strike ended.[61] The problem of determining a fair figure for proceeds, raised by Wright in relation to the entire industry,[62] was shown by Berry and colleagues to be even more acute for individual collieries because of interdependence within the industry, for example through blending coal of different qualities from a group of collieries. Moreover, the relative contribution of a pit was difficult to appraise, particularly given the way in which budgeted production influenced profitability (volume determined overheads but over-production was penalised). The accountants supported the economists' criticisms of the inclusion of subsidence, early retirement and headquarters overheads and suggested that a capital charge would be more relevant to decision-making than depreciation. Finally, they criticised the use of the F23, an historic account, to predict the future when geological conditions implied inherent uncertainty and development work often had long lead times.

Many of the criticisms of NCB accounts were subsequently incorporated in alternative economic assessments of NCB areas.[63] The accountants' work was most influential in the preparation of cases against the closure of particular collieries (see Chapter seven). Their arguments demonstrated how the NCB created financial results to justify closure decisions, but this was insufficient to build a case *against* closure since it failed to confront the *processes* underlying the crisis (see below). To a degree, these criticisms could also be levelled at the NUM case; as O'Donnell argues, it could be seen as 'a negative case for coal'.[64] Glyn's paper, as the major articulation of the NUM case during the strike, was naturally subject to much scrutiny.

An inherent assumption of the NUM's opponents was that an extension of the Glyn case 'to all industries implies a permanent locking-in of resources to particular economic activities', a view which Towers noted neglects the special features of the coal industry (an allusion to the physical resource argument) and the context of high unemployment.[65] Others claimed that 'the central assumption in Glyn's work was that those miners made redundant remain so'[66] and that Glyn's conclusions:

rest heavily upon the assumption that workers, displaced from the industry, would be unable to secure alternative employment . . . The contingent nature of Glyn's defence can be demonstrated with reference to the rundown of the industry during the 1960s. Between 1960 and 1969, 420 pits were closed and approximately fifty per cent of the workforce (about 322,000 miners) left the industry. In the more buoyant economic conditions of that time, however, many miners were able to find alternative employment. Under these circumstances Glyn's case against closures collapses.[67]

From Glyn's account, however, whether or not a particular individual leaving the pit becomes re-employed is largely irrelevant. In a situation of demand deficiency for labour, taking workers out of the economy merely adds to unemployment. If a cohort study of 100 redundant mineworkers found them all to be re-employed within three months, the case would not be weakened so long as at least an equivalent number of school leavers or other workers were unemployed; the miners would merely have displaced others in the labour market. While the context of high unemployment obviously strengthens the Glyn case, even the conventional wisdom that the 'major run down of the mining industry in the fifties and sixties was against a backdrop of full employment'[68] neglects the contribution which pit closures made to unemployment during that period. Between 1957 and 1971 almost 417,000 jobs (fifty-nine per cent of the 1957 labour force) were lost from coal-mining. Over the same period, unemployment nationally increased from 297,000 to 724,175, a rise of 427,175 almost exactly the number of jobs taken out of the coal industry. Moreover, the view that pit closures will not only release manpower from an inefficient industry but will also generate additional employment depends upon *increasing* unemployment in the first instance. The additional jobs are assumed to be produced either by the entrepreneurial activity of redundant miners, which is implausible, or through the depressive effect on wages of adding to the unemployed. Once such additional jobs are produced and unemployment falls, the resulting rise in wages will presumably negate any temporary gain in employment.

The case against pit closures can easily be advanced against redundancies in any industry. The premise that unemployment incurs opportunity costs of employment as well as government subsidy and social costs is central to Glyn's programme for planning full employment.[69] When elaborated in the case of the coal industry *alone*, however, it appeared to be a sectional plea for mining which, as O'Donnell noted, 'was hardly likely to succeed in establishing the case for coal at a time when industries in general were undergoing a massive programme of rationalization'.[70] The grounds for the exceptional treatment of coal-mining were not made central to the case against pit closures.

Had the question of physical resources been given more prominence, then perhaps the NUM case could have overcome its apparent sectionalism and more effectively have challenged bourgeois market economics. Of the various academic contributions supportive of the NUM during the strike, few prioritised the resources issue. The second WERG report to the NUM, prepared during the first three months of the strike, emphasised that coal reserves would be sterilised by the NCB policy of concentrating production in the central coalfield,[71] a view later supported by the British Geological Survey.[72] The WERG case rested upon Schumacher's argument that:

> the concepts of 'economic' and 'uneconomic' cannot be applied to the extraction of non-renewable resources without very great caution . . . To eliminate the losing factory means the elimination of waste. But to close the losing colliery means merely to change

the time sequence in which finite resources are being used . . . It is a policy of doubtful wisdom and questionable morality for this generation to take all the best resources and leave for its children only the worst. But it is surely a criminal policy if, in addition, we wilfully sterilize, abandon, and thereby ruin such relatively inferior resources as we ourselves have opened up but do not care to utilize.[73]

The sterilisation problem arises because when a coal mine is abandoned, roadways, headings and coal faces are liable to collapse under the pressure of the strata which forces the roof to fall (crush) and the floor to rise (creep). In addition, the build-up of gases and water in a closed colliery increases the costs and hazards of attempting to reopen it. As former NCB Chairman, Lord Robens explained: 'it is easiest and cheapest to conserve coal by continuing to mine it . . . because it is not feasible to 'mothball' mining capacity . . . underground roadways and workings, if left for any length of time, will be crushed out of existence and whole areas of coal will become virtually impossible to work'.[74] While 'impressed by the tenacity with which coal experts held to this argument', the economists responsible for the PEP report, *Fuel Policy*, found it difficult to accept that the reserves would be lost forever and were 'more prepared to gamble upon improvements in mining technology, or different ways of getting the energy in coal out of the ground if the time comes a decade or two later'.[75] Two decades of significant developments in coal-mining technology have not affected the veracity of the sterilisation argument.

The fundamental conflict at the root of the 1984–85 strike against pit closures was between the long-term conservation of natural resources and the short-term pursuit of financial objectives in accordance with the 'logic' of the market. The contradiction between the two is epitomised in the sterilisation question rather than in the question of 'uneconomic capacity', which is merely a consequence of allowing the market mechanism to prevail. Barratt-Brown stressed the inadequacy of allowing the price mechanism to dictate energy policy: 'the possibility of oil or natural gas running out in 30 years time does not at all enter into its current price. Scarcity is only measured by price as a result of a gap between supply and demand *today*, or in the near future, not in the distant future'.[76] Aage's analysis of economic arguments on the sufficiency of natural resources similarly concluded: 'economics, particularly in the form of neo-classical economic theory, is of practically no use for the evaluation of the sufficiency of natural resources . . . technological and political considerations not only cannot be ignored but are of overwhelming importance'.[77]

The MMC Report on the NCB, in detailing the problems with the concept of colliery exhaustion, contrasted the NCB view with that of commercial mining companies. While the NCB took into account the value and reserves of coal as well as its costs of extraction: 'commercial mining companies, when judging whether a mine should be regarded as exhausted, would concentrate on whether it offers any potentially profitable mining opportunities rather than the quantity of mineral known to remain physically in the ground'.[78] The MMC inquiry legitimised the Government's imposition of such commercial criteria upon the NCB, thereby shifting debate away from the question of conservation of natural resources. Glyn's case against pit closures acknowledged Schumacher's argument but prioritorised other aspects of closure.[79] The *Aberystwyth Report on Coal* similarly accepted the resource arguments but focused upon NCB financial calculations.[80] During the 1984–85 strike the resource issue was not re-established on the agenda and the argument was conceded to the Government by default. Six months after the strike ended, Alistair Goodlad, Under-Secretary of State for Energy, was able to tell a TCPA conference: 'I therefore do not believe it is right to treat the question of "*the* appropriate energy mix" as one to which it is Government's business to find a single, optimum, answer. We ought not to be in the business of this sort of central planning.'[81]

1.3 Markets for coal

On the demand side, the economic recession has so far been the most important single cause of contraction of coal markets. Total UK energy demand is largely a function of the level of economic activity so that, even though coal more or less maintained a constant market share from the mid-1970s, coal demand collapsed with the contraction of manufacturing industry. Estimates of primary fuel demand made by the NCB, Department of Energy (DEn) and CEGB fell as the recession deepened. *Plan for Coal* forecast 1985 energy consumption of 400–490mtce,[82] while in 1978 the CEGB forecast for 1985 was 352–420mtce.[83] Actual energy consumption in 1983, the last undisturbed year before the strike, was 313mtce. The projection of *Plan for Coal* was therefore twenty-eight to fifty-seven per cent too high, and even the CEGB's later estimate was twelve to thirty-four per cent too high. Projections to 2000 have dropped even further. *Coal for the Future* forecast 500–650mtce total energy demand for 2000,[84] whereas by 1982 the DEn forecast for 2000 had fallen to 308–413mtce.[85] In five years authoritative predictions of energy demand for 2000 ranged from 650 to 308mtce.

In its *Review of Energy Policy*, the TUC's Fuel and Power Industries Committee, then chaired by Frank Chapple of the EETPU, produced even more optimistic demand projections than the DEn, in excess of 500mtce by 2000.[86] In effect, the TUC projections add together the supply targets of coal, nuclear, oil and gas as though they are not in competition, ignoring the fact that energy demand is declining, in order to obscure the conflict of interest between the NUM and the AUEW and EETPU. Because of the difficulties in assessing future energy requirements, the DEn no longer publishes energy demand forecasts. However, estimates of coal demand can be synthesised from existing patterns of consumption according to assumptions made about the future.

Non-CEGB coal demand

Coal demand other than for electricity generation represented 30m tonnes before the 1984–85 strike. The major markets were coke works (for the steel industry), other industries and domestic demand, each accounting for about

8m tonnes. With the extension of the Coal Firing Scheme, the NCB anticipated increasing industrial use to 14m tonnes by 2000, while the introduction of a parallel scheme for public-sector coal-firing could provide another 5m tonnes demand. Coking coal requirements are unlikely to rise beyond 10m tonnes, and this is probably also the ceiling for domestic demand. The most optimistic demand scenario would rise to 45m tonnes by 2010, while the worst demand is unlikely to fall below 20m tonnes. A constant demand of 30m tonnes is the most likely outcome. In addition, electricity generation in Scotland (SSEB) is expected to add a further 5m tonnes coal demand per year to the end of the century, 6m tonnes thereafter. These scenarios are produced as Table 1.

Table 1 *Non-CEGB coal demand (mt)*

Year	*1990*	*1995*	*2000*	*2005*	*2010*
High	40	40	46	46	51
Medium	35	35	36	36	36
Low	35	30	31	26	26

CEGB total fuel requirements

Since almost three-quarters of Britain's coal is consumed by the CEGB, electricity demand has the greatest impact upon coal demand. For the Sizewell Inquiry, the CEGB forecast electricity demand to 2010 based on four economic scenarios. Two scenarios assumed high economic growth rates (2·6% p.a. GDP) based alternatively on service industries (A), and manufacturing (B), one (C) assumed medium growth (1% p.a. GDP), and the fourth (E) assumed unstable low growth (−0·4% p.a. GDP). Table 2 shows CEGB fuel requirements under the extreme scenarios (B and E) and the medium growth scenario (C), widely assumed to be the most likely outcome.

Table 2 *CEGB fuel requirements (mtce)*

Year	*1990*	*1995*	*2000*	*2005*	*2010*
Scenario B	127	138	155	169	182
Scenario C	105	112	115	117	121
Scenario E	86	85	84	81	80

Source CEGB Evidence to Sizewell.

The divergence between the extreme estimates, which increases from 127–86mtce in 1990 to 182–80mtce in 2010, demonstrates the effect of economic activity upon energy demand. The primary fuel mix decided upon by the CEGB (and SSEB in Scotland) further affects the proportion of these fuel requirements to be met from coal.

Nuclear power
Nuclear power represents the major threat to coal demand in the long term. Nuclear power has been justified to meet increased energy demand but this deception has been committed in order to make room for nuclear power at the expense of coal. Leaked secret minutes of a cabinet meeting held on 23 October 1979 stated: 'a nuclear programme would have the advantage of removing a substantial portion of electricity production from the dangers of disruption by industrial action by coal miners or transport workers'.[87]

The CEGB similarly argued at the Sizewell Inquiry that nuclear power stations were necessary to avoid the risk of discontinuity of supply of coal. By the end of March 1984 nuclear generation capacity represented approximately 14mtce; a further 5mtce of nuclear capacity will come on stream by 1990 with Hartlepool, Heysham I and II, and Dungeness B. The majority of Britain's nuclear reactors are Magnox, fuelled by natural uranium, cooled by carbon dioxide under pressure and controlled by neutron-absorbing rods. About one-third of the stations are AGRs, advanced gas-cooled reactors, like the Magnox but fuelled with enriched uranium. Although the Thatcher Government had approved the construction of new AGRs at Heysham and Torness without debate, its more controversial programme of building a family of Pressurised Water Reactors (PWRs) prompted the establishment of a public inquiry into the proposed siting of the first PWR at Sizewell in Suffolk. The leaked cabinet minute of 23 October 1979 noted that the inquiry should be 'wide ranging' but expressed fears that this 'would arouse prolonged technical debate between different representatives of scientific opinion'. Even before the inquiry opened, the CEGB began site clearance work and had ordered the pressure vessel for the reactor.[88]

There are three major arguments against the introduction of further nuclear generating capacity in Britain: it is unnecessary, expensive and dangerous. The first point is self-evident since the CEGB have to inflate energy demand projections to justify the nuclear programme. Questions of relative generating costs are more contentious. The conventional view is that nuclear plants are costly to build but entail low fuel costs so their unit costs over the life of the plant are lower than for coal-fired power stations. Such calculations depend upon assumptions about the capital costs and the reliability of nuclear plant. Evidence from the USA suggests that nuclear plants are more often out of service and that capital costs are frequently under-estimated by as much as one hundred and thirty per cent. In 1975 the CEGB claimed that total costs (capital and fuel) of electricity generation were 0·85p/kWhr for coal and 0·53p/kWhr for nuclear. However, the cost of uranium has increased significantly and capital costs were held down by historic cost accounting, so Sweet suggests a more realistic comparison would be 1·75p/kWhr for coal and 2·8p/kWhr for nuclear.[89] While the capital costs of AGRs were assumed to fall to the level of coal-fired stations, in fact they rose to *seven times* the costs of coal-fired stations. Moreover, relative generating costs based upon performance favour nuclear power stations because they are kept on base load, supplying electricity all day throughout the year, whereas coal-fired stations are operated at a lower load factor.

The major issue for most objectors to nuclear power is that of safety. The disasters at Windscale in 1957, Three Mile Island in 1979 and Chernobyl in 1986 serve as powerful reminders of the risks associated with every nuclear power station and reprocessing plant.[90] The dangers from radiation leaks, accidents and explosions are, however, only part of the health and safety argument against nuclear power. Equally important is the risk of terrorist attacks upon nuclear installations, or of the creation of crude nuclear weapons from the spent fuel. The measures society must adopt to protect against these risks are almost as serious in themselves as the possibilities of criminal or terrorist action. The 'nuclear state' serves to legitimise the extension of repressive police powers of surveillance and detention, particularly in relation to protection of nuclear plants once decommissioned, since these would be under armed guard for a hundred years.

Despite powerful arguments against nuclear power, the Government is clearly seeking to expand nuclear capacity. Apart from the obvious advantages of further weakening the miners as the vanguard of organised labour, capital has much to gain from the *construction* of nuclear plant, although there has been little interest in private ownership of plants once built. The *scale* of the nuclear programme depends upon how far capital is able to pursue its objects and the effectiveness of anti-nuclear campaigners in mobilising public opinion against nuclear energy.

Fothergill has estimated the nuclear capacity with and without alternative PWR programmes up to 2010, which forms the basis of Table 3.[91] In addition to nuclear power generation within the UK, there is also 'imported nuclear electricity' via the cross-channel 2GW cable link.[92] Electricity supplied from French nuclear power stations could displace between 2m and 4m tonnes of coal.[93] The low scenario assumes that no new nuclear plants are introduced beyond those already completed or under construction, and that no 'nuclear electricity' is imported. The medium scenario assumes Sizewell B and four other PWRs are commissioned in the 1990s in accordance with the minimum nuclear plans of the Electricity Council, with 2mtce of imported 'nuclear electricity'.[94] The high scenario represents an average of the 'medium nuclear background' and 'high nuclear background' options outlined by the CEGB at the Sizewell

Table 3 *CEGB nuclear options (mtce)*

Year	*1990*	*1995*	*2000*	*2005*	*2010*
Low	19	19	15	12	12
Medium	21	24	31	28	28
High	23	30	41	55	74

Inquiry, with 4mtce of imported 'nuclear electricity'. Such a programme would entail at least ten PWRs by 2000 and thirty-one in total by 2010 (compared with thirty-nine in total under the high nuclear background).

While nuclear power represents the major element of CEGB diversification strategy, power stations typically take a decade from planning to operation and may remain in service for fifty years. Changes in power station infrastructure therefore invariably reflect out-dated market considerations. In the short term, the fuel mix cannot be radically altered, but since 1979 the CEGB has been encouraged to establish multiple sources through importing coal.

CEGB oil burn

Oil is not generally considered a viable alternative, although there are occasions when oil prices and the US dollar exchange rate permit oil to be substituted for coal. A year after the 1984–85 strike ended, the London Business School predicted a further contraction of the coal industry with the collapse in world oil prices.[95] It has been estimated that, if the price of oil falls below $15–$16 a barrel, some 25m tonnes of coal could be displaced in the power-station market.[96] While the additional cost of increasing the oil burn during the 1984–85 strike suggests that oil is not a serious alternative to coal (see Chapter six), the CEGB will maintain oil-fired capacity as part of its diversification strategy and to take advantage of short-term oil price movements. It is unlikely that both the highest nuclear *and* oil options would be pursued, although forecasts which have put CEGB coal demand below 40m tonnes in 2000[97] must have assumed this in the context of low unstable growth. The size of oil-fired capacity which is developed into the next century will depend upon the rate of economic growth and the scale of the nuclear energy programme. From CEGB evidence to the Sizewell Inquiry, maximum, minimum and most likely scenarios have been projected and are summarised in Table 4. The high scenario assumes high economic growth and reduced nuclear capacity. The low scenario is associated with low economic growth and high nuclear capacity. The medium scenario assumes moderate growth and a minimal nuclear programme.

Table 4 *CEGB oil options (mtce)*

Year	*1990*	*1995*	*2000*	*2005*	*2010*
High	8	12	12	14	14
Medium	5	7	7	6	6
Low	5	4	4	6	6

Oil is an important part of CEGB diversification strategy, which must be understood in the context of the economic relationship between the CEGB and the NCB. This relationship is politically determined since both are, at present, within the public sector. The CEGB has been used as a lever on the NCB, giving the appearance that the 'market' is responsible for the coal crisis. As the major consumer, the CEGB enjoys a monopsonistic advantage which the NCB, as the major supplier, ought theoretically to be able to counter with its near-monopoly of coal production. Diversification of fuel supply reinforced the CEGB's monopsonistic advantage, and resulted in increasingly aggressive bargaining with the NCB over coal prices. In October 1979 in a 'Joint Understanding' the CEGB agreed to purchase 75m tonnes of coal, provided its price did not increase faster than the retail price index. The CEGB increased its use of imported coal from 2m tonnes in the year ended March 1980 to 3·75m tonnes in the year ended March 1981, but following the February 1981 coal crisis (see Chapter two) the CEGB reduced coal imports. In November 1982 the Joint Understanding was amended so that the CEGB would only purchase coal in excess of 70m tonnes at the 'world market price', effectively the spot price of coal. In November 1983 a further 5m tonnes tranche of coal was to be at the world price and the price of the basic 65m tonnes was to increase *less than* the rate of inflation over the previous year until the end of October 1987. After the strike the CEGB imposed further demands on British Coal (see Chapter seven).

While the CEGB exploits competition between alternative sources of coal and other fuels, the electricity supply industry enjoys a monopoly. Electricity experiences no competition from other fuels for lighting and the area boards of the Electricity Council cannot compete with one another since they operate in geographically discrete markets. Nevertheless, the privatisation lobby would doubtless claim that area boards could compete, as has been suggested in the case of water authorities! In contrast with the NCB's financial difficulties, CEGB profits grew from £190m in the year ended March 1981 to £635m in the year ended March 1984, while the Electricity Council increased its profits from £303m to £901m over the same period. Since all are in the public sector, the allocation of profit between electricity supply enterprises and the NCB is arbitrary. The total annual profits of electricity supply, increasing from £493m to £1,536m over 1981–84, could have been used to maintain indigenous reserves of coal. Effectively, the CEGB has been encouraged to extract coal profits behaving like part of the private sector even before electricity privatisation was officially on the agenda.

Imported coal

The key to the CEGB's bargaining power over the NCB is the availability of cheap coal on the international market. As with the damage inflicted to Britain's coal industry in the 1960s, multinational capital is at the centre of the coal crisis in the 1980s. While much of the internationalisation of capital is associated with US hegemony, Radice reminds us of the integration of capital from France, Germany and Britain: 'In Britain, in particular, this process has gone so far, together with a rapid centralization of capital, that the only group of large business enterprises that are plainly national are those in the public sector: *there are no longer any significant nationally-limited capitals*.'[98]

The World Energy Conference (1980) estimated recoverable reserves of coal in the world of 882,000m tonnes, approxiately six per cent of total known reserves.[99] The largest recoverable reserves of hard coal (anthracite and bituminous coal with a calorific rating greater than 23·76 GJ/tonne) are in the Soviet Union, the USA and China, each amounting to approximately 100m tonnes, while Britain has the fourth largest with 45,000m tonnes. Although Australia, South Africa, Poland and Colombia have only around 25,000m tonnes each, these reserves have become particularly significant. Wright and Rutledge have extensively analysed the role of the oil companies in the development of the international coal market.[100] Limited reserves of hard coal are being exploited by modern opencast operations, offering enormous price advantages over deep-mined coal. However, the huge export-orientated opencast operations, entail enormous capital expenditure by the Trans-National Corporations (TNCs), both on the mines and associated infrastructure: 'This infrastructural expenditure represents a huge fixed cost to the operator and requires very high production levels (typically from 3m tonnes to 15m tonnes per year) in order to maintain a profitable operation. These changes in the mode of production have contributed significantly to the present crisis in the World coal market'.[101]

Even without the slave labour conditions of mining under apartheid, the geological advantages of South African strip mines are enormous. Technological improvements have enabled strip mines to be worked profitably with a ratio of overburden to coal as high as 35:1.[102] The Grootegeluk mine in the Transvaal, however, enjoys an overburden to coal ratio of 0·7:1. Beneath a 19m overburden lies 92·7m of usable coal interspersed with three dirt bands totalling only 13·4m.[103]

Optimistic demand forecasts for energy and steam coal consumption led the TNCs to open up new reserves of coal, which they have to exploit to recoup overheads but which results in dumping of excess production and a fall in the market price of coal.[104] During the 1984–85 strike, Louis Turner advocated further increases in coal imports in view of the attractiveness of spot prices.[105] Almost a century ago, Jeans had exclaimed: 'it will be an evil case for England should the day ever arrive when coal can be mined upwards of 3,000 miles from our shores, and landed at our very doors in successful competition with our own'.[106]

CEGB evidence to the MMC showed that for the Thames power stations the delivered costs of overseas coal were lower than indigenous coal shipped from Durham pits on the collier fleet.[107] Since overland transportation costs add significantly to coal costs, Turner argued that future coal-fired power stations should be located on the coast near deep-water facilities capable of handling high volumes of coal. Such a policy would increase the transportation costs of indigenous coal and tie the CEGB to imported coal in the future irrespective of relative costs.

The continued availability of cheap coal on the international market is questionable on four counts. Firstly, most of the coal has been dumped on the international market because of overproduction; as capacity and surplus production is eliminated, prices will rise. Already Australian coal production has been cut back because it cannot be profitably sold at prevailing prices. Any major increase in demand from the CEGB for internationally-traded steam coal 'would drive prices up towards their breakeven levels'.[108] Secondly, reserves of high-quality (low sulphur, high calorific rating) opencast coal are limited and could be exhausted relatively soon. Thirdly, since coal is sold in US dollars, the cost of imported coal is very sensitive to the exchange rate. With an exchange rate of £1 = $1.50 and below, international coal prices are generally above the average price of British coal. Fourthly, as more coal production and shipping comes under the control of TNCs, the cartelisation of the coal market becomes increasingly likely. Despite these substantial grounds for ensuring independence from reliance upon coal imports, the TNCs could well succeed in increasing their share of the coal supplied to the CEGB, particularly as part of a strategy to privatise coal-mining and electricity generation within the UK. Robinson and Marshall, for example, stress the role of unrestricted coal imports in their proposals to 'liberalise' the coal industry, and suggest that imports could be between ten and twenty million tonnes per year in the next decade.[109] The American Coal Investment Company recently predicted UK coal imports of 20m tonnes in 1990 and between 30m and 40m tonnes by 2000.[110] Gerald Manners, adviser to the House of Commons Select Committee on Energy, assumed that the 'political position' of the NCB and NUM would prevent coal imports rising above 10m tonnes in 1990 and 15m tonnes in 2000.[111] These various estimates are translated into scenarios in Table 5. The low scenario assumes a total ban on all coal imports in order to promote energy self-sufficiency and improve the balance of payments. The medium scenario is seen as the most likely outcome, reflecting a hybrid of the Manners and Robinson/Marshall projections. The high scenario is the worst case, assuming the strategy of the TNCs succeeds.

Table 5 *Coal imports (mt)*

Year	*1990*	*1995*	*2000*	*2005*	*2010*
Low	0	0	0	0	0
Medium	10	15	20	25	30
High	20	25	30	35	40

Indigenous opencast production

It is evident that the range of possible demand for indigenous coal into the next century is wide. It remains to quantify the volume of coal likely to be supplied from UK opencast operations before drawing together the above scenarios to produce a range of demand for indigenous deep-mined coal. Opencast production steadily increased from around 10m tonnes in the mid-1970s to almost 15m tonnes just before the 1984–85 strike. Because opencast mining entails the greatest environmental destruction, the Flowers Report recommended opencast output should continue to be kept at under 15m tonnes per year by government and local authority action.[112] The NCB Opencast Executive is responsible for establishing production targets but private operators run the opencast operations and workers are members of the TGWU rather than the NUM. On major sites the private companies act as contractors but sites likely to yield less than 25,000 tons were licensed for private operation under the Opencast Coal Act 1958. In 1981 the limit was raised to 35,000 tonnes for individual sites and 50,000 for adjacent sites, while the MMC recommended a further increase to 100,000 tonnes.[113] Robinson and Marshall applaud the Government's decision of 27 February 1984 to allow the level of opencast production to be determined by the market subject to local planning approval:

> This move away from a centrally-determined target for opencast output to a more decentralized system seems to us entirely appropriate for a liberalized coal market. Relatively low-cost, high-quality opencast production should not be held down – as it evidently has been on occasions – in order to avoid cuts in deep-mined coal . . . there seems no reason why opencast coal production should not be fully privatised.[114]

The parent companies of the major opencast operators include Taylor Woodrow, Charter Consolidated, Derek Crouch and Trafalgar House, further evidence of the involvement of multinational capital in the UK coal industry. Capitalist opencast operators have developed a powerful parliamentary lobby, particularly through Marcus Fox MP, whose Shipley constituency includes no mining.[115]

Table 6 *Opencast production (mt)*

Year	*1990*	*1995*	*2000*	*2005*	*2010*
Low	10	10	5	5	5
Medium	15	15	15	15	15
High	20	30	35	40	45

It is difficult to quantify the expansion of opencast production which could arise if restrictions were removed, but some 100m tonnes of reserves were being worked at forty-three sites by major contractors in 1984.[116] The high scenario assumes that the output of these sites could be doubled within a decade and increased by as much again in the first decade of the next century through opening further sites. The medium scenario assumes that environmental considerations will prevent opencast production from exceeding the 15m tonnes level advocated in the Flowers Report.[117] The low scenario represents the outcome of a policy of reducing opencast operations to a minimum, restricting their use to situations where coal can be removed as part of a programme of land reclamation or where special coals can be obtained for blending purposes. These are combined in Table 6.

Indigenous deep-mined coal demand

From the above scenarios, ranges of indigenous deep-mined coal (IDMC) demand can be generated according to assumptions about the future. Taking all the scenarios most favourable to IDMC (high economic growth with no new nuclear power, no coal imports and minimal opencast and oil), would produce IDMC demand of 130m tonnes in 1990, rising to 169m tonnes in 2000 and 202m tonnes in 2010, which represents Scenario I in Table 7. Similarly, taking all the scenarios least favourable to IDMC (low economic growth, maximum nuclear, coal imports and opencast) would produce demand of only 53m tonnes in 1990, falling to 5m tonnes in 2000 and zero thereafter. However, Scenario III in Table 7 assumes that the high nuclear strategy with increased imports and opencast are unlikely to arise in a context of low economic growth. The central Scenario II assumes moderate economic growth, minimal nuclear, and the middle assumptions of oil, imports and opencast. Scenario I is seen as the highest IDMC demand conceivable under strong Labour governments committed to a socialist energy programme based on coal. The minimum scenario would be expected to be closest to what will occur if Conservative monetarist policies continue. The central IDMC scenario corresponds to the future for coal under the labour movement energy policy of 1984.[118] Since the strike, LP policy has moved to approximately mid-way between the high and medium IDMC demand scenarios (see Chapter seven).

Table 7 *Indigenous deep-mined coal demand (mt)*

Year	*1990*	*1995*	*2000*	*2005*	*2010*
Scenario I	130	137	169	184	202
Scenario II	89	86	78	79	78
Scenario III	70	57	45	19	27

1.4 Restructuring the coal-mining industry

On the supply side, the crisis of the coal industry arises from the introduction of new capacity. In a situation of falling demand, new capacity creates economic disequilibrium which has necessitated an acceleration of restructuring and the closure of collieries before exhaustion.[119] New capacity is being introduced which, under the 1974 *Plan for Coal*, was intended to replace capacity lost through exhaustion. Sir Derek Ezra explained the philosophy of *Plan for Coal*:

without a substantial and continuing investment programme, the output from our deep mines would decline to 85m tonnes by 1985 . . . The solution to this problem had three elements:

1. the expansion of capacity at existing longlife collieries with sufficient reserves to justify large scale investment
2. the extension of life of collieries that would otherwise exhaust by the provision of access to new areas of reserves
3. the construction of new mines

Our aim was to provide 42m tonnes of new capacity from our deep mines by 1985; *this new capacity would be used to replace exhausted reserves rather than be incremental, and was intended to enable us to secure an annual deepmined output of 120m tonnes by 1985.* The Plan also envisaged an increase of opencast production from the 1974 output of 10m tonnes to 15m tonnes. Our minimum target was, therefore, to achieve a total output of 135m tonnes by 1985 . . . *Coal for the Future* [1976], looked forward to the likely demand for coal in the year 2000 . . . producing a total of 170m tonnes of coal by the end of the century – 150m tonnes from deep mines and 20m tonnes from opencast. [Emphasis added.][120]

In addition to providing replacement capacity the investment programme under *Plan for Coal* aimed to increase efficiency, particularly through the application of systems engineering, which was expected to repeat the productivity gains made under mechanisation. The actual increase in output per manshift has been less than projected, although face output per manshift increased signficantly from 1977 following the introduction of area incentive schemes.

With falling demand for coal, the introduction of new capacity causes the elimination of corresponding capacity, not through exhaustion but because it becomes *surplus capacity*. The 'current grave over-capacity situation' was recognised by Ken Moses when writing of the Belvoir project.[121] Indeed, Manners opposed the Belvoir development largely on the grounds that it was unnecessary.[122] The restructuring has led to the closure of surplus capacity which the NCB describes as 'uneconomic pits'. The restructuring entails a massive loss of jobs deriving from three processes: (a) the introduction of new capacity from new developments and from the reorganisation of existing collieries; (b) the elimination of human activity through automation, and (c) the closure of pits representing surplus capacity. These processes are affecting every NCB Area to a differing degree so that each area has a unique restructuring profile. In order to assess the scale and location of the industry's contraction, each process must first be considered separately.

The most dramatic example of capacity being introduced through the construction of new mines is the Selby complex in the NCB North Yorkshire Area. Selby was planned to produce 12·5m tonnes per annum by the end of March 1988.[123] The N.E. Leicester Prospect, in the Vale of Belvoir, was originally to have entailed three new mines producing 7·2m tonnes of coal with 3,800 miners.[124] Local opposition forced the NCB to make more modest proposals for a 2·2m tonnes per annum mine at Asfordby with 1,100 miners,[125] and production is planned for 1994.[126] The South Warwickshire prospect will come on stream next, towards the end of the century when the Kenilworth colliery will be expected to produce 3·3m tonnes and employ 1,100 miners. The East Nottinghamshire Prospect is likely to be brought ahead of East Yorkshire in order to sustain levels of employment in the right-wing coalfield. Witham, between Newark and Lincoln, will probably produce 2·2m tonnes, perhaps by 2005, and is likely to require at most 1,000 miners. Snaith, in East Yorkshire, may not produce before 2010, with an estimated output of 2m tonnes and 1,000 miners. Similar developments at Banbury (Oxfordshire) and Ouse (North Yorkshire) are unlikely to be in production before 2010 and it is impossible to speculate on capacity or employment in these cases. Although new developments will only occur in some areas, they will have a profound effect upon the whole industry.

The reorganisation of existing collieries is expected to occur in all areas where sufficient long-life pits exist. The objective of reorganisation is to create colliery complexes by linking pits underground and centralising surface operations. The Longannet complex in Scotland was an early example; more recently the process has begun in North Derbyshire and Nottinghamshire. In the Barnsley Area, sixteen pits have been grouped into three complexes centred on automated coal preparation plants at Woolley, Grimethorpe and South Kirkby. The reorganisation enables an optimisation of resources so that output was to be increased by twenty-one per cent and manpower reduced by twenty per cent according to the Chief Mining Engineer,[127] although media reports based on NCB press releases indicated productivity levels implying job losses of the order of thirty per cent.[128]

The effect of MINOS on colliery manning levels can be assessed from the colliery models used by the NCB for planning purposes.[129] Table 8 shows that the planned manning levels for fully-integrated MINOS pits are fifty-three per cent of manning levels at conventional mines. Job losses could be even greater, depending upon the outcome of the Advanced Technology Mining programme which has the 'objective of a two-man face line in 1987'.[130]

As the restructuring proceeded without the increased demand for coal it was ostensibly undertaken to satisfy, pit closures on grounds other than exhaustion became inevitable. In February 1981 the NCB issued a list of twenty-three collieries it planned to close but withdrew this after the threat of a national strike.[131] Later that year the NCB undertook an 'intensive review' of about twenty pits and by October 1981 had implemented half of the closure programme withdrawn earlier.[132] In July 1982 the NCB admitted to a 'searching financial review' of some thirty pits, having by then closed or merged fifteen of the twenty-three pits on the original list through the local joint review procedure.[133] In November 1982 documents leaked from the NCB suggested that seventy-five 'short-life pits', employing 50,000 miners, could close in the next decade.[134] In June 1983, after the Monopolies and Mergers Commission Report defined 141 of the 198 pits then operating as unprofitable, the NCB announced plans to cut up to 70,000 jobs over the next five years.[135] On 30 September 1983 the NCB reply to the 1983 pay claim referred to the need to eliminate 'uneconomic pits'.[136]

Table 8 *NCB colliery model*

MINOS Package	*Manning levels per shift* Pre-MINOS	*MINOS*	*Jobs lost*	*Wage class*
Coal Clearance	20	4	16	U5
Environmental Monitoring	3	0	3	U3
Face Deployment	71	48	23	U1
Coal Preparation	6	6	3	U5
Fixed Plant	4	0	4	U4
Maintenance	10	5	5	CII
Total	114	60	54	

Source Burns *et al.*, *An Interim Assessment of MINOS*, Report No. 4, Working Environment Research Group, 1982, 4.9.

At the centre of the crisis which led to the 1984–85 strike was 25m tonnes of new capacity to be introduced by the end of March 1990: 12·5m tonnes from Selby and 12·5m tonnes from the reorganisation of existing collieries. This new capacity was to replace some 9m tonnes of exhausted capacity but would also result in the elimination of 16m tonnes of capacity from high-cost pits. The scale of job loss with this substitution can be appreciated from the operating statistics for Selby compared with those for South Wales and Scotland. Selby will produce 12·5m tonnes per annum with a workforce of 3,500, while the two areas together produced 13·5m tonnes with a workforce of 39,000. Similarly, as the automation programme increases output from MINOS pits, so more high-cost pits become classified as surplus capacity.

Under the restructuring, every area will suffer job losses, but pit closures will be most widespread in what the NCB regard as 'peripheral' areas. Like new mines, the application of MINOS is heavily concentrated in the central coalfield: while Scotland, Durham and South Wales comprise thirty per cent of NCB collieries, only ten per cent of MINOS applications are found there, and only four per cent of major projects investment reached these areas. In the last financial year NCB expenditure per capita on major capital projects in South Wales was under three per cent of the figure for Doncaster and under one per cent of the figure for North Yorkshire which includes the Selby development.[137] The geological problems of the peripheral areas have influenced NCB investment strategy to create the self-fulfilling prophecy of 'uneconomic pits'.

Glyn estimated that the elimination of high-cost capacity to make room for new capacity would result in the closure of seventy pits and the loss of about 70,000 jobs by 1988–90.[138] The continued application of MINOS with an associated increase in output could jeopardise more high-cost pits and will entail job losses through computer control in the central coalfield. The rationale behind the restructuring is the elimination of high-cost capacity ('uneconomic pits') and the concentration of production at low-cost collieries. If the NCB were able to expand production and reduce manpower at low-cost pits as well as closing high-cost pits, then it was estimated that the industry would contract to the central coalfield with 79,000 miners working ninety-nine pits (counting Selby as six mines) to produce 93·6m tonnes of coal per annum.[139] However, the restructuring is more complex so the collieries under threat cannot be ascertained simply from an examination of operating costs.

One complicating factor is that some low-cost collieries are nearing exhaustion or running into geological difficulties so cannot be considered long-life collieries. It is therefore necessary to look beyond operating costs and to identify collieries that are liable to close for such reasons, regardless of operating costs. Conversely, where a new face is being developed or a new drift being excavated, or when a temporary fault is encountered, a long-life colliery may pass through a high-cost phase. In such cases collieries receiving significant investment in new technology and other developments must be identified, because these ought, *ceteris paribus*, to be long-life collieries.

To estimate the numbers of pits and jobs which could be lost from each area, the geological conditions, technology installed, location and operating costs of each colliery must be considered. When such an assessment was undertaken for the Coalfield Communities Campaign, it was concluded that from a 1984 pre-strike situation of 181 collieries producing 96·7m tonnes per year, the industry would decline by 1990 to 101 collieries producing 93·4m tonnes.[140] Manpower was predicted to fall from 184,000 to 123,000 through closures and to as low as 76,000 with the further introduction of new technology. Between March 1981 and the onset of the strike 41,000 jobs disappeared from the industry. Since the strike the industry has contracted further; employment fell by another 22,000 between March 1984 and July 1985. These losses occurred before it was announced that planned production was to be reduced further to 90m tonnes per year.[141] Actual production is normally about ninety-six per cent of budget so 90m tonnes represents a likely annual production of 86·4m tonnes saleable. Taking these developments into account, the industry was expected to be further reduced to ninety-one pits producing 86·4m tonnes with between 61,500 and 109,000 miners.[142] Ian Lloyd, then Chairman of the Energy Committee, confirmed the mid-point of these estimates. When asked whether the number of miners would be reduced to 80,000 within five years, Lloyd replied 'yes, somewhere of that order'.[143] Table 9 summarises the estimated reduction in collieries and manpower between 1984 and 1990, by NCB area.[144]

Table 9 *Collieries, output, investment and employment projections by NCB Area, 1984–90*

	End March 1984				End March 1990					
Area	*Pits*	*Output (m tonnes)*	*Employment*	*Investment (£m)*	*Pits*	*Output (m tonnes)*	*Employment Max.*	*Employment Min.*	*Job losses Min.*	*Job losses Max.*
Scotland	10	5·30	13831	70	3	3·68	6431	3408	7400	10423
N.East	16	12·33	23482	27	6	6·71	12680	4588	10802	18894
N.Yorks.	32	15·99	25903	1510	18	24·55	15830	11262	10073	14641
S.Yorks.	25	12·92	28105	676	16	11·42	19441	10869	8664	17236
N.Derbys.	11	7·19	10961	30	6	5·44	7032	5095	3929	5866
N.Notts.	14	11·58	16909	84	10	11·67	12290	6918	4619	9991
S.Notts.	11	7·43	13732	11	5	4·62	7516	4304	6216	9428
S.Mids.	16	6·78	13647	24	8	5·91	7367	5256	6280	8391
Western	17	10·32	18776	153	10	9·08	13821	6537	4955	12239
S.Wales	28	6·72	18758	20	9	3·32	6259	3294	12499	15464
Total	180	96·7	184104	2605	91	86·4	108667	61531	75437	122573

Source J. Winterton and R. Winterton, *The Implications of NCB Restructuring Policies for West Yorkshire*, Report to West Yorkshire Metropolitan County Council, November 1985.

1.5 The scale of the crisis

By combining the above analysis of restructuring with the demand forecasts developed in the first section, it is possible to estimate the size of coal industry likely to exist under the three political scenarios. Scenario I in Table 7, representing the possibilities of a socialist energy strategy, confirms the folly of closing existing capacity before exhaustion. However, the destruction of the coal-mining industry is already so far advanced that IDMC demand projections under Scenario I could not be met from existing capacity and planned developments, even with an immediate moratorium on all pit closures on grounds other than exhaustion. Under these conditions by 1990 the industry could comprise 113 pits with a total capacity of 100m tonnes, employing about 86,000 miners. Coal demand of 130m tonnes could not be met from IDMC, given the colliery closures that have already taken place. Although new capacity will be introduced up to 2010 this represents less than 10m tonnes and will therefore be inadequate to raise IDMC production to anything approaching the 200m tonnes projection even if no collieries were to exhaust. In fact, the new capacity will not even compensate for exhausting capacity so 100m tonnes IDMC has to be regarded as a ceiling. The possibility of introducing a socialist coal-based energy policy has effectively already been sabotaged.

On the basis of 1984 LP/TUC energy policy, a Labour election victory would not, in any case, have resulted in the adoption of a socialist energy strategy. Even so, IDMC demand Scenario II, based upon LP/TUC policy of 1984, would have been significantly more attractive than Scenario III based on a continuation of Conservative policies. The nature of the restructuring of the coal-mining industry under Scenarios II and III can be compared at quinquennial intervals. In the following analysis, minimum employment levels, assuming the continued application of new technology, are cited since employment levels had fallen below the maxima cited in Table 9 within two years of the end of the strike.

Even by 1990 the different effects of Labour and Conservative policies on the coal industry become apparent. Under Scenario II IDMC demand of 89m tonnes implies a loss of 11m tonnes of existing capacity by 1990; at least 4m tonnes is expected to have effectively exhausted by then. Scenario III, however, with IDMC demand down to 70m tonnes, entails a loss of 30m tonnes, 26m tonnes of which will represent viable collieries and entail the sterilisation of reserves. The effects of restructuring under the two scenarios are shown in Table 10. In national terms the difference between them is one of twenty-two collieries and fewer than 20,000 miners, roughly the annual contraction of recent years. However, these twenty-two collieries represent twenty-two per cent of the capacity under Scenario II, over four times the proportion which a similar number of pits represented before the strike. Scenario II represents a contraction of 62 per cent of pre-strike employment, while under Scenario III seventy-three per cent of (1984) mining jobs will have disappeared by 1990.

The differences between the Labour and Conservative scenarios are most noticeable in the North-East, South Yorkshire and South Nottinghamshire. In the North-East eighty-eight per cent of pre-strike jobs would be lost under Scenario III (compared with sixty-four per cent under II), while in South Yorkshire the comparable figure is seventy per cent under III (compared with sixty-six per cent under II). Under either scenario, seventy-five per cent of Scottish mining jobs will be lost by 1990, while South Wales will contract by between eight-two per cent (II) and eighty-nine per cent (III).

Table 10 *Colleries, output, and employment projections by BC Area, 1990*

	Scenario II (Labour)			*Scenario III (Conservative)*		
Area	*Pits*	*Output*	*Employment*	*Pits*	*Output*	*Employment*
Scotland	3	3·84	3511	3	3·84	3511
N.East	7	9·08	8357	3	3·30	2764
N.Yorks.	16	23·75	12186	13	20·24	9456
S.Yorks.	18	13·10	12901	13	10·14	8624
N.Derbys.	6	5·56	5197	5	3·96	4021
N.Notts.	10	11·37	7798	9	10·55	6856
S.Notts.	5	4·84	4602	3	3·23	2971
S.Mids.	8	5·36	4632	7	5·20	4448
Western	9	8·56	6788	7	7·19	5447
S.Wales	9	3·89	3414	6	2·35	2021
Total	91	89·35	69386	69	70·00	50119

Note Discrepancies between Scenario II above and projections in Table 9 reflect changes in the circumstances of individual collieries in the interval between the two analyses.

The divergence between Scenarios I and III by 1995 results in a sharp contrast in the impact of restructuring. Under 1984 LP policies, IDMC demand would fall slightly to 86m tonnes and the introduction of new capacity at Asfordby would necessitate a further capacity reduction between 1990 and 1995 of 5·5m tonnes; capacity losses due to exhaustion may, in fact, prove higher than this. Under Conservative policies, however, IDMC demand is expected to fall to 57m tonnes, requiring a capacity reduction of 15m tonnes; as much as 10m tonnes of capacity could be closed at viable collieries with proven reserves, entailing further sterilisation. The effects of these capacity reductions are shown in Table 11. Scenario III entails the closure of thirty-four collieries more than Scenario II, and the loss of 27,000 more jobs. By 1995 under Conservative policies there will be fifty-eight per cent of the mining jobs that would have remained under Labour policies and no coal industry left in Scotland. In South Yorkshire, South Nottinghamshire and South Wales, coal-mining employment under the Conservatives is estimated to be half of the level expected under

Labour, while the North-East would be left with a single pit and one-quarter of the employment expected under Labour.

Table 11 *Collieries, output and employment projections by BC Area, 1995*

	Scenario II (Labour)			*Scenario III (Conservative)*		
Area	*Pits*	*Output*	*Employment*	*Pits*	*Output*	*Employment*
Scotland	3	3·84	3511	0	0	0
N.East	4	5·70	5543	1	1·69	1525
N.Yorks.	16	23·75	12186	10	18·14	7596
S.Yorks.	17	12·26	11198	9	7·32	5659
N.Derbys.	6	5·56	5197	5	3·96	4021
N.Notts.	10	11·37	7798	9	10·55	6856
S.Notts.	5	4·84	4602	2	2·36	1880
S.Mids.	9	7·56	5732	6	5·82	4104
Western	8	8·09	6281	6	6·01	4463
S.Wales	7	3·02	2404	3	1·35	1108
Total	85	85·99	64452	51	57·20	37212

By the turn of the century IDMC demand under Scenario II would fall to about 78m tonnes and, with the introduction of Kenilworth, capacity would need to be reduced by a further 11m tonnes. It is impossible to calculate what capacity is likely to be lost through exhaustion during this period, although 5m tonnes per year would not be regarded as an excessive estimate. Under Scenario III, however, with IDMC demand down to 45m tonnes by 2000 a capacity reduction of 15·5m tonnes is more likely to involve the closure of viable collieries and the sterilisation of reserves. Total employment under Scenario III is only forty-five per cent of that under Scenario II. Table 12 shows the effects upon BC areas of restructuring under the two sets of assumptions. Under 1984 Labour policies the North-East suffers the greatest reduction, followed by South Yorkshire, but the losses in these areas are largely accounted for by exhaustion and the remainder of the contraction is fairly even across other areas. Conservative policies, however, have a devastatingly uneven effect upon the areas. Coal-mining is eliminated from both the North-East and South Wales, like Scotland, so that by the turn of the century the peripheral areas and their reserves of coal are abandoned. Employment in South Yorkshire, South Nottinghamshire and Western areas will be about one-third of what would have remained under 1984 Labour policies, while North Derbyshire's one remaining colliery will employ six per cent of the miners who would have remained under Labour.

Table 12 *Collieries, output and employment projections by BC Area, 2000*

Area	Scenario II (Labour) Pits	Output	Employment	Scenario III (Conservative) Pits	Output	Employment
Scotland	3	3·84	3511	0	0	0
N.East	1	1·69	1525	0	0	0
N.Yorks.	14	21·36	10364	10	18·14	7596
S.Yorks.	14	10·88	9408	6	4·99	3644
N.Derbys.	6	5·56	5197	1	0·68	336
N.Notts.	9	10·55	6856	7	8·96	5462
S.Notts.	4	3·84	3518	1	1·42	1075
S.Mids.	9	10·70	6648	6	8·62	4817
Western	7	7·19	5447	2	2·33	1628
S.Wales	6	2·34	2021	0	0	0
Total	73	77·95	54495	33	45·14	24558

The reliability of any energy scenario projected into the next century is questionable, but, on the basis of the market assumptions outlined in the first section, the effects of Labour and Conservative policies on the coal industry would continue to diverge. Under Scenario II IDMC demand remains at about 79m tonnes to 2005 when Witham will be on stream. Only one million tonnes capacity would need to be shed by 2005 and it is probable that exhaustion would eliminate 5m tonnes, leading to a deficit supply of IDMC. Under Scenario III, however, IDMC demand of 19m tonnes necessitates the closure of all capacity except Selby, Asfordby, Kenilworth and Witham, with extensive sterilisation of reserves. By 2005 Conservative policies would result in twelve per cent of the mining workforce which would remain under Labour policies. Table 13 contrasts the restructuring anticipated under the two scenarios. Whereas the industry alters little in five years (assuming minimal exhaustion) under Scenario II, coal-mining is confined to four sites in three BC areas under Scenario III.

While it is necessary to reiterate the low degree of confidence which can be attached to either scenario by 2010, *if* current policies were continued by each party the respective IDMC demand would be 78m tonnes under Labour and 27m tonnes under the Conservatives. New capacity will by then be available at Snaith and under Scenario II a modest 3m tonnes of capacity (again, possibly *less than* natural exhaustion) would be shed. Table 14 shows that, under Labour policies, coal-mining would continue in every BC area until 2010. In contrast, under Scenario III coal-mining is confined to three BC areas, and the total workforce represents fifteen per cent of that which would remain under Labour policies. Moreover, by 2010 the folly of pit closures on grounds other than exhaustion becomes apparent even with all the conservative assumptions and

Table 13 *Collieries, output and employment projections by BC Area, 2005*

Area	*Scenario II (Labour)* *Pits*	*Output*	*Employment*	*Scenario III (Conservative)* *Pits*	*Output*	*Employment*
Scotland	3	3·84	3511	0	0	0
N.East	1	1·69	1525	0	0	0
N.Yorks.	13	20·24	9456	5	12·5	3500
S.Yorks.	14	10·88	9408	0	0	0
N.Derbys.	6	5·56	5197	0	0	0
N.Notts.	10	12·75	7856	1	2·2	1000
S.Notts.	4	3·84	3518	0	0	0
S.Mids.	9	10·71	6648	2	5·5	2200
Western	7	7·18	5447	0	0	0
S.Wales	6	2·34	2021	0	0	0
Total	73	79·03	54587	8	20·2	6700

Table 14 *Collieries, output and employment projections by BC Area, 2010*

Area	*Scenario II (Labour)* *Pits*	*Output*	*Employment*	*Scenario III (Conservative)* *Pits*	*Output*	*Employment*
Scotland	3	3·84	3511	0	0	0
N.East	1	1·69	1525	0	0	0
N.Yorks.	14	22·24	10456	6	14·5	4500
S.Yorks.	13	10·13	8624	0	0	0
N.Derbys.	5	3·96	4021	0	0	0
N.Notts.	10	12·75	7856	1	2·2	1000
S.Notts.	3	3·23	2971	0	0	0
S.Mids.	9	10·71	6648	2	5·5	2200
Western	7	7·18	5447	0	0	0
S.Wales	6	2·34	2021	0	0	0
Total	71	78·07	53080	9	22·2	7700

policies *least favourable* to IDMC: making no allowances for natural exhaustion, demand exceeds supply by almost 5m tonnes, part of the *56m tonnes of viable capacity which will by then have been closed in pursuit of short-term objectives.* This situation is reminiscent of the coal industry during the 1970s oil crises, the history of which has been re-written by those advocating a further contraction of coal. Robinson and Marshall claim that British coal did not revive after the oil shocks because of 'the inherent intertia of energy markets . . . the monopolistic position of the NCB . . . [and] disruptions to coal supplies in 1971–72 and 1973–74'.[145] They neglect the more straightforward explanation, which justified the 1974 *Plan for Coal*, that coal production could only be slowly increased in the wake of the enormous capacity reductions of the 1960s, since this might suggest the need to restrain the restructuring underway in the 1980s.

It is difficult not to conclude that Conservative policies have already been ruinous for the coal industry and that their continuation into the future would be disastrous for Britain's energy needs. On the basis of current information and assuming no radical change of energy policy by either party, the prognosis is very different under the Tories from under Labour. Perhaps the most important conclusion to be drawn from the above analysis is that the effective destruction of the British coal industry will be largely completed in the next decade if Conservative policies continue.

Privatisation

The question immediately arises as to *why* the Conservatives wish to destroy the coal-mining industry. Sweet commented during the strike on Mrs Thatcher's uncharacteristic reticence to publicise Tory objectives and suggested: 'Her objectives are first to undermine the coal industry, and to do that she needs to promote the nuclear alternative; then to reorganise it, and that includes breaking up the NCB . . . Privatization may well be the name of the game'.[146] Coal-mining has long been on the Conservatives' privatisation agenda although it attracted little attention because the NCB was an apparent loss-maker and more lucrative enterprises like British Telecom and British Gas took precedence. In 1981, when John Moore, now leading the Tories' privatisation programme, was at the Department of Industry, he described opencast mining as 'more akin to quarrying, a privately-owned industry, than to deep mining'.[147] In December 1983, when Financial Secretary to the Treasury, Moore's Conservative Party pamphlet *Why Privatise?* declared the intention of looking at 'further ways of increasing competition and introducing privatisation into the energy sector'.[148] By then the NUM was already alerted to the risk as the Union's *Campaign for Coal* Briefing Booklet No. 2 spelt out:

> There is already disturbing talk of the peripheral coal fields being allowed to go to the wall with private capital invited for stakes in the lucrative central coal field. It is not at all inconceivable that Selby, paid for and developed at tremendous cost to the taxpayer, could be sold off at a knock-down price to private buyers. Precedents have already been set for the sale of valuable natural resources, developed by the taxpayer, to private speculators.[149]

During the early part of the 1984 strike the Government was careful not to reawaken fears of a return to private ownership. The press seized on a comment which Norman Tebbit made at a Commons luncheon in April that there would have been no strike if the coal industry had been denationalised: 'it is a thought, is it not, for the future?' he asked.[150] Later the remark was explained away as mischievous so as not to alienate the scab areas into supporting the strike. Within a month, however, by which time it was clear that production would continue in Nottinghamshire, a Cabinet sub-committee was reported to have 'agreed that the NCB should be allowed to seek private capital to develop a new generation of highly-profitable pits'.[151] The Cabinet sub-committee was reluctant to countenance outright denationalisation, but suggested agreements involving companies like Taylor Woodrow investing in coal for use in privatised power stations. The Tories' reticence to discuss privatisation during the strike was also explained: 'it must also be a prerequisite that the political power of Mr Arthur Scargill and the National Union of Mineworkers should be broken. It is unlikely that private investment could be attracted if Mr Scargill was able to stop the current programme of pit closures. If he won that battle he could prevent privatisation.'[152] When it became apparent that the strike was going the Government's way, the right wing launched its coal privatisation offensive. In September 1984 the Adam Smith Institute's report on *Energy Policy* under the Omega Project recommended that the NCB should cease to be the licensing authority for independent opencast operators and independent deep mines, arguing that both of these should be allowed to expand.[153] The report also recommended that the NCB should be divided into at least twelve independent units which would become autonomous corporations. The MMC Report had similarly recommended that NCB areas 'be operated, as far as possible, as separate business units'.[154] MacGregor had already declared his intention to decentralise the industry and claimed that the ability 'to tap private capital' was an indication that 'the industry has a future';[155] on 5 December 1984 he enthused from the pulpit of St Lawrence Jewry that it would be 'wonderful' to sell 'uneconomic' pits to miners.[156]

Once the strike was over, the chorus of support for coal denationalisation was more vocal. As the strikers returned, The *Times* editorial proclaimed: 'the year-long dispute has reinforced the lesson of history that a state monopoly is a wholly unsuitable structure for the British coal industry'.[157] The *Times* suggested that if a Trust were established to let out extraction rights, opencast operations could be privatised immediately, and the more profitable areas privatised. Peter Rost, writing in *Economic Affairs*, added parliamentary approval claiming that 'the most significant outcome' of the strike was: 'a belated wider consensus that nationalisation had proved itself a disastrous failure. The industry should, and could, now be privatized. What has been privately acknowledged throughout the NCB and in the Government should now be openly debated.'[158] Academic credibility for the right's arguments was supplied by Robinson and Marshall's Hobart Paper (also published by the Institute for Economic Affairs). The importance of decentralisation was made clear in the preface to this paper: 'Decentralisation is not the least important if the Government is to build on Mr Scargill's work in weakening the monolithic power of a politicised NUM, which the authors acknowledge remains the major concern for potential private investors'.[159] Robinson and Marshall argued the case for 'liberalising' the coal industry: privatising small, competitive units rather than selling off the NCB as an enterprise. Restrictions on imports, opencast and licensed mines could immediately be lifted, said the Hobart Paper, since these were more profitable and would promote deep-mine efficiency if allowed to compete. As a first step towards privatising deep-mined coal production, they recommended joint ventures involving private capital in the development of super-pits and looked forward to the sale of packages of pits and power stations: 'the offer of pits *and* nearby power stations in the central coalfield areas should be very attractive to private investors'.[160] It was even claimed that the high-cost pits could have a better future 'if they were run by private companies prepared to invest in technologies appropriate to such mines'.[161]

The *Financial Times* had put forward similar arguments during the strike, noting that under private ownership workers might have a choice between closure and keeping their colliery open by forfeiting pay increases. Those familiar with the misery caused by the coal owners between the wars will recognise other ways in which marginal collieries could remain competitive at the workers' expense. However, to extract the maximum profit under private ownership, high-cost collieries would certainly be shed: 'many parts of the coal industry are profitable . . . even though the NCB with its trail of high-cost pits is running at a loss'.[162] It is equally significant that many units in *isolation* are less profitable and that super-profits would acrue to the low-cost collieries once freed from the burden of cross-subsidy. This need to be free from maintaining costly infrastructure explains why the Conservatives plan to destroy so much of the coal industry.

While the significance of privatisation should not be underrated, this last step is not inconsistent with the ways in which the nationalised coal industry has been subordinated to capitalist interests in the past. Over-compensation of the coal owners enabled them to continue to bleed the industry as mining machinery manufacturers, while the 1960s contraction facilitated the TNCs' domination of energy markets and held wages down as redundant miners entered the labour market. Privatisation, enabling capital to extract the profit from massive state investment in new coal-mining technology, is an interim phase in the destruction of the coal industry which will allow further profits to be made from importing coal and building nuclear power plants. Defeating the NUM was widely viewed as a prerequisite for the Conservatives to carry out their policy for the coal industry, but this can also be seen as an objective of that policy serving wider capitalist interests. Whatever damage is inflicted on the coal industry will at least weaken the NUM in proportion and this has profound implications for the whole labour movement. The NUM's image at the outset of the 1984 strike was still one of the vanguard of the labour movement, even if economic reality was somewhat different, so that psychologically, if not strategically, the entire movement is weakened along with the NUM.

Chapter two

Strike mobilisation

The coal crisis has been traced back over the decade before the 1984 strike (see Chapter One) but it is less clear when the *conflict* began. March 1984 marked an escalation in the conflict rather than its origin; a strike had been inevitable since the first Thatcher Government took office. The failed attempts to launch a major confrontation earlier illustrate the problems of mobilisation in general and help explain much of the character of the 1984 strike. This chapter examines those false starts, the mobilisation of the 1984 strike in Yorkshire, and the problems encountered in spreading the strike to other areas.

2.1 False starts

January 1980, South Wales

After the 1979 election, Conservative policies had their most immediate impact upon the collieries of South Wales, because British Steel had been encouraged to increase imports of coking coal and had simultaneously reduced capacity. Although the South Wales Area NUM had vacillated over resisting the closure of Deep Duffryn, by the end of 1979 two-thirds of South Wales pits were at risk. Newport dockers agreed not to handle imported coal which was, according to the United Mineworkers of America, being produced by non-union labour, and a Welsh steel and coal strike was planned for 21 January 1980. Events in South Wales were closely watched in Yorkshire where earlier that month seventy three per cent had voted to reject the NCB pay offer, compared with only forty nine per cent nationally.[1] The NEC of the NUM would not support the Welsh strike and the TUC succeeded in diluting it into a Welsh day of action on 28 January. In any case, these preparations were overtaken by a national steel strike which started on 2 January 1980 in support of a pay claim. Although a South Wales delegate conference on 20 February called for an area strike from 28 February, the South Wales membership overwhelmingly rejected the call. This result has been attributed to complacency on the part of the leadership, who did little campaigning in the pits.[2]

Gormley's right-dominated NEC was able to isolate the conflict in 1980 because only South Wales appeared to be at risk. The NCB did not immediately embark on a closure programme and denied the existence of a hit list, making it more difficult to raise membership consciousness. However, the implications of the Coal Industry Act 1980 (see Chapter one) demonstrated the need for a membership campaign. This campaign was launched on 10 November with the publication of *The Miners and the Battle for Britain*. The sixty-page booklet outlined the problems facing the coal industry owing to imports and inadequate financial support, at the same time attacking the Government's monetarist policies and anti-union legislation.[3] Although 100,000 copies of the pamphlet were circulated,[4] its presentation was unimaginative and its effectiveness questionable. The pamphlet argued that the solution to the industry's problems lay in the return of a Labour government committed to the expansionist policies of the Alternative Economic Strategy (AES), and failed to analyse the roots of the crisis. Its content was further evidence of the complacency of the leadership, while its presentation showed a lack of professionalism. The NEC's success in avoiding conflict was reaffirmed in the November 1980 ballot when nationally fifty six per cent endorsed their recommendation to accept the NCB wage offer.[5]

January 1981, Yorkshire

The campaign against pit closures was carried on in Yorkshire with more enthusiasm than at national level. The left had come to power in Yorkshire largely because of the Union's failure to confront closures in the 1960s, and it was necessary for the new area leadership to demonstrate more effective resistance.[6] Moreover, judging from Scargill's speeches at the 1979 and 1980 area demonstrations, the left had already identified the need for industrial action to prevent pit closures as the central theme of the campaign for the national presidency. In anticipation of NCB proposals to reduce capacity by closing Orgreave and other Yorkshire collieries, an area ballot was held towards the end of January 1981, which authorised, with a majority of 85·6%, industrial action 'to stop the closure of any pit unless on the grounds of exhaustion'. The ballot result demonstrated the success of the Yorkshire campaign and, in particular, showed that voting on the *principle* of a strike against pit closures offered a number of advantages. Most significantly, because no collieries were named miners could be expected to vote as if it were their own colliery at risk and not take a sectional view. While the preamble noted that *Yorkshire* collieries were at risk and many miners believed that they were voting to resist area closures, the ballot referred to *any* pit, and so could be used to sanction solidarity action with another area.[7] There was also less likelihood of interference by the media or NCB in a ballot about possible action at some indefinite future date. Moreover, it was easier to win support for a hypothetical strike than to obtain a definite commitment to forfeit earnings, although it was not long before this resolve was tested.

February 1981, national

At the regular CINCC meeting in February 1981, the NCB outlined the industry's overcapacity problems and announced the intention to reduce output by closing anything up to fifty pits (although twenty three was the eventual number). Specific collieries were to be named at later area review meetings. The Board's approach fragmented the *timing* of the response, but brought an unusual degree of unanimity for action from the NEC because all areas could have been at risk. The NEC meeting on 12 February agreed to hold a ballot on a strike to prevent the closures, but first to arrange a meeting with the Government. However, the meeting between the NUM, NCB and Ministers concerned was arranged for 23 February, after the ballot was to go ahead. As the NCB held its area review meetings, spontaneous, unco-ordinated strikes swept through the coalfields. On Saturday 14 February a mass meeting at the threatened Coegnant colliery voted for an immediate strike; pickets were deployed throughout South Wales and on Monday 16 February half of the coalfield was at a standstill. The Kent Area joined the unofficial action on the Tuesday over the proposed closure of Snowdown. By Wednesday 18 February, Scottish miners were on strike to resist three closures in their area. Three closures were also announced in Yorkshire, including Park Hill, whose workforce immediately struck. George Hayes, South Yorkshire NCB Area Director, issued a statement refuting 'persistent rumours' that Brookhouse 'features on some closure list'.[8]

The right of the NEC were obviously being pushed by the rank-and-file into leading a strike that they would rather have avoided. Gormley and Daly went to the House of Commons on 17 February to discuss the situation with Michael Foot and Len Murray, during which time Gormley telephoned Lord Ezra to propose a tripartite meeting with the Government in order to 'calm things down a bit'.[9] At the meeting, held the next day, David Howell, Secretary of State for Energy, accepted the need to reduce imports and relax financial constraints on the NCB, while Ezra agreed to withdraw the list of twenty three collieries to close. Gormley had also visited James Prior's flat to discuss 'reactions of various union leaders to the proposed Trade Union Bill' and there was speculation that the Department of Employment had stitched up an agreement to end the strike without consulting the Department of Energy.[10] It has also been suggested that the closure plans were intentionally leaked by the NCB to subvert Government policy.[11]

Wide media coverage was given to the Government's 'climb-down' and, as the NEC met on Thursday 19 February, miners were returning in some areas, although the strike had spread to almost a quarter of Yorkshire's collieries. The NEC voted 15–8 for an immediate return-to-work but left-wingers agreed to continue the strike because the Government had merely given promises. Gormley's action had sucessfully undermined the strike movement, and the left was unable to exploit the best opportunity that was to arise for resisting the destruction of the industry.

Despite a commitment within the Forum to maintain the action, the South Wales Area delegate conference decided after the NEC meeting to call off the strike to preserve area unity. There was no co-ordination with Scotland and Yorkshire where delegates met on the Saturday to extend the action and then agreed to stop the strikes once the South Wales decision was known. Left unity was damaged and the representatives from South Wales played a less significant role in the national Forum thereafter. The experience of 1981 profoundly affected the left's 1984 mobilisation strategy, yet the essential lesson went unheeded. An unofficial strike could not be sustained once solidarity was weakened without threatening the Union itself.[12]

Naturally the Government's 'U-turn' proved to be nothing of the sort; the twenty three pit closures were merely taken through the local Colliery Review Procedure. In Yorkshire the strike had been called off after the NCB gave assurances about Orgreave colliery but, at the general colliery review meeting on 23 April 1981, Hayes denied that the proposed closure of Orgreave was related to the February 1981 closure programme, while Scargill referred to a recorded telephone message in which James Cowan 'said quite clearly that Orgreave was no different to the rest of the pits on that list'.[13] Orgreave closed soon after with branch agreement, as did Brookhouse after the 1984 strike. By October 1981 half of the national closure programme was implemented and by July 1982, when a further thirty pits were being investigated, fifteen of the twenty three pits on the original list had been closed through the local Review Procedure. In most cases the NCB was able to obtain branch agreement after softening the workforce with redundancy terms, although isolated cases raised hopes of resuming the February 1981 action.

The 1981 pay claim

The next opportunity for strike mobilisation arose out of the 1981 pay claim. A Special Delegate Conference on 18 December 1981 agreed to put the NCB wage offer to the membership in a ballot with a recommendation for strike action. The day before the ballot Gormley placed a full-page newspaper advertisement urging acceptance of the NCB offer.[14] Nationally, only forty five per cent voted for strike action, while in Yorkshire the figure was 66·5%.[15] Wage militancy was clearly no substitute for raising consciousness over pit closures.

October 1982 linked ballot

With Scargill's election to the national presidency, the campaign against pit closures was taken up by the National Union. Of the 1982 pit closures, two, Kinneil in Scotland and Snowdown in Kent, were contested but resistance remained local. NUM officials estimated that forty six pits were candidates for full or partial closure and when they met the NCB over Snowdown on 31 August, Deputy Chairman, James Cowan referred to 'technological unemployment' and the need to abandon inefficient pits. Hoping to combine the support for action over closures in the militant areas with the wage militancy assumed to exist in the moderate areas, the NEC agreed to link the issues of pay and pit closures in a ballot which a Special Delegate Conference on

4 October decided to hold on 28–9 October 1982.[16] An overtime ban was imposed from 11 October 1982.

NUM officials spoke at several rallies in the coalfields, and the NCB reputedly feared that the ballot would support action. Before the ballot result was officially announced, press reports predicted a sizeable majority against industrial action and interpreted this as licence to accelerate the closure of up to forty pits employing over 2,000 miners.[17] The NEC's ballot tactic backfired as nationally only thirty nine per cent voted in favour of action and in Yorkshire the strike majority had fallen to fifty six per cent. McGahey identified the inherent weakness of the national campaign: 'The lesson of the miners' campaign is to recognize that the Union must be a campaigning, crusading union, not only in periods of wage negotiations or fighting in the conditions decided by the Coal Board. The Union must campaign every day in educating our members about the situation at the base level, namely at the pits.'[18]

Taylor commented on the 'confusion' throughout the industry and suggested that the varying status of collieries gave miners different, but equally compelling, reasons for not striking:

> Four broad groups of pits can be identified: first pits with a long life but which are facing temporary problems (e.g. geological). Strike action in this instance might be seen as jeopardising long-term investment. Second, those pits with a short life but which were doing well. Here strike action might contribute to speedy closure. Third, pits with a long life facing no immediate problems. Again, there was little reason to strike as the future seemed secure; equally, it could be argued that this group of pits could strike with little risk. Finally, pits with a short life doing badly and facing severe problems; here strike action might result in rapid closure.[19]

Within Yorkshire, the deterioration in support for action between the 1981 and 1982 ballots was generally greatest at collieries on the periphery of the coalfield. There was no clear link between voting and real or perceived security.[20] The cohesiveness of the mining community appears to offer a better explanation of voting behaviour than the status of the colliery.

On 11 November 1982 the NUM severed all national consultative procedures with the NCB and threatened to withdraw from area and local consultations unless provided with full details of planned closures at talks scheduled for 23 November. At the meeting, the NCB told the NUM of its intention to close sixty pits over the next eight or nine years owing to exhaustion, but indicated that overproduction might necessitate further closures in the short term. The case for Kinneil was widely used in the campaign before the ballot and the LP Conference in Blackpool passed an emergency motion of support. After the ballot, the overtime ban was called off and the NCB formally announced Kinneil's closure, prompting an immediate strike at the colliery. Pickets were deployed throughout Scotland, where the vote in the October ballot had shown sixty nine per cent in favour of striking, but there were angry exchanges at some collieries. The Scottish Area Executive recommended accepting the closure rather than risk divisions within the area or at best isolation from the rest of the NUM. Kinneil miners staged an unofficial stay-down strike over Christmas and McGahey lost some support among activists. On 9 December the NEC decided to withdraw from consultation with the NCB at all levels unless provided with more information about proposed pit closures by the end of the month.

January 1983, South Wales

As the capitulation over Kinneil was taking place, another struggle was developing in South Wales. The NCB announced at an area review meeting that development work would cease at Blaengwrach, Britannia and Tymawr-Lewis Merthyr. A South Wales Area delegate conference on 8 December 1982 voted for strike action from 17 January unless the NCB would provide the necessary investment, and the decision was ratified by pit meetings. In the week before the strike was due to begin, Emlyn Williams spoke of the area's determination to use industrial action 'to safeguard what remains of South Wales coalmining after two decades of investment starvation and managerial vandalism'.[21] The NCB agreed to joint monitoring of investment needs and, although it was feared that this could be at the expense of up to eight other high-cost collieries in the area, the strike was called off on 11 January, largely because of the failure of the Scottish strike. Ceri Jones noted: 'The policy of individual pit closures will now continue. It remains to be seen whether there will be isolated resistance at these pits. But even the militants are becoming disheartened by the Duke of York tactics – called up to the top of the hill only to be marched down again'.[22]

When the NCB announced the closure of Tymawr-Lewis Merthyr and Blaengwrach in February 1983, twenty eight Merthyr miners, including Des Dutfield, began an immediate stay-down strike. Seven neighbouring collieries joined the strike but a stormy delegate conference on 23 February decided to ballot with a strike recommendation. Strikers interrupted a press conference given by Area President Emlyn Williams, asking 'why he had not implemented a mandate to strike once the Board confirmed plans to close another pit'.[23] By the end of the month the coalfield was at a standstill and delegations were speaking in the militant areas. Although Yorkshire consistently voted in favour of strike action, against the national trend, paradoxically, as the threat to Yorkshire collieries became clearer, the margin in support of strike action narrowed. In August 1982 a list of forty five pits 'in jeopardy' compiled by Paul Routledge included twelve in Yorkshire,[24] yet in the ballot at the end of October 1982, linking pay with pit closures, support for action in Yorkshire fell to fifty six per cent (compared with thirty nine per cent nationally). By the end of the year commentators were doubting whether the South Wales strike would be supported in Yorkshire where 'jobs and wages are more secure'.[25] Nevertheless, on 28 February 1983 Yorkshire Area Council pledged support for

South Wales using the 1981 area ballot as its mandate.

On 3 March 1983 the NEC considered officially sanctioning the strike in South Wales and the Yorkshire recommendation under Rule 41 but decided instead upon a national ballot, a decision for which the right-wing caucus claimed responsibility.[26] The March 1983 ballot on pit closures returned the same result nationally as in the linked issues ballot (thirty nine per cent), while Yorkshire support for a strike fell slightly to fifty four per cent. South Wales was the only area where the majority in favour of strike action increased in the ballots from January 1982 to March 1983. The coal crisis in South Wales had already been more severe than elsewhere because of the destruction of so much of the steel industry: eleven pits with 8,400 jobs had disappeared since 1979.

The Triple Alliance of unions in the coal, rail and steel industries, established in October 1980, had produced a pamphlet outlining the effects of Government policies on these industries,[27] and in April 1983 published their case for alternative policies to prevent further job losses.[28] The 1983 pamphlet documented the job losses, market contraction, withdrawal of state finance and privatisation common to all three industries. The 'alternative approach' amounted to a reassertion of the need to maintain public industries through state subsidies and a commitment to planned economic recovery through the AES. The 'nine point plan' comprised a list of tactics to extend the influence of the Triple Alliance. Expansion of coal demand through the adoption of combined heat and power schemes was the only element of a strategy to deal with the coal crisis, and, significantly, no mention was made of the technological changes affecting coal-mining.

The overriding conclusion to be drawn from the experience of the first Thatcher Government's handling of the coal industry is that, despite the deepening coal crisis, the Union had not prepared adequately for the inevitable conflict. The Union's failure to prepare was manifest both in a lack of awareness of the dimensions and scale of the crisis (see Chapter one) and in the absence of a coherent strategy. The repeated inability to mobilise the NUM membership indicated the need for greater co-ordination of area and national officials in the pursuit of a strategy developed within the Union and agreed at NEC level. It was crucial for the strategy to be underpinned by an educational campaign involving the entire membership, rather than just activists, with propaganda materials professionally prepared and marketed.

2.2 Raising consciousness

By June 1983 an alternative strategy for the coal industry still depended upon the election of a Labour government committed to expanding coal, but the return of the Conservatives with an increased majority demonstrated that the responsibility for devising such a strategy could no longer be delegated to a future Labour government. Following the publication of the MMC Report which defined 141 of Britain's 198 pits as unprofitable,[29] the pit closure programme continued with renewed vigour. The NCB told NUM leaders of plans to axe between 65,000 and 70,000 jobs over the next five years.

Once the NUM had learnt of the June closure plans, the Scottish Area NUM conference determined that a wages offensive be launched to combat closures. The widespread myth that pit closures would cease when the miners rediscovered their former wage militancy reflected frustration and impotence rather than any coherent strategy. The June NEC meeting agreed to put an emergency resolution to conference seeking approval for a pit-head campaign against closures. The Perth conference supported the wages offensive and the campaign, but decided that another ballot should be held on strike action to halt pit closures. After the conference the NCB closed three more pits, removing 4,000 jobs in total, and within a month threatened to close the Coedely coke works and Wyndham Western colliery, both in South Wales. The closure of Cardowan was announced on 16 August. On 26 August the Cardowan membership agreed to closure, the day after the NCB had announced Brynlliw was to close.

Management offensive

At the same time as these closures, the first signs of a management offensive over discipline and productivity were emerging in Scotland. Development work was halted at Monktonhall and miners were threatened with dismissal for alleged restriction of output. The Board then provoked a strike at Monktonhall by sending home men who were a few minutes late for work after attending a union meeting on redundancies.

Questions of managerial prerogative featured in more disputes after MacGregor took up his appointment as NCB chairman on 1 September 1983. On 19 September George Marsh, a faceworker at Dodworth colliery, was dismissed for allegedly hitting an overman and the pit struck for the fifth time that month. Although summary dismissal is normally accepted as automatic in cases of fighting underground, the dispute spread throughout the Barnsley Area to involve 14,000 miners by 24 September, demonstrating the intensity of feeling over discipline and working practices. Yorkshire Area Council instructed miners to return to work but Dodworth stayed out for another two days. MacGregor visited Bilston Glen on 20 September, arriving two hours early to avoid demonstrators from Monktonhall, now on strike against closure, and warned miners there 'perform and you have a future; don't and you have no future, it's as simple as that'.[30]

On 27 September 1983 the NUM submitted a claim for a substantial increase in wages, the consolidation of average incentive pay into basic rates, and the introduction of a shorter working week.[31] The NCB rejected the claim on 30 September with a 'final offer' of 5·2% in return for co-operation in eliminating high-cost collieries.[32] A special Executive meeting on 1 October decided that the NCB reply should be discussed at an SDC on 21 October 1983. The pay claim, the NCB reply and the resolution from the special NEC meeting were circulated to the areas on 3 October. Coalfield campaign briefing notes, prepared by officers of the Union and distributed in September, also influenced the character of the special conference.

Although conflict over closures and the new management style continued, it remained localised. A strike over altered shift times at Westoe lasted a week, while 560 miners at Cronton struck for the day when the pit's closure was announced on 10 October. Monktonhall miners, still on strike against closure after a month, were supported by a one-day strike throughout the coalfield. MacGregor visited South Wales, warned the NUM there was no hope of £400m investment and refused to comment on pit closures.[33] Although attempts to save the Coedely coke works failed, the coking plant at Fishburn was saved from closure following a local struggle, which the leadership urged miners at nearby Herrington to replicate in order to resist the pit's closure.

Campaign for Coal

The SDC imposed an overtime ban from 31 October 1983. Although the ban arose out of wage militancy, it formed the basis of an emerging strategy against pit closures. In November the campaign notes were reproduced as an information pack, which was distributed to NUM branches, sponsored MPs and others.[34] The *Campaign* notes began with a catalogue of the redundancies which had occurred in coal-mining during the recession. Unemployment in coal-mining was shown to be particularly concentrated and it was argued that the costs of closures outweigh the costs of retaining existing capacity. Privatisation was opposed on the grounds that this would weaken NUM organisation, worsen conditions and lead to profiteering at public expense. The reduction in capacity sought by the Conservative Government was put at ninety five pits and up to 100,000 jobs over the next eight years; 'uneconomic pits' were shown to be an artefact of NCB accounting. Nuclear power was discussed as a serious threat which could eliminate 55m tonnes of coal demand and 100,000 jobs by the year 2000. The *Campaign* notes argued that increased coal demand should be generated with the application of coal utilising technologies already available, such as combined heat and power. After outlining the further threat to jobs from the application of new technology, the *Campaign* notes argued for the implementation of a new technology agreement so that miners would derive benefits from their acceptance of technological change. Finally, the notes criticised the AIS for distracting miners from the pursuit of an adequate basic wage and for creating inequalities and divisions within the industry.

Campaign for Coal themes were publicised through five rallies organised by the Triple Alliance at major cities through November and December. In April 1983 other unions in the three industries had joined the Alliance formed by the NUM, NUR and ISTC in 1981, and in November transport unions, notably the NUS and TGWU, became involved. On 20 October 1983 all three unions in the coal industry held a historic meeting to co-ordinate their campaign against pit closures. Three major deficiencies with the *Campaign* made it unlikely to convince anyone. Firstly, the *Campaign* literature gave the appearance of having been produced quickly and cheaply. The package was dull and unattractive as if its authors believed that the importance of the issues would overcome deficiencies of presentation. The arguments were all conveyed textually without any supporting diagrams or illustrations. Even highly-motivated activists, such as those attending the University day-release courses, were reluctant to use the *Campaign* material. Few copies were distributed among the rank-and-file; some branch officials in Yorkshire even claimed not to have seen it. Secondly, the rallies affected only the minority of activists; the majority of members were never exposed to Scargill's unquestionable skills as orator and agitator. Many militants boasted of having gone to several of the rallies but few could claim to have persuaded uncommitted members to go with them. Moreover, in preaching to the converted, the national officials obtained no feedback concerning the effectiveness of the *Campaign* on the typical inactive member. Thirdly, the campaign began years too late. Too much had been left even for the indefatigable Scargill, whose schedule of pit visits and rallies during the overtime ban put an enormous strain on the Union's research and industrial relations officers. The deficiencies of the *Campaign* established a pattern which was repeated throughout the strike. Activists were subjected to morale-building rallies instead of being equipped to take the arguments to the membership.

Reactions to the overtime ban

The overtime ban, and mangement's reaction to it, proved more decisive for strike mobilisation than *Campaign for Coal*. When the overtime ban was announced, MacGregor appeared on television expressing disappointment and arguing that the miners would suffer reduced income just before Christmas, whereas the action would not affect coal stocks. Before the effect of the ban could be assessed, the NCB claimed that disruption was 'minimal', although work was delayed at many Yorkshire collieries on the first day.[35] At the start of the second week, work was delayed at about half the country's pits.[36] As the backlog of essential maintenance, normally undertaken at weekends, built up, more miners were sent home on Monday and, by the fifth week, Tuesday. Scargill reported that the action was more effective than the Union had expected, while the NCB claimed that it was having less effect than at the outset.[37] Claims and counter-claims continued throughout the ban, with the NUM estimating NCB lost production of £10m per week and the media reporting that South Wales miners had lost £1m in the first four weeks of the action.[38] In November output was reported to be down by fourteen per cent[39] and Scargill put lost production at 2·4m tonnes just before Christmas.[40] The action was obviously having an impact and if it could be maintained long enough the NEC believed coal stocks would fall to a level sufficient to make a strike effective. In retrospect, these assumptions about the level and rate of consumption of coal stocks further demonstrate the Union's lack of preparation for the strike (see Chapter six).

The 1973 overtime ban had raised the miners' consciousness, committed the membership to further action in order to recoup the necessary investment of lost earnings, and prepared miners for the forthcoming deprivation. The 1983 ban was similarly expected to develop into a strike but the NEC were reluctant to organise a national ballot after the recent votes against action. The overtime ban was popular with most of the membership as a form of 'cut-price' action, while the *de facto* three- or four-day working week stimulated demands for reduced hours and demonstrated the inadequacy of the basic wage. A small minority, mostly those accustomed to working extensive overtime, were opposed to the ban, while some activists criticised the strategy. In reducing current earnings, some feared the ban would make it more difficult to sustain a strike: 'if we're going to have a strike let's get on with it while we can afford it'. Pits executed the ban with a degree of enthusiasm which varied with the quality of local leadership. Whereas some branches clearly sought to restrict output, others undermined their own action to maintain bonus pay; such variations could have seriously damaged unity within Yorkshire had they been more widely known. The NEC were reluctant to organise a national ballot in the light of recent rejections, but sought other ways of intensifying the action. As the ban continued, pressures grew from some sections to call off the ban and from others to escalate it into a strike.

Management increasingly resorted to provocation in an effort to break the overtime ban. At Westoe, 200 overmen struck when managers took over safety checks because of the NUM action.[41] By December leading left-wingers on the Yorkshire Area Executive were advising people to make preparations for a strike. Although Kellingley delegate Davy Miller was to remind members 'there'll be other Christmases', the approach of the holiday gave activists less scope to organise action out of the growing discontent. MacGregor appealed for a 'truce' over Christmas,[42] but at its 20 December meeting the NEC voted to continue the overtime ban and to withdraw from the CINCC Joint Policy Advisory Committee.[43]

Following the NEC decision, colliery winding enginemen in North Staffordshire threatened to defy the ban, claiming that their sixth shift was part of normal working hours, not overtime.[44] Roy Ottey, General Secretary of the Power Group, the craftworkers' Area which organises Staffordshire winders, wrote to the National President at the end of the year outlining the dispute and calling for a ballot on the overtime ban.[45] After a week, when the backlog of maintenance had caused severe disruption to production, the Staffordshire winders attempted to work on Saturday 7 January 1984. Midlands NUM and Power Group members mounted pickets at the five collieries and only two winders went in. The winders retaliated with a twenty-four-hour strike which effectively laid off miners at five pits on the Monday. The following day miners at Silverdale walked out because a winder who crossed the weekend picket line was operating the winding gear; they remained on strike that week and voted to walk the three miles underground each day rather than ride in the cage operated by this winder. More miners were laid off on Monday 16 January than on any previous day that the ban was in force. The following week 18,000 miners were sent home and craftsmen at Solsgirth held a twenty-four-hour strike over weekend picketing by miners.[46]

The election of the NUM Secretary also provided a focus for internal conflict over the overtime ban, but opinion polls continued to show support for the action among the rank-and-file.[47] John Walsh, the right's candidate from North Yorkshire, campaigned against the overtime ban and only narrowly failed to defeat Heathfield. Although the media interpreted support for Walsh as oppostion to the ban, activists attributed it to the parochialism of Yorkshire miners.

Major maintenance work provided management with a powerful lever against the ban. Kellingley miners were locked out for a week for refusing to allow rope-capping at the weekend.[48] At Wearmouth colliery miners threatened to picket the pit to a standstill unless craftsmen would work round the clock to avoid a six-week shutdown and management argued that rope-capping would necessitate a three-day week.[49] MacGregor's hostile reception at a number of collieries was further evidence of the deterioration in labour-management relations. Brodsworth miners stopped work when MacGregor arrived on 2 February and held a noisy demonstration.[50] On 22 February, in a similar reception at Ellington colliery, the NCB Chairman lost his balance when angry Northumberland miners pushed forward to voice their opposition to pit closures.[51]

Management tactics during the overtime ban shifted the attention of the rank-and-file from the pay issue to questions of control and managerial prerogative. FIDO reports (see Chapter one) were seen as a crucial weapon for management attempting to maintain production, while computerised monitoring of machine conditions was used to alter the work of craftsmen, who were most affected by the ban. As the central role of technology in the coal crisis became more widely acknowledged,[52] rank-and-file support for the new technology agreement increased. The NCB refused to consider the draft agreement and Rodney Dunn, Director General of Mining, described discussion with the NUM over new technology as 'inappropriate'.[53] The experience of the rank-and-file at colliery level and the lack of progress in Grading and Technology negotiations caused the NEC to agree at its February meeting to give 'further consideration to the banning of all new technology in the absence of a new technology agreement'.[54] At the March NEC meeting, the emerging strike took priority and technology was dropped from the agenda.

2.3 The trigger

Yorkshire was the key to the mobilisation of the 1984 miners' strike for strategic reasons, because of earlier area decisions and because action was already underway. The strategic importance of Yorkshire outweighs the size of the area, even though it represented almost one-third of national output and manpower before the strike. While being a part of the central coalfield extending through the moderate Midlands areas, Yorkshire was transformed into an ally of the militant peripheral areas in the run-up to the strikes of the 1970s. Any attempt to spread a strike to other areas, as South Wales miners had in 1983 over Tymawr Lewis Merthyr, was expected to be more effective from Yorkshire. The decision of the Yorkshire ballot of January 1981 legitimised area resistance to pit closures without recourse to a further ballot. Most significantly, disputes in Yorkshire arising out of the overtime ban were growing and increasingly developing into strikes.

In the last few weeks of 1983 several disputes had erupted through management tactics in response to the overtime ban. The Yorkshire Miners' Forum which had been a crucial instrument in articulating opposition to pit closures in the 1960s, and which brought the left to power in the 1970s, was reawakened. After a number of informal meetings a consensus view of a left strategy to intensify the effectiveness of sanctions began to emerge. Each minor dispute was assessed for its potential to be generalised to make a stand against the new-style authoritarian management and to express opposition to pit closures. In the Barnsley Area a dispute at Denby Grange over the transfer of men from neighbouring Bullcliffe Wood was too parochial and divisive; men returned to work during the week which saw the mobilisation of the entire South Yorkshire Area. A long-running dispute at the East Side complex (South Kirkby) over the use of a private washery was more directly related to the overtime ban and stood more chance of winning the support of the Barnsley Panel. In Doncaster Area a strike began on 20 February following the merger of Goldthorpe and Highgate collieries and the discovery that Highgate's miners enjoyed higher bonus earnings than Goldthorpe's, even though the colliery was supposedly a loss-maker.

Manvers Main

The dispute chosen as the basis for widespread mobilisation came from the unlikely quarter of South Yorkshire, with a reputation for moderation rivalled only by North Yorkshire, and over the even more unlikely issue of snap times. At Manvers Main the twenty-minute meal breaks were staggered and daywagemen were paid at time-and-one-half to work through snap time to prevent production from being interrupted for an hour. Although this practice had caused serious problems during the overtime bans before the 1972 and 1974 strikes, management had not attempted to introduce uniform snap times. From the onset of the 1983 overtime ban, men did not work through snap time and the colliery stood for an hour on each shift. In February 1984 management responded to the effectiveness of this action by deducting pay for idle time, even though meal breaks were allowable under the collective agreement, changed the time of shaft inspections unilaterally, and attempted to introduce a four-shift system without agreement. Manvers miners walked out spontaneously on Monday 20 February and at a union meeting that evening voted unanimously to strike and call upon other South Yorkshire pits for support. The next South Yorkshire Panel meeting recommended strike action and Silverwood proposed an all-out stoppage at Yorkshire Area Council but this was rejected. Nevertheless, Barnburgh, Wath and Kilnhurst collieries in the Manvers complex were affected by the dispute and followed the strike call. Cadeby and Silverwood came out on 5 March, while Kiveton Park's miners were locked out, owing to supporting action by the winders.

Cortonwood

A situation comparable with 1981 and 1983 emerged at the South Yorkshire Area general colliery review meeting on 1 March 1984. The review meeting was scheduled as usual to follow a CINCC meeting, but the national meeting was rearranged to 6 March and the review was not postponed. George Hayes outlined the budget reductions decided by the NCB (and which would be made known nationally at the CINCC meeting) and announced the intention to meet these through the closure of Cortonwood. This announcement provided a focus for the separate strands which constituted the roots of the strike. According to the review procedure, Hayes should have made Cortonwood the subject of a reconvened review meeting, allowing the area union time to investigate the colliery and providing for the attendance of Cortonwood branch officials. Goodman claims Cortonwood was deliberately chosen to provoke a stoppage over the question of the closure of 'uneconomic' capacity, and cites both Yorkshire Area leaders and people 'inside the Whitehall machine' as adherents to this view.[55] Several facts are inconsistent with such a conspiracy theory. Area Directors would not relinquish lightly their discretion to decide which colliery to recommend for closure or investment, even though the budgetary constraints are established nationally. Hayes chose Cortonwood because its output matched the required capacity reduction, the market for its high-quality coking coal had been severely reduced and its reserves were expected to be exhausted within five years. Had Hayes rescheduled the review meeting and kept within procedure, Cortonwood's closure would have been overshadowed by the NUM reaction to capacity reductions overall. In the 1982 ballot on pay and pit closures, forty five per cent of Cortonwood miners supported action, compared with fifty six per cent for Yorkshire as a whole. Given the moderate reputation of the branch, and the age structure of the membership, the NCB probably expected little resistance to the closure of Cortonwood. The timing of the review meeting before the expected announcement of widespread capacity reductions and the fact that a stoppage was already gaining momentum in South Yorkshire brought Cortonwood to prominence as the 'trigger' of the 1984 strike.

After the review meeting on 1 March, Arnie Young, Area Agent for South Yorkshire, conveyed the news to the Cortonwood branch. On Saturday 3 March Jack Taylor and Scargill had addressed a meeting at Askern originally arranged to spread the Manvers strike; they argued that the existing strikes should be extended into an area-wide stoppage in support of Cortonwood. The Cortonwood branch meeting the following day had also been convened originally to discuss the Manvers dispute, but news of the threatened closure caused over 500 people to crowd into Brampton Parish Hall. After an emotive address from Branch Secretary Jack Wade, the meeting voted unanimously to take immediate strike action, picket other South Yorkshire collieries, and to lobby the Area Council meeting the next day for the implementation of the 1981 ballot decision.

Cortonwood miners were joined by strikers from other South Yorkshire pits and activists from the other Yorkshire areas on their lobby of the Area Council meeting in Barnsley on 5 March. As Council met, the Manvers and Cortonwood issues had brought every South Yorkshire colliery except Manton to a standstill. Seven of the fifteen South Yorkshire collieries voted to work but six were picketed out on 5 March. Pickets were deployed from Wath, Manvers and Silverwood over the Manvers issue, while Cortonwood pickets brought out Treeton and Brookhouse. In Doncaster the Goldthorpe stoppage, which also entailed lay-offs at Highgate, was still in progress and two others had begun. At Yorkshire Main, unsafe face conditions prompted a stoppage and, when Edlington pickets stoned staff entering by a side gate, management threatened to close the colliery. Askern colliery also came out on the Monday over management's proposal to abolish chock fitters as part of the reduction in the number of craft workers.

Inside the Area Council meeting, however, discussion focused not upon the separate stoppages already underway but upon the principle of the proposed closure of Cortonwood and Bullcliffe Wood collieries. The threatened closure of Yorkshire pits before exhaustion, without agreement and in breach of procedure, contained the essential ingredients to overcome colliery parochialism and trigger a general stoppage. Mick Carter, Cortonwood delegate, described how the NCB had guaranteed the pit a life of five years at the previous review meeting, how men had been transferred from Elsecar colliery when this closed only weeks earlier, and how over £1 million had recently been invested in Cortonwood. The review procedure had been avoided and Mick Carter argued that if this closure was not resisted then no pit in Yorkshire would be safe; the Cortonwood miners were unanimous in their determination to fight and they sought area backing. Delegates in support of an all-out strike recognised a branch making a stand against pit closures as the long-awaited opportunity. Taylor emerged to announce that the Area Council had voted to implement the 1981 ballot decision and instruct branches to cease work from the last shift on Friday 9 March. National approval for the action would be sought under Rule 41 at the next meeting on 8 March.

Four more Yorkshire collieries stopped work before 9 March. Miners at Sharlston were locked out from 6 March following a dispute in connection with the overtime ban. Manton was eventually picketed out on 7 March, a small number of blacklegs, including Robert Taylor (see Chapter five), having left the pit on the previous day under heavy police protection. South Kirkby miners refused to work a single-entry face because of the reduced manning levels and came out on 8 March, as did Kinsley Drift because Fitzwilliam miners 'saw no sense in waiting another day'. Many individuals stopped work early and at several collieries the last shift worked was 'afters' on the Friday.

Perceptions of the strike issues

Although the strike was ostensibly 'against pit closures', the arguments used by activists during the early mobilisation were more complex and not without confusion. The cause of a strike is no more susceptible to unambiguous interpretation than any other social phenomenon. All strikes involve both latent and manifest issues and just because a particular issue is chosen as the rallying cry does not even mean that this is the major one. The social, legal, political and economic contexts may be more relevant than the manifest issue which has proved to be the catalyst for a particular stoppage. Neither need the central issue remain constant as a strike progresses. The question is further complicated by the observation that management and workers are liable to define a strike situation differently. What management may define as the exercise of legitimate authority, workers may perceive as the arbitrary use of power. In such an instance, to classify the strike as over 'discipline' sterilises the actual conflict of ideologies. In the strike process three elements need to be considered: the fundamental objectives of those involved; the arguments summoned in pursuit of the objectives, and the institutional channels available. The reasons formally given for striking may bear little resemblance to the fundamental 'cause' or objectives of the participants but may be more informative of the arguments used and the channels available. The act of striking stimulates an articulation of motives; as Hyman notes, it 'obliges those involved to specify their grievances; negotiations for a settlement are otherwise impossible'.[56] While there is no guarantee that the fundamental grievance will be specified or the objectives presented in negotiations, there is little risk in asserting that 'there is some relationship between what people involved in a strike say a strike is about and what it is about'.[57] Given the complexity of the issues underlying the miners' strike (see Chapter one), the participants' views of its major causes and objectives are of particular interest.

Overall, there was little concordance in the ranking of strike issues by the respondents to our questionnaire, but three issues were more often ranked as the most important reason for the strike: the appointment of MacGregor (twenty nine per cent); Conservatives seeking revenge for the defeat of 1974 (twenty seven per cent), and the introduction of new capacity (twenty three per cent). By contrast, new technology (eleven per cent), deficient demand (nine per cent) and colliery reorganisation (five per cent) were rarely cited as major causes of the strike. The responses did not vary systematically with occupation or type of community. There was considerably more agreement over the objects of the strike, the two major objects being no pit closures (fifty eight per cent) and no job losses (thirty three per cent). A new *Plan for Coal* was ranked first by slightly more respondents (ten per cent) than the other three objects: the four day week;

new technology agreement, and the consolidation of bonus earnings (three per cent in each case). The objectives of avoiding pit closures and avoiding job losses were to some extent treated as alternatives. Those in pit communities tended to emphasise opposition to pit closures (sixty eight per cent) more than opposition to job losses (twenty seven per cent), whereas outsiders ranked avoiding job losses first marginally more often (forty six per cent) than preventing pit closures (forty two per cent). Those living within a pit community clearly associated job losses with the symbolism of pit closure. The ranking of objects showed no variation between different occupational groups. There was only a small correlation between the ranking of issues and objects.[58]

The response to the strike call across Yorkshire was more enthusiastic over the pit closure issue than in respect of Manvers, but not without equivocation in some quarters. Branch meetings were held throughout Yorkshire between 5 and 12 March to receive delegates' reports of the Area Council meeting. Many meetings were held over the weekend and all were exceptionally well attended. In South Yorkshire, where the strike was already effective, six branches were unanimous in support of action and five others were virtually unanimous. At Brookhouse, resentment was expressed over having been picketed out, while at Manton objectors to the strike succeeded in passing a vote of no-confidence in the area officials. In Doncaster, where three pits were already fully stopped and one was partially out, eight branches were unanimous in supporting the strike, and at the other three dissent was minimal. Similarly in Barnsley, nine pits were unanimous in support for the strike and at the remaining seven objectors were very few in number. In North Yorkshire, however, only four branch meetings expressed unanimous support, although at six others supporters overwhelmed the small minority of objectors. At Kellingley, Gascoigne Wood and Prince of Wales there were substantial numbers opposed to the strike, which was attributed to Scargill and political activists. At Glasshoughton, where the branch had already agreed to closure before exhaustion to facilitate transfer to Selby, the delegate was mandated to call for a national ballot. Similarly, at Allerton Bywater a resolution was sent to Area Council calling for a national ballot and a minority threatened at the outset not to observe the strike instruction. Reactions to the strike call at pit level correlate significantly with voting patterns in the 1982 ballot, and broadly with the *drop* in support between the 1981 and 1982 ballots. In neither case, however, could voting behaviour be said to be a good predictor of membership reaction. The status of the colliery appears to have had no influence upon membership reaction, again suggesting the primacy of sociological 'community' explanations over economic ones.[59]

2.4 The Domino Strategy

The Yorkshire decision was almost immediately overtaken by national events. At the CINCC meeting on 6 March the context of the Cortonwood closure plan was revealed. The NCB intended to reduce the production budget for 1984–85 to 97·4m tonnes, a cut-back of 8·2m tonnes on the previous year's budget (4m tonnes lower than *actual* production in 1983–84). Area-by-area budget reductions had already been established and Area Directors were to determine the collieries to be closed to fulfil these. Polmaise had already been on strike for three weeks without winning the support of the rest of the area, but in response to the capacity reductions the Scottish Area Executive called Scotland's miners out from Monday 12 March. The CINCC meeting and the strikes in Yorkshire and Scotland were discussed at the NEC meeting on 8 March 1984 which faced the same situation as it had a year previously. After five hours debate the Yorkshire and Scottish strikes were endorsed under Rule 41[60] and, on an amendment from South Wales, approval was extended to any area joining the action. The day before the NEC met, McGahey had outlined the philosophy which underlay the decision to use Rule 41: 'we are not dealing with niceties here. We shall not be constitutionalized out of a defence of our jobs. Area by Area will decide, and in my opinion it will have a domino effect.'[61] At the NEC a succession of speakers from the right argued for a national ballot under Rule 43,[62] which required a fifty five per cent majority in a membership ballot to declare a national strike. Sid Vincent presented a call for a ballot from the Lancashire Area Executive but recognised it was 'now or never' and the endorsement under Rule 41 was supported 21–3.[63] In contrast to the support of most Yorkshire branch meetings, only six pits out of twenty eight voted for strike action in South Wales, and miners at Bilston Glen rejected the Scottish Executive call for a strike without a ballot.[64]

The ballot question

The decision not to hold a national ballot effectively abrogated responsibility to the areas for spreading the strike. Therein lay the weakness of the domino strategy because areas could resist action and legitimise their continued strike-breaking by the absence of a ballot. In retrospect some have seen the decision not to hold a ballot as a 'tactical mistake' because it undermined support and because of evidence that the vote could have been won.[65] A MORI opinion poll, for example, estimated that sixty two per cent would vote for action; even with the five per cent who were uncertain, this result would produce fifty seven per cent, sufficient to authorise a national strike, while another poll showed sixty per cent in favour of a strike.[66] By mid-April, however, a Harris poll estimated fifty five per cent for a strike with twelve per cent uncertain, while the *Sunday Times* suggested sixty eight per cent in favour with six per cent uncertain.[67]

The response in a poll to a hypothetical ballot may be very different from *actual* voting behaviour, so it is impossible to estimate what would have been the outcome of a ballot. The ballot question dogged the strike to the end, but a number of factors influenced the decision against holding one in March 1984. Foremost among these reasons was the failure to obtain a majority for action in the earlier ballots; there was no reason to suppose that conditions were any more favourable to win support whatever the polls might claim. Secondly, the considerable success which pickets had in spreading the action in the early days suggested that the object could be achieved without recourse to a ballot, campaigning over which might take momentum out of the strike movement. Thirdly, it was clear that the supporters of a ballot were generally opposed to a strike and knew that the media would mount a 'vote no' campaign. The power of the media to influence decisions made by trade-unionists was demonstrated in the AUEW elections of 1978, where a campaign against Bob Wright cost him the presidency. Moreover, the right's calls for a ballot were made in a favourable climate generated by discussion of the Conservatives' Green Paper on *Democracy in Trade Unions*, published in preparation for the Trade Union Act 1984.[68] It was also noticeable that the NEC members who called for a ballot represented the areas which in 1977 *overturned* a national ballot and conference decision by negotiating local incentive agreements. The incentive scheme contributed to the crisis by stimulating over-production and fragmenting unity over national pay claims. There was no guarantee that Notts miners would have joined a strike which the area voted against, even if a national ballot had been held and a majority attained overall; their defiance of a national ballot would doubtless have been supported by the courts. However, the central argument against a ballot remained that pit closures were a divisive issue and that no person should have the right to vote another out of a job, a theme repeated by miners in media appearances during the mobilisation phase.

Calls for a ballot, from whatever quarter, were given a high media profile. Trevor Bell, COSA, appealed on television for a ballot, and Durham Area Executive members wanted the NEC recalled to discuss the issue.[69] MacGregor claimed 'our people are shocked by the failure of the NEC to call a ballot'. The more Thatcher, MacGregor and Energy Secretary, Peter Walker advocated a ballot, the more it became identified with attempts to undermine the action and *not having* a ballot became a principle among activists. One protested he had 'heard enough of this ballot nausea: middle-class housewives, journalists, Barnsley shopkeepers all saying we should have a ballot.[70]

As area ballots rejected strike action and picketing intensified, the divisions over the ballot deepened. By 20 March none of the ten areas which had balloted showed support at the fifty five per cent level required, although Northumberland and Derbyshire came close. Over the next few days area leaders in Nottinghamshire, Lancashire and the Midlands appeared to attempt to overcome the ballot results. In Lancashire, where forty one per cent voted for strike action, the Area Executive instructed the members to strike from Monday 26 March 'in order to maintain unity', after most pits had been picketed out. Lancashire delegates, however, voted to end the strike a week later.[71] At best NEC moderates were ambivalent towards the strike. Ken Toon called on pickets to leave South Derbyshire and declared his members would work normally until a national ballot was called.[72]

On 27 March a caucus meeting of seven NEC right-wingers was organised by Ottey and Jack Jones (Leicestershire) to co-ordinate their campaign for a ballot. Ottey apparently advised Chadburn of the meeting via Ned Smith, head of NCB Industrial Relations, and the 'secret' gathering was filmed by a TV crew.[73] The meeting itself was less significant than the fact that this unconstitutional behaviour was publicised, which is probably explained by the erosion of right influence on the NEC. After the meeting Ottey claimed a majority of NEC members would support the ballot call and urged miners who had voted against the strike to return to work individually. Heathfield was 'astonished' that NEC members should encourage miners to cross picket lines,[74] yet this was quite consistent with the popular view of 'trade-union democracy' cultivated by the Government. The image of unrepresentative union leaders calling strikes to serve their own political ends was grafted on to Scargill even though the strike grew out of rank-and-file action, so media reports repeatedly misrepresented the strike as Scargill's doing (see Chapter five).

It was against this background that the Leicestershire Area representative proposed at the NEC meeting on 12 April a motion which sought to reverse the previous decision of the NEC to endorse the Scottish and Yorkshire strikes. In accordance with the NUM constitution, the President ruled the motion out of order and his ruling was upheld when challenged. Two days earlier the Durham Area Executive had moved its position from calling for a ballot to demanding an SDC to resolve the issue, which Billy Stobbs then put to the NEC. The meeting agreed to convene a delegate conference to consider the ballot question and a recommendation from the NEC to reduce the majority required from fifty five per cent to a simple majority.[75]

Problems with Nottinghamshire

Opposition to the strike centred on Nottinghamshire where sufficient production was maintained to prevent a victory. Under twenty seven per cent of Notts miners voted for strike action on 16 March, although this was *higher* than in the ballots of the previous two years. Henry Richardson had warned at the NEC meeting on 8 March that Yorkshire pickets could cause attitudes to harden in Nottinghamshire: 'Calling us scabs will not help us. I have been called that already outside. If Notts are scabs before we start, Notts will become scabs. You will make our jobs [as Area officials] impossible.'[76] On 21 March Chadburn and Richardson strongly urged members not to cross picket lines but it was not until 3 April that the Notts Executive voted to offer the same advice, and delegates rejected the call on 5 April. Before the SDC on 19 April, the Notts delegates voted to reject the NEC recommendation to lower the majority required for action and agreed to continue campaigning for a ballot. It appeared that the Notts miners would support any barrier to strike action, which led Taylor to comment 'we don't really trust you' at the SDC. By this time Chadburn and Richardson were exhorting Notts miners to join the strike regardless of any ballot. Militants among the 6,000 lobbying the SDC in Sheffield adopted the somewhat paradoxical slogan 'no ballot, simple majority' which, in the event, summarised the actual outcome. The opportunity for winning a ballot, if it ever existed, was gone before the SDC so the ballot became irrelevant except as propaganda for opponents of the strike. By 21 May the *Times* was claiming that a survey showed miners would vote for a return-to-work if a ballot was held, a view which Walker apparently held throughout the strike.[77]

The problems with Nottinghamshire were not resolved by the SDC. As Chadburn and Richardson fell into line with the NEC, Lynk and Prendergast began to re-establish right-wing control in Nottinghamshire. Scargill continued to campaign for those working in Nottinghamshire to join the strike; on 26 April he called for a national demonstration in Nottinghamshire and on 28 April at a rally in Cardiff emphasised the importance of Notts miners joining the strike. Two days later, with little change in the situation, Scargill declared that the strike could be won *without* Nottinghamshire, although it would be easier with their support.[78] The statement may have been a gross miscalculation or designed to raise morale; more likely it represented frustration at the growing realisation that Notts miners would not be mobilised.

The right-wing backlash in the Notts Area became evident on 1 May 1984 when about 7,000 working Notts miners demanding the resignation of Chadburn and Richardson clashed with half as many strikers from several areas at the Berry Hill offices in Mansfield. Lynk, formally unauthorised to speak, snatched the microphone and encouraged the mob haranguing Richardson and Scargill. In the week that followed, Yorkshire pickets concentrated on key Notts targets, but with little success (see Chapter three). About 15,000 strikers and supporters attended the national rally in Mansfield on 14 May. Scargill claimed the strike would be won even if they had to 'stay out till Christmas' but that, with the support of Notts miners, victory was only eight or nine weeks away. The inaccuracy of this forecast was disastrous for NUM strategy and for the credibility of the leadership (see Chapter six). The rally had no impact on the scabs safely out of reach in their collieries and later that night Bentinck President Neil Greatrex warned that the attitudes of Notts strikers were hardening against national and Notts officials.[79] Nevertheless, speakers like Tony Benn and Dennis Skinner had an enormous effect upon strikers' morale.

Other events demonstrated the irrevocable polarisation of Nottinghamshire and the National Union. The Notts Area's defiance of the SDC decision to call all miners out on strike without a ballot was condemned by the NEC, who warned that 'any Officer . . . who urges men to cross an official picket line will be acting contrary to the established practice of the NUM and Trade Union movement generally . . . and the National Executive will have to consider taking action against such officials in accordance with the Rules of the Union'.[80] The NEC then discussed a proposed new disciplinary rule which was recommended to an Extraordinary Annual Conference to be held in July in place of the planned Annual Conference. The new Rule 51 would establish a National Disciplinary Committee to consider complaints against members covering misdemeanours such as embezzlement, disclosure of Union documents to outsiders and 'any act (including any omission) which may be detrimental to the interests of the Union'.[81] The NEC also agreed to postpone branch elections for the duration of the dispute.[82]

The proposed Rule 51 was clearly designed to make the Notts Area more accountable to the NEC, while postponing elections was intended to prevent branch officials, most of whom were sympathetic to the strike, from being removed by the Notts anti-strike feeling. In response, three working miners in Nottinghamshire initiated legal action. On 23 May Sir Robert Megarry decreed the NEC ruling void and instructed the Notts Area to go ahead with elections, thereby ensuring that the right would gain control of the Union apparatus. On 25 May the High Court delivered a second blow in the form of an injunction upholding the Notts miners' 'right to work' in defiance of Union instructions. This action marked the first intervention of the law, which was to play a decisive role in restricting the Union's freedom to manoeuvre (see Chapter five).

The NEC agreed to hold the EAC concerning Rule 51 on 10 and 11 July 1984.[83] The Notts Area Council was expected to mandate its representatives to vote against any disciplinary rule change, but Notts strikers occupied the committee rooms to prevent the meeting taking place. Again working miners used the law and a High Court injunction on 10 July sought to prevent the EAC unless the Notts Area Council was allowed to meet. The NEC decided to defy the High Court[84] and on 11 July the rule change was accepted with a two-thirds majority, the Notts representatives having voted against. In his presidential address to the EAC, Scargill accused the Vice-Chancellor of 'state interference' in the strike, following his legal ruling that the strike in Nottinghamshire was unofficial. The High Court ruled the EAC rule change null and void because the Notts Area Council had not been convened although the Notts representatives 'did vote against the proposed Rule changes and the question in reality became academic'.[85] The NEC therefore resolved that the issue be placed on the agenda of an SDC to be held on 10 August 1984, thereby giving the Notts Area opportunity to discuss the proposals.

The NEC was under pressure from Notts strikers to introduce the new Rule 51. Strikers, who suffered great hardship and abuse from former workmates, threatened to return to work unless branch officials who were scabbing were disciplined.[86] The rule change was agreed at the SDC on 10 August, which was boycotted by delegates from Nottinghamshire, Leicestershire and South Derbyshire. This withdrawal from the democratic processes of the Union was further evidence of the widening rift and of the intention of the Notts Area leadership to become completely autonomous of the NUM (see Chapter seven).

Whether or not the strike was lost in the first six weeks, as Goodman appears to suggest, the failure to mobilise the Notts miners was a decisive factor in the eventual outcome.[87] Moreover, it is important to investigate the reasons for this failure because, for all that Nottinghamshire is a 'special case', the problems of strike mobilisation encountered there were present to a lesser degree throughout the industry. Rather than a complete aberration, the Notts experience highlights weaknesses in NUM strike strategy *in general*. Several theories have been advanced to explain why Nottinghamshire did not support the strike. Analysts from the right favour the excuse used by the scabs that 'the failure to win over Notts flowed essentially from the refusal to countenance a national ballot'.[88] However, ballots are not the only mechanisms for strike mobilisation and, as Fryer notes, they 'have never been the cornerstone of union government or democracy'.[89] The Notts miners called for a ballot as a way of stopping the strike; the ballot issue was a *result* of their disaffection for the strike rather than a *cause* of it. More fundamental explanations for why the Notts miners wanted to vote against strike action therefore need to be considered.

Part of the explanation for the behaviour of most Notts miners lies in the relatively late development of the Notts coalfield and the social relations of the Dukeries.[90] Only three of the North Notts collieries were sunk in the last century and miners were drawn from other areas during the depression and the post-war contraction. While it may still be true that 'most Notts miners are related by blood to those who work in the pits of Scotland, Durham and Wales',[91] a historical sense of mining *community* is virtually absent from most of Nottinghamshire. The environment in which Spencer's company unionism flourished was never eradicated. Easy coal meant less rank-and-file pressure for militant leadership and fewer opportunities for militants who succeeded in gaining branch positions to demonstrate their skills. Although the left 'captured' certain area and branch positions, the membership remained unpoliticised with no more than an instrumental attachment to the NUM. Henry Richardson's election was seen as a victory for the left but most of the campaigning was undertaken by *Yorkshire* miners, which underlines the inadequacy of left support in Nottinghamshire.

The politics of the Notts Area was exemplified in their sectional approach to pay bargaining. Notts miners played a full part in the wage militancy of the 1970s, but this was a period when they were *dependent* upon national negotiations under the NPLA. In Spencer's time local negotiation allowed Notts miners to take advantage of easier mining conditions. The Notts Area first attempted to reintroduce local incentives in 1970,[92] and eventually helped to engineer the AIS, fragmenting pay determination and guaranteeing higher earnings to the higher productivity areas (see Chapter one). Under local incentives Notts miners had less to gain from national pay disputes and applied the same logic to the dispute over pit closures. This attitude depended crucially upon the perpetuation of two myths which were accepted by the majority of miners in Nottinghamshire *and the other areas*.

The first myth was that Notts miners were bought off from striking by the bonus earnings of high productivity pits. Certainly Nottinghamshire enjoyed higher than average bonus earnings, which Phil Basset showed to vary by +/− 18 per cent between areas.[93] Although the scheme is claimed to reward effort, not output, differences in effort cannot possibly explain the disparity between bonus earnings in South Wales and Nottinghamshire, concealed in the censored paragraphs of the MMC Report.[94] However, several Yorkshire pits average bonuses higher than many in Nottinghamshire and bonus earnings *within* Yorkshire ranged from +36 per cent to −15 per cent of national average earnings in 1983. Variations in the bonus showed little conformity with the patterns of strike mobilisation in Yorkshire.

Bonus variations within Yorkshire mean that canteen workers at a high productivity pit may receive more bonus pay than face-workers working in bad conditions in a pit nearby. These inequities between effort and reward have been compounded by local variations like the 'Kellingley option' and tonnage agreements contrary to the original AIS.[95] The NCB was well aware that the AIS was in disarray[96] yet, in contrast with the orthodox view that pay systems in decay invariably cause a deterioration in industrial relations,[97] nothing was done, suggesting that the bonus was used 'to facilitate labour mobility between pits and speed up the closure programme'.[98] Since its introduction in 1978 the AIS has degenerated from a *proportional* scheme, where increased effort was rewarded with proportional increases in bonus, to a *regressive* scheme, where a twenty five per cent increase in effort is rewarded with a bonus payment representing less than nineteen per cent of standard earnings. The average U1 faceworker was losing £15.51 per week in the year up to the strike, while those on a bonus of £90 were losing £25 per week. The longer the AIS remains, the more success the NCB will have in cheapening labour; the miners have been robbed through the bonus and the Notts miners robbed most of all.

The bonus myth was crucial in divorcing Notts miners from national pay disputes, but the myth that Notts jobs are safe was even more important in divorcing them from the struggle over pit closures. In 1983 a report produced by the Notts Area NUM Education and Research Department on possible job losses in the area recognised that 'pits hitherto regarded as "long life" pits could become vulnerable virtually overnight to the threat of closure'.[99] The report identified short-life collieries, and showed the risk to Notts pits from an expansion of nuclear power and the threat to jobs from new technology. In April 1984, it was shown that pit closures and automation could remove forty four per cent of the jobs in the Notts Area.[100] In February 1985 the suggestion that the area's manpower would be reduced over the next five years by at least 10,000 was dismissed by the NCB as 'politically-motivated mischief'.[101] The Coalfield Communities Campaign prediction that by 1990 manning levels in the two Notts Areas would be reduced by between 11,222 and 19,806[102] was incorporated in July 1985 in a Report prepared by the Nottinghamshire County Council Director of Planning and Transportation.[103]

Relatively few Notts pits face closure in the near future compared with, say, Scotland or South Wales, but the process of concentrating capacity on high productivity pits will inevitably mean closures, particulary in South Notts where Moorgreen and Pye Hill have already closed. Babbington, the largest loss-maker in the whole of Nottinghamshire and Prendergast's power base, seems certain to be closed. Continual geological and water problems at Annesley could jeopardise Newstead and Bentinck which are linked in the same complex. South Nottinghamshire could be left with under five pits within five years. Fewer pits are expected to close in North Nottinghamshire, although the future of several is uncertain. Sutton, Lynk's stronghold, will close owing to a combination of exhaustion and subsidence damage, in between two and seven years. If output is increased from long-life pits like Ollerton, to which £30m investment was recently earmarked, then Sutton's life is probably nearer to two years. Blidworth is a consistent loss-maker and may face closure through geological problems. Mansfield has geological problems, Rufford suffers from water and even Sherwood could be at risk if its history of subsidence and geological problems is repeated.

The good mining conditions of Nottinghamshire and its relatively late development have made it the most *mechanised* area of all but *automation* and the creation of super-pit complexes is less developed than in Yorkshire. Since the Notts Area offers the best conditions for advanced technology mining, job losses from MINOS could be particularly high (see Chapter one). The Notts Area is also vulnerable on the demand side because seventy four per cent of all Notts coal is supplied to CEGB power stations. Over fifty per cent of the area's production is from collieries supplying over ninety per cent of their coal to the CEGB. For six Notts collieries, including Bilsthorpe and Cotgrave among the five long-life pits producing over a million tonnes per year, the CEGB is the only customer. Above all other areas, Notts is at the mercy of the CEGB's monopsonistic advantage. The introduction of nuclear power and use of imported coal will directly jeopardise Notts miners' jobs.

The message that jobs are on the line in Nottinghamshire never got through to sufficient Notts miners.[104] In the first month of the strike Alan Moss, President at Moorgreen, claimed: 'North Notts is a young coalfield, with very little chance of closures.'[105] This view was never successfully challenged because of the silence of the Notts leadership and the ineffectiveness of the National Union's campaign. It was in the interests of the area leadership not to publicise the evidence of future job losses in Nottinghamshire because these were concentrated in the collieries where Lynk and Prendergast enjoyed most support. Their bid to take over the area may even have been prompted by the knowledge of the evaporation of their power bases. The ineffectiveness of the National Union's campaign derived both from its message and its medium. The campaign emphasised job losses in peripheral areas, so served to *reinforce* the myth that jobs were safe in Nottinghamshire. The argument against a national ballot was similarly couched in terms of the immorality of voting another person out of work, rather than demonstrating the need for Notts miners to defend their own jobs. Heathfield elaborated at the 19 April SDC that MacGregor was effectively saying to miners in the central coalfield 'turn your backs on your comrades in Scotland, Durham, South Wales and Kent'. The National Union was unable to overcome area autonomy and establish direct communication with Notts miners, so picketing became the medium in place of campaigning. By its very nature, picketing encourages sloganising and simplification, so preventing the presentation of complex arguments. In any case, pickets were usually prevented by the police from communicating with Notts miners, but the widespread acceptance of the 'Notts myths' ensured that such exchanges as took place were misinformed.

A campaign which demonstrated the threat to jobs in Nottinghamshire might have succeeded in persuading Notts miners of the conformity between their own sectional interests and those of the rest of the NUM. This, after all, was the situation in 1972 and 1974. Freeman's claim that 'there was a good chance that the Notts miners could be persuaded to support a strike that was presented as being in the interest of the working class',[106] however, is inconsistent with the whole history of the Notts Area. The failure to campaign in Nottinghamshire put more emphasis on picketing, distorted the tasks of strike organisation (see Chapter three), and ultimately allowed the strike to be undermined (see Chapters five and six).

Chapter three

Strike organisation

This chapter focuses upon the major preoccupation of strike organisation, picketing, in which 'union tactics are designed not only to deter the strike breaker, but also to maintain discipline over the pickets while they are dissuading him'.[1] Activities such as seeking solidarity action from other unions and raising finance to sustain picketing or alleviate hardship are considered separately under strike maintenance in Chapter four.

The chapter is organised into four sections: in the first we describe the evolution of strike organisation and its key features; the second section analyses the problems posed by key strategic issues which illuminated serious organisational deficiencies; the third section addresses the administration of the picketing strategy in Yorkshire by the panels and the branches, and, finally, this chapter analyses patterns of activism within the Yorkshire coalfield to determine what proportion of miners became involved in various aspects of strike organisation, and to examine whether activism was influenced by factors such as age, occupation, or community.

3.1 The evolution of strike organisation

Immediately the mobilisation of a strike is underway, a union is confronted with an extraordinary range of organisational problems which highlight any deficiencies in union government and administration. Although the union provides the resources and machinery of strike organisation, 'formal organisation is both an aid and an obstacle to striking'[2] because of an inherent conflict between bureaucracy and activism. Moreover, there is a conflict between the need for increased cohesiveness and higher membership participation during a strike: 'All the while it is necessary for leadership to retain control over strategy and tactics for the sake of power and unity, whilst maintaining the legitimacy of their position through the support of membership'.[3]

The NUM had not prepared the machinery of strike organisation in advance of the mobilisation of the strike in March 1984. Strike organisation evolved with the strike and therefore illuminated both the dynamics of the strike process and organisational features of the NUM. Successive levels of the Union attempted to co-ordinate and control the action. In the mobilisation phase the strike was led from below; branch officials had difficulty in restraining pickets and neither

Yorkshire Area nor national officials appeared to be in control. To regain control and co-ordinate the picketing, a Strike Co-ordinating Committee (SCC) was established as a sub-committee of the Yorkshire Area Executive. The federal government of the NUM intensified the organisational problems of the strike. Power within the NUM resides firmly in the areas, and this feature was particularly evident in Yorkshire. From its first meeting on 2 April 1984, the Yorkshire SCC represented the gravitational centre of strike organisation which, together with other key areas, distorted attempts to develop the national co-ordination agreed on 19 April. The National Co-ordinating Committee (NCC) was eventually operational from 14 June 1984, but was constrained throughout the strike by area autonomy. The resultant organisation of the strike mirrored the hierarchical government of the NUM, as shown in Fig. 4. At each level strike organisation was delegated to a committee and communication within the hierarchy was primarily downward.

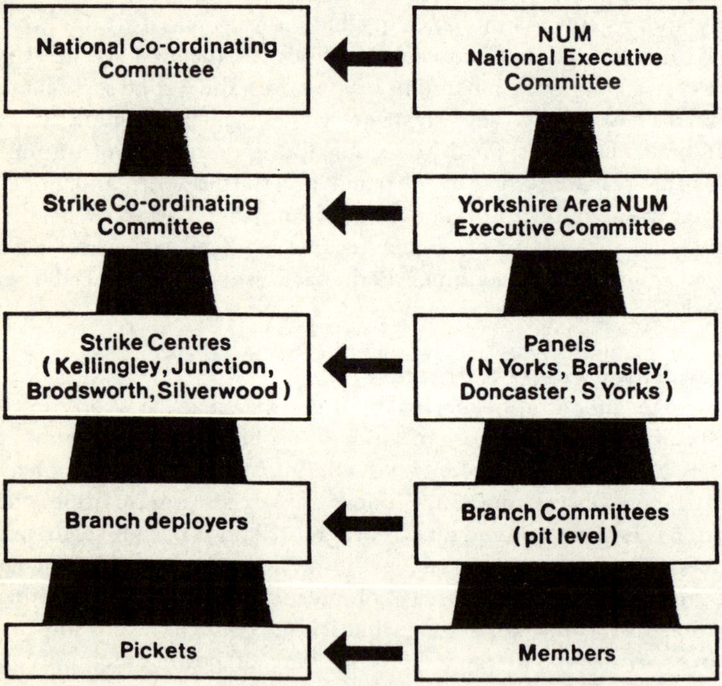

Fig 4 Formal strike organisation

The structure of this section is dictated by the levels of strike organisation and by the chronology of their evolution. The strike and strike organisation began in Yorkshire and the area continued to play such a central role that it is considered first. The NCC was established three months after the SCC, and in Yorkshire the SCC continued to determine strategic questions of strike organisation, although efforts were made to reconcile major policy differences with the NCC.

Yorkshire pickets lead the strike

From 12 March 1984, the organisation and control of picketing became of primary concern throughout Yorkshire. Meetings were held at local clubs and welfare halls where branch officials called for volunteers for picketing duties. The response was so high that resources were not available to involve all of those who had volunteered which some have claimed weakened initial enthusiasm among the rank-and-file.[4] However, the euphoria of these early meetings was not the forum to test people's later commitment to getting up in the middle of the night to risk being arrested and beaten. Pickets were deployed from panel strike centres at Kellingley, Barnsley, Brodsworth and Silverwood.

On the first day Doncaster pickets went to Harworth, the nearest Notts colliery which had been the scene of violent struggles to oust the Spencer union in the 1930s. On 13 March Harworth, Bevercotes and Thoresby pits were picketed out. Later that day after the flying pickets were endorsed by an Area Council meeting, Taylor declared 'we are asking our workmates to support us in attempting to save our industry'. By the Wednesday pickets from across the four Yorkshire Areas were at every colliery in Nottinghamshire and Lancashire, and succeeded during the day in stopping one-third of the Notts pits and half of those in Lancashire, despite the increased effectiveness of police operations. On the same day the NCB was granted an injunction by the High Court ordering Yorkshire Area to withdraw instructions to members to picket other areas (see Chapter five). Taylor announced that pickets would be withdrawn from Nottinghamshire pending the decision of an Area Executive meeting the following day. Pickets were defiant, however, and told the media 'we'll continue whatever'; 'when we've got the pits out then we'll get the power stations out and stop the blacklegs who're moving coal by road'.[5]

Officials despatched by the Executive to call the Yorkshire pickets back were largely unsuccessful. National and area leaders met to discuss the injunction that evening, while many pickets ignored instructions and returned to Nottinghamshire. During the evening David Jones, an Ackton Hall miner picketing with others from Frickley, died after being crushed between the police front line and a picket van and struck by a brick thrown at pickets by youths opposed to the strike in Ollerton.[6] As news of the incident spread, pickets converged on Ollerton and only dispersed in the early hours after Taylor and Scargill had been summoned to appeal to them to go home.

As the Area Executive meeting on 15 March unanimously agreed to defy the court injunction, pickets from South Yorkshire clashed with police at Thoresby. The picketing of Nottinghamshire was suspended when Notts Area officials instructed members to strike from 6 p.m. that evening for forty-eight hours while a ballot was organised. The early picketing of Lancashire mainly by North Yorkshire miners, contrasts with that of Nottinghamshire involving the other Yorkshire areas and particularly Doncaster pickets. The approach at several Lancashire collieries was to have a few Yorkshire pickets, often accompanied by members of the local branch, talking to men as they arrived for work. The issues were put across more effectively and many Lancashire miners joined the strike. This approach depended upon limiting the numbers picketing and required the co-operation of the local branch and even the police at the colliery. Where these conditions were not met, such as at Agecroft, the situation was much as it was in Nottinghamshire.

While Notts picketing was suspended, Yorkshire pickets went to Lancashire and the West Midlands in greater numbers and attitudes seemed to harden at pits where men were still working. Yorkshire pickets at Lancashire collieries admitted 'our tactics need rethinking: the mass picketing has lost us some support'. New picketing targets proved more fruitful: a private wharf at Scunthorpe was prevented from moving imported coal for the steelworks; collieries in Staffordshire were brought out. In Lancashire, Staffordshire, and to a lesser degree Nottinghamshire, local miners joined the pickets after witnessing police violence. By the end of the week Lancashire leaders recommended joining the strike and several Staffordshire pits voted to strike.

Divisions reappeared at the weekend when the result of the Nottinghamshire ballot was announced. Picketing was again concentrated on Nottinghamshire, but police operations involving up to 8,000 officers prevented all but 1,000 pickets from reaching their targets. In the third week, after the Nottinghamshire ballot, police action intensified and fewer pickets reached Nottinghamshire targets. More pickets reached Derbyshire collieries and attention was also paid to power stations. Out of frustration, pickets converged on Coal House in Doncaster and blocked roads with slow-moving convoys of cars when turned back from the Nottinghamshire border. Moreover, on the Tuesday, Taylor announced that the £100,000 Industrial Action Fund was virtually exhausted, necessitating fund-raising for further picketing.

The numbers attempting to picket during this initial phase had fluctuated with the success of the overall operation and the effectiveness of the police action, although the position had stabilised somewhat by the end of March. The area leadership, panels and branch officials had experienced difficulties in controlling the pickets and there was some disagreement between the different levels of the Union over the strategy adopted.

Yorkshire Strike Co-ordinating Committee

To co-ordinate branch deployment of pickets, channel resources coming from supporters to fund this, and resolve some of the organisational difficulties, the Yorkshire Area Executive established the SCC, comprising the four Yorkshire officials, the agents of the four areas, and a representative of each panel. The SCC first met on 2 April 1984 and each morning thereafter. The SCC's first concern was the optimal deployment of pickets, but its terms of reference increased to include financial and legal questions, inter-union relations, parliamentary lobbies, hardship and social services, propaganda and police intelligence.

The SCC had two major objectives in relation to picketing: 'to prevent the movement of coal' and 'to picket collieries continuing to work'.[7] Initially the picketing of power stations and wharves to prevent coal movement was far more successful than efforts to stop the pits working, especially in Nottinghamshire. Tactical considerations were important in the pursuit of these objectives, hence the SCC decided that picketing in Lancashire would be suspended until that area made a decision on the strike, to avoid alienating potential support. As soon as the Lancashire position was clear, the SCC resumed the deployment of pickets to the area and arranged mass-pickets of Parkside and Agecroft collieries where large numbers had continued working. The picketing of Nottinghamshire was a priority for the SCC since it was believed that 'a victory in Notts . . . would be a bigger morale booster than public meetings'.[8] The efforts made to get pickets into Nottinghamshire involved co-operation with other NUM areas. Yorkshire officials met leaders of North Derbyshire Area and Richardson from Nottinghamshire to arrange for Yorkshire pickets to be deployed to Nottinghamshire pits from Derbyshire strike centres. When Nottinghamshire miners were advised by Chadburn and Richardson not to cross picket lines, the SCC organised an immediate push on Nottinghamshire.[9] The SCC directed propaganda at their own pickets to rally support for an all-out campaign to stop Nottinghamshire and counter the effect which the media barrage and the police operation was having upon morale: 'It was agreed that a circular be sent to all branches informing them that sixty per cent of Notts miners were now on strike. Notts Area was contacted and it was agreed that a letter signed by their officials be forwarded to us containing this information.'[10] More accurate information than this was needed to deploy pickets effectively and Nottinghamshire officials were asked for 'a written statement of the numbers of men on strike and pits affected'.[11]

Minority political groups proved an irritant to the SCC during the strike. On 22 May it was reported at the SCC meeting that the SWP had established a fighting fund for the miners 'but none of the money was coming back to the NUM', and the Union had no control over how the money was being used. The SCC considered this so serious a problem that members suggested making a public statement concerning the behaviour of the SWP and decided that a 'circular could go out to all branches telling them that money is not coming into the NUM' from this source. In its efforts to curb independent action by political groups within the Yorkshire membership, any report regarding minority group activity was followed up. For example, it was reported on 25 May that York University had sent a sum of money to SWP members at South Kirkby branch. The Union was also keen to discipline members who ignored the area's authority over the organisation of the strike, or those who demonstrated more allegiance to directives of political groups than to those of the SCC. SWP members at South Kirkby branch were told 'in no uncertain terms that they must not speak or collect without the authority of the branch', which those involved reluctantly accepted.[12]

The SCC even experienced problems controlling its own members and officials: 'Any decision made by the Strike Committee can be criticised by the Executive Committee or absent members at the appropriate time and place but . . . no decision can be countermanded. The Committee expresses serious concern that decisions are being countermanded before being referred back to the Coordinating Committee'.[13] Statements made by national officials did not always coincide with Area policies and this made them difficult to maintain. For example the question of dispensations became problematic for the SCC 'particularly in view of statements by the national president that none were to be given'.[14] The SCC were worried that if they complied with this request, they could endanger the support received from other unions locally and impede the progress of the strike.

The relationship of the area with other unions was also sometimes strained. In June, Owen Briscoe wrote to the Regional TUC asking them to circularise their affiliated organisations with a number of requests: firstly, that donations should be sent to area office rather than to individual branches; secondly, that speakers should only be requested from area level, and thirdly that area should be informed of donations of food so that distribution could be equalised among the branches. Inequalities in the support received by different branches was a result of organisations taking independent action and dealing directly with branches. The importance of links with other unions was demonstrated in April when the SCC arranged for pickets to be mounted on rail bridges around Fiddlers Ferry power station after informal discussions with ASLEF and the NUR. At a later stage NUPE offered to arrange for two van-loads of nurses to join the NUM picket line. Although the support of other unions was crucial in the strike, the SCC only requested such support in times of crisis, such as Orgreave, when 'it was suggested that Sheffield and Rotherham Trades Councils might be persuaded to help our pickets'.[15]

National Co-ordinating Committee

The SDC on 19 April 1984 had decided that the NUM head office should co-ordinate all future deployment of pickets and requests for solidarity action. Many activists expected this organisational change to have a profound impact on the progress of the strike but, in the event, area autonomy proved more robust, although the National Union was to present a major challenge to the authority of the Yorkshire SCC. On 24 April the SCC meeting learnt that a National Co-ordinating Committee (NCC) had been established and a request had been sent to areas for officials to represent them on this Committee: 'It was agreed that whatever was decided [by the NCC] would not affect this Committee's right to deploy pickets. The main aim of National Coordination would be to avoid duplication of picketing by areas'.[16] Yorkshire Area officials feared that the NCC was an attempt by national office to take control of Yorkshire pickets out of the hands of the SCC. The following day, Taylor reported that: 'he had had words with national office and they had no authority whatsoever over this Committee [the SCC]. They were there to advise as to where pickets should be sent and then it was left to this Committee to agree or disagree to send pickets as they think necessary'.[17] The area officials' scepticism about the intentions of the national office was reinforced before the NCC was established. On 27 April it was reported that requests had been received from national office for Yorkshire pickets to relieve COSA who had been picketing Garston docks the previous day. The Yorkshire strike centre receiving the request had complied with it and only *later* referred the matter to the SCC: 'It was decided that this Committee deploy pickets and that any request from national office or any other area be discussed. National office were informed that they were not to request the deployment of pickets. A notice was produced to that effect for display in Area Strike Centres.'[18] The SCC decision did not prevent national office from attempting to influence picketing in Yorkshire nor did the SCC thereafter feel constrained to respond to national office requests favourably. For example, when Mike McCarthy, National Research Officer, suggested that Yorkshire's target for the next day's picketing should be Sherwood colliery, 'after a long discussion it was agreed that North Yorkshire cover Bickershaw with the other three areas covering Nottinghamshire . . . It was agreed to cover Annesley from 5 p.m.'.[19]

The National Union struggled to formulate a national plan to which areas would adhere but made little progress because there was no democratic representative body directing the strike at national level in the first three months of the strike. Animosity appeared between areas and the National Union, and between the areas themselves. A meeting in Sheffield broke up without agreement and the SCC learnt that 'our representative walked out after Mr Jack Collins [Kent] had been complaining about what Yorkshire was doing'.[20] Lack of communication between national office and the areas left the National Union ill-informed of progress being made in the strike. There was 'very little feedback of picketing successes and it was hoped that this information could be sent to national office'.[21]

The NEC only agreed to establish the NCC on 14 June:

> It was further AGREED:
>
> To appoint a 'National Co-Ordinating Committee' to co-ordinate Industrial Action in accordance with the decision of the Special Conference held on the 19th April, 1984, and that the following members be appointed to serve: Messrs A. Scargill, M. McGahey, P.E. Heathfield, O. Briscoe (to act as Chair in the absence of a National Official), E. Clarke, G. Rees, W. Chambers, W. Stobbs, L. Morgan, J. Colgan, G. Butler, D. Murphy and S. Vincent.
>
> Also that the Sub-Committee meet as frequently as required.[22]

The NCC first met on 19 June 1984, three months after the strike began and the day after the battle of Orgreave (see below). Before then, formal co-ordination of picketing at national level had been minimal. The NCC agreed that 'future arrangements for picketing which involved more than one area be coordinated by the Committee'.[23] The NCC was to meet each Wednesday at

2.30 p.m. and invite other NEC members to attend 'on specific matters in which they had particular knowledge or interest'. The NCC was to maintain contact with other unions who were affected by the dispute and agree upon relevant action.

The NCC attempted to co-ordinate the TUC lobby and asked all areas to appoint stewards on coaches travelling to Brighton. Members were requested 'to act in a disciplined and organised way outside the conference hall'.[24] Following rumours that a return-to-work was being organised to coincide with the TUC, the NCC also asked areas 'to mount a mass-picket at every pit, cokeworks, workshops, depots etc., on Monday 3 September 1984'. Many areas turned out record numbers of pickets during that week, fearing that the strike was about to crumble.

3.2 Challenges to strike organisation

The deficiencies in strike organisation became apparent in the Union's failure to resolve key strategic questions, particularly with respect to the steel industry and power stations.

Steel

The effect of the dispute upon steel production and relations with the ISTC was to prove a major problem. Although picketing was at first directed at stopping collieries, the strategic importance of denying coal to steelworks and power stations was immediately recognised in the areas. If industry could be starved of steel and electricity, pressure would increase for the Government to settle the dispute. On 20 March pickets prevented coal moving into the Scunthorpe steelworks from its private wharf.[25] On 27 March the Triple Alliance agreed to picket ports to prevent coal imports, to picket Ravenscraig steel complex, and to plan a series of stoppages in all industries. On 29 March the transport unions offered to black the movement of all coal by rail, sea, road, and on the docks. The following day the ASLEF Executive instructed members not to cross NUM picket lines and not to move any coal. Workers in BS had suffered a decisive defeat in their 1980 wage strike,[26] employment had fallen dramatically by March 1984 although the rationalisation programe was not yet complete.[27] It was rumoured before the miners' strike that another steelworks was to close, but BS and the Government would not name the plant. Bill Sirs, ISTC General Secretary, warned that the steel workers' support for the miners would not extend to risking steel jobs, while BS Chairman, Sir Robert Haslam simultaneously announced that coal shortages could cost more jobs in steel than pit closures would in coal.[28] Although BR claimed they were able to move coal normally, BS laid off 160 workers at Scunthorpe because of coal shortages.[29] On 3 April the NUS implemented their decision to black all imported coal but the following day Ravenscraig steelworkers agreed to use coal carried by private hauliers and reneged on the Triple Alliance agreement with the miners.[30] Miners were angered by the justification Sirs offered: 'the miners' fight is not our fight . . . In our own strike we didn't ask the miners to support us, even though they did give us moral and other support'.[31]

Area unions attempted to constrain the violence which erupted on the picket lines at steelworks following the steel-workers' acceptance of scab coal. Scottish Area representatives met workers from Ravenscraig to agreee a level of coal dispensation sufficient to protect the future life of the steel plant. Violence at Port Talbot resulted in an agreement to limit pickets to forty and police to three. BS threatened to stop production at Scunthorpe entirely,[32] and when Ravenscraig steel-workers failed to persuade the Scottish Area to increase the coal dispensation, they agreed to accept ninety lorries of scab coal driven in convoy from the coast. The South Wales Area agreed to a special delivery of coal to Llanwern but cokemen rejected the agreement and refused to let coke at Nantgarw be moved.[33] Scottish miners were also critical of the Union's agreement to allow a train-load of coal into Ravenscraig, prompting McGahey to state 'there is maximum unity in this movement today . . . no one has been defeated'.[34] Three days later the Ravenscraig agreement was destroyed when ISTC members unloaded a boat of American coal at Hunterston. The vacillations of the area unions over steel plants in their regions reflected a wish to avoid alienating workers in the same locality by putting their jobs at risk. In seeking to impose *uniform* treatment of steel plants, the National Union sought both to increase pressure on the steel industry and ensure no plant could be singled out and the miners' action blamed for its closure. Kim Howells, South Wales Area Research Officer, blamed the dispute with the steel-workers on National Union efforts to establish a uniform agreement with the steel unions in place of established area channels of inter-union negotiation.[35] Certainly, the confusion which arose following the intervention of national office served BS purposes. Both the South Wales and Scottish Areas agreed to uphold the NEC decision to halt coal deliveries to steelworks until Sirs met the NUM, even though Port Talbot steel-workers had agreed to the NUM terms.[36]

Orgreave demonstrated how the Yorkshire SCC was not always free to pursue what the majority of its members believed to be the most effective means of winning the strike. At a meeting on 25 May it was noted that fuel from Orgreave coke works was reaching Scunthorpe steelworks. The SCC was eager to 'come to some settlement of the Scunthorpe situation very quickly as this would with Orgreave tie up all our pickets'.[37] After Scargill was bruised in a police charge at Orgreave, he called for NUM members to 'build a picket line like that at Saltley' and to come to Orgreave in such numbers that it would be impossible to move coke from the plant.[38] The SCC had wished to avoid focusing resources on Orgreave and resented the pressure that had been brought to bear on them by the National Union: 'the lads on strike in Notts are under more pressure that we in Yorkshire; for us to win Orgreave it would not mean we would win the dispute. It was necessary to stop the Notts area.'[39] In accordance with the policy of the SCC the pickets were deployed to Nottinghamshire on 29 May, although Orgreave was designated the fallback target. There was to be no change in picketing strategy until the Yorkshire Area Executive had fully discussed the subject. Nevertheless, that morning all the SCC and Area Officials had gone to Orgreave in support of their National President. Over 2,000 pickets assembled at Orgreave on 28 May; the following day 5,000 pickets reached the target and over the next few days clashes between police and pickets intensified. 'Today's scenes represent the most violent in any dispute since the war', claimed the BBC news.[40]

At a meeting with Yorkshire Area officials on 31 May, Scargill reiterated his opinion that Orgreave coking plant ought to be picketed every day. Area officials accepted the strategic importance of Orgreave and agreed to picket it after picketing Calverton Colliery in Nottinghamshire from 5 a.m. The meeting of the SCC on 1 June, however, was divided over picketing Orgreave, as a number of Yorkshire pickets had been detained by the police on serious charges. South Yorkshire and Doncaster felt that Orgreave should continue to be picketed and it was decided by the SCC to contact National level 'and clear it from them about us picketing Orgreave'. National Office was unable to resolve the SCC's dilemma since 'it was not up to them to direct pickets to particular places'.[41] National officials had succeeded in influencing picketing strategy but recognised the limitations of their authority over the areas. It was eventually agreed that no mass-picket be held but members who wished to go to Orgreave or Scunthorpe over the weekend should be allowed to do so. The SCC now felt that 'politically' it would be difficult to withdraw from picketing Orgreave and 'the best solution was to hit Orgreave on an irregular basis rather than every day'.[42] The SCC had to take control of organising the picketing in Orgreave because 'the national president had declared Orgreave our target and most of the rank and file went anyway so we had to protect those well-meaning pickets who keep going there'.[43] By appearing to endorse the NCC decision, the SCC also avoided advertising the fact that the National President had more influence over Yorkshire activists than they did themselves. On 5 June a mass-picket was organised by the SCC at Orgreave. Members of the SCC were still unhappy about this picketing target, 'Orgreave could become a whirlpool which draws us deeper in and [pickets would] be unable to go elsewhere'.[44] After it was decided to restrict Orgreave to a token picket on 7 June, the struggle for control between National Office and the Area SCC came to a head: 'It was thought that too many people were trying to coordinate this strike – there was a definite rift between AS and our officials, particularly about Orgreave. It was felt that an early meeting between our officials and the three national officials be arranged'.[45]

Strategically the Orgreave mass-pickets were a failure, as coal continued to be brought out and hundreds of pickets were injured and arrested. However, the NEC reiterated the need to 'blockade the movement of all coke/coal into steelworks' and agreed to continue picketing Orgreave: 'until such time that the ISTC were able to accept the invitation to meet with the NUM and reach an agreement which would provide sufficient coke/coal to keep their furnaces intact and the industry safe but not allow any production'.[46] On 18 June 5,000 pickets from several areas returned to Orgreave to prevent coke being taken to Scunthorpe. The 'battle of Orgreave' represented the peak of police-picket violence and can be regarded as a turning point in the strike. The mass picket failed to prevent the convoy removing the remaining coke; at least 100 miners, including Scargill, were injured, and ninety were arrested, many on serious charges.[47] Later that day the SCC agreed to a total withdrawal of pickets as BS had said 'they are getting no more fuel out of Orgreave for the moment'.[48]

When the NCC first met on 19 June the steel question took priority. Members reported that 'meetings had been held locally with Regional Triple Alliance leaders to effect the decision of the NEC', but it was agreed that 'a national agreement should supersede all previous local agreements relating to coke and coal deliveries to steel works'.[49] The NCC was available to negotiate such an agreement with the ISTC but if none was reached, 'the transport unions would be urged to blockade in addition to coke and coal, iron ore and all other supplies in and out of the works'.[50] From 20 to 25 June Ravenscraig and Llanwern continued to receive all the coal they needed to operate. On 25 June pickets prevented iron ore leaving Llanwern and the following day Scargill asked the NUR to stop all movement of iron ore. After a coke oven collapsed at Llanwern,[51] the NCC learnt that Sirs had agreed to meet the Union on 29 June,[52] to 'discuss solidarity action to be taken by them as members of the Triple Alliance'.[53] Before the meeting with Sirs the NEC convened at 10.30 a.m. when Scargill emphasised that the transport unions' offer on 21 June to blockade the movement of coke and coal into steelworks was entirely their own decision. At the meeting later that morning Sirs told the NUM that the ISTC was: 'prepared to give moral and financial support but did not feel they could take what amounted to industrial action. He did, however, state that he was willing to put the matter before the Steel Committee and his own Executive, and would convey their decisions to the Union'.[54]

The NEC agreed that 'a further meeting with the transport unions should be arranged to consider the steel Union's response'.[55] The TUC Steel Committee endorsed the ISTC stance in refusing to support the NUM, which Scargill described as 'disappointing but not unexpected'.[56] The NCC challenged the decision conveyed by Sirs: apparently the TUC Steel Committee had not met to discuss the issues beforehand, and of the fourteen members of the Committee only eight were involved in the decision, three of whom had dissociated themselves from the ISTC position. The NCC agreed: 'that the secretary be requested to communicate with all members of the TUC Steel Committee on an individual basis, putting forward the NUM case and that a campaign document be considered that could put over the Union's point of view regarding the steel industry'.[57]

British Steel simultaneously applied pressure to steel-workers; BS Chairman Haslam sent a letter to every employee, in which he argued that the miners were putting steel jobs at risk and denied that steel-workers owed any debt to the miners for 1980.[58] On 7 July BS organised a convoy of lorries to take iron ore from Immingham to Scunthorpe, which brought Immingham and Grimsby dock-workers into the conflict. On 10 July dockers struck over the use of unregistered labour and the following day the TGWU voted to extend the action to ports outside the dock labour scheme. The disruption to ferry services during the peak season boosted rank-and-file morale and Scargill was able to claim 'for the first time since the strike began even the experts are saying the pendulum is swinging in favour of the NUM'.[59] While the number of dockers on strike was still increasing, the Immingham workers accepted an offer from BS. The action continued, however, because the TGWU wanted an agreement from the employers that there would be no further breaches of the Dock Labour Scheme.[60] On 15 July ACAS intervened in the dispute and the following day Scargill met the transport unions' leaders to co-ordinate the coal-ore blockade. Lorry drivers blockaded channel ports *in protest* at the dockers' action, which finally ended on 20 July; the NUM was once more on strike alone.

Trouble flared on the docks again on 23 August when steel-workers unloaded the *Ostia*, which the dockers had blacked, at Hunterston. Scottish dockers struck but Sirs claimed that unloading the coal was ISTC work and that 'steelmen will not be intimidated by the miners'.[61] Ravenscraig was reportedly desperate for coal. At a delegate conference on 24 August TGWU dockers' representatives voted 78–11 to strike from the weekend. Hull, Liverpool, Middlesbrough, and Tilbury dockers came out in support of the Hunterston workers, and Nicholas Ridley accused them of fomenting 'a political strike . . . to support the crumbling miners' strike'.[62] Meetings of dockers produced a variety of decisions on whether to take strike action, with Immingham, Grimsby and Belfast dockers resisting. Nevertheless, the situation was serious enough to warrant a cabinet meeting on 28 August to discuss both the docks and coal strikes. On 6 September the TGWU decided to send flying pickets to ports still working and to attempt to talk to lorry drivers. On 10 September the docks strike was described as being 'fairly solid', and the BBC reported growing numbers of dockers joining the action. Once dockers and steel-workers agreed the volume of coal to be sent from Hunterston to Ravenscraig, the strike was effectively over. Scargill voiced dissatisfaction with the agreement,[63] and Eric Clarke told the next NCC meeting: 'at no time had the NUM Scottish Area been involved in discussion regarding such arrangements nor had the Area been approached to attend any meeting and in any case, had consequently maintained that were such an agreement reached, it would have to be at national level in accordance with national policy'.[64] Despite these setbacks, the NUM still attempted to limit steel production, but when Scargill met ISTC representatives on 21 September he was unable to persuade them to support the miners. The support of dockers and rail-workers failed to halt steel production. Nor was the TUC any more effective when it resolved to ban all coal movement as part of a pact to assist the miners with action in other industries (see Chapter four). The failure to halt steel production was emphasised by reports that steel output was slightly above the 1983 level.

Power stations

The NCC's dependence upon area co-operation to organise picketing was clear when they attempted to co-ordinate the picketing of power stations which areas had undertaken from the outset. After NEC discussion of 'opencast working, power stations, commercial and industrial users of coal and other fuels and domestic requirements', it was agreed that 'the National Co-ordinating Committee draw up national guidelines for areas, in order to obtain uniformity of arrangements'.[65] The NCC received reports of the level of coal stocks at power stations from 'sources within the CEGB';[66] the reliability of the information provided cannot be assessed but it must have formed the basis of any picketing strategy. After learning that the CEGB planned to seek parliamentary authority for power cuts before the end of the summer,[67] the NCC agreed that: '24 hour picketing of power stations be maintained and that no fuel or other materials be permitted to cross our picket lines. Where there were difficulties, the National Control Centre would be available for assistance. Furthermore, the secretary be requested to communicate with all unions representing power workers seeking solidarity action'.[68] When the Scottish Area found it difficult to prevent oil entering Blyth power station the NCC sought assistance from the transport unions, although many of the bargees were not NUS members and possibly non-union. The TGWU instructed members in power stations not to use any coal or additional oil brought into the power stations through the miners' picket lines,[69] but by the end of the month it was clear that many TGWU members ignored these instructions and picketing was proving ineffective.

Guidelines to improve the effectiveness of picketing power stations were issued to the areas,[70] and when the position failed to improve a 'co-ordinator' was appointed to direct power-station picketing from National Office.[71] Despite twenty-four hour picketing, power stations were still receiving coal and oil, particularly in the Midlands, so the NCC despatched 'J Weaver to the Trent Valley to arrange the coordination of power station picketing'.[72] The volume, location and potential use of pit-head stocks were difficult to ascertain so areas were requested to supply details in order that pits holding coal suitable for power stations could be targeted. By 31 October the responsibility for co-ordinating the picketing of the Trent power stations was back in the hands of the area union; direct NCC intervention in the organisation of picketing had been unpopular.

The NCC called for assistance from TGWU national officials to prevent oil being moved in the Midlands area,[73] but a week later substantial amounts of oil were still entering Blyth power station and the NUM sought the co-operation of shop stewards.[74] Other areas had more success; shop stewards at Bold power station refused to handle scab coal and threatened strike action if any man was sent home.[75] On 5 December the NCC learnt that there were no coal stocks held by the NCB in North Nottinghamshire and only a small amount of coal at Bentinck colliery in South Nottinghamshire, but lack of information from the areas prompted a request for 'replies to circular 436/84 . . . from other areas'.[76] When the oil burn at Fiddlers Ferry power station was found to have increased dramatically a week later, the NCC sought the help of the TGWU in investigating the cause.[77] NCC efforts to co-ordinate power-station picketing were limited by the co-operation they could command from the areas. Information on pit-head stocks and the movement of coal, coke and oil was deficient; often intelligence was so late in getting through to the NCC that the picketing response organised was by then inappropriate. The NCC's attempt to use the TGWU to implement NUM policy and to restrict power-station performance was a further limitation, which caused delay, confusion and conflict when TGWU members refused to comply with TGWU advice.

Other problems

In the absence of effective national strike organisation, the area unions had to foster local relationships with other trade-unions, but this created further difficulties. In Yorkshire, a Regional Triple Alliance was established to co-ordinate supportive action, but this simply fragmented what support existed. Thus when National Office requested help in picketing Toton rail yards where coal was being transported, they were told that 'these drivers were not within the Yorkshire and Humberside Region. They were not subject to the Area's Triple Alliance'.[78] Similarly, whereas the TGWU had agreed at national level to support the miners, TGWU drivers crossing Yorkshire picket lines claimed that they had received no instructions from their union *not* to do so! Pickets often vented their frustration at this situation on those they saw undermining the strike. At Glasson dock, for example, coal boats docked on every tide from East Germany and Poland, to be unloaded directly onto lorries queueing at the dock gates. The TGWU agreed to 'turn back some of these lorries if the picket was removed'.[79] Once only a token picket remained, the lorries returned and the pickets retaliated by smashing windscreens.

Coal imports were clearly a matter for national action, but before the NCC was established the Yorkshire SCC found it necessary to suggest that national office contact the relevant ambassadors.[80] Scargill contacted the Polish Ambassador, but coal continued to arrive from Poland so the SCC called for a picket to be mounted on the Polish Embassy and tourist office.[81] When Kent Area refused to picket the Polish Embassy, the SCC decided to mount a token picket from Yorkshire and arranged this by the end of June.[82] To make the picket effective, the NUM's International Department was asked to supply pickets with details of the amount of Polish coal being imported.[83]

One of the most powerful weapons at the disposal of the NCC was its power to decide dispensations for the supply of coal, over which the NEC was determined to maintain control. The first NCC meeting ruled that 'no industrial dispensations be granted without the sanction of the National Union'.[84] In practice, however, dispensations continued to be granted at area level. Some requests for dispensations took the form of threats to use European coke suppliers, as in the case of GEC Foundries.[85] When Yorkshire Area requested a dispensation for coke from Smithywood coking plant to British Rail Engineering Ltd, Doncaster, the NCC deferred its decision 'pending further enquiries through the national officials'.[86] The NCC was inevitably not in a position to make an informed decision compared with the areas on whether to allow dispensations, so removing responsibility to national level caused further delays and worsened relations between the miners and other trade-unionists at area level. On occasions, after discussion at the NCC, requests for dispensations were referred back to the area in question for a final decision, as when South Wales requested a dispensation for anthracite beans to be supplied to Lincoln and Aylesbury for the elderly and sick.[87] When there was less ambiguity over a dispensation the NCC took the necessary action. Thus a request from TGWU Region 5 for a dispensation to transport coal to hospitals and prisons was immediately agreed by the NCC.[88] The dispensation cases referred back to the areas outnumbered those agreed at national level, most of which occurred towards the end of the strike.

The NCC was formed too late to formulate an effective strategy on key national questions such as steel, coal imports, and the picketing of power stations. The National Union's intervention with respect to the steel industry and dispensations disturbed existing area arrangements and caused conflict with the area unions. The strategy for picketing power stations and preventing coal imports, however, should have been developed at national level but remained ineffective because responsibility was abrogated to the areas. The weakness of the two-tier organisation was that there was no agreement on the jurisdiction of the national and area unions.

Throughout the strike the NCC was circumscribed by area autonomy. The democratic government of the NUM depends upon the power of the area unions, but the NCC attempted to ignore this and operate a system of centralised control. The NCC terms of reference of 19 June 1984, to co-ordinate 'future arrangements for picketing which involved more than one area',[89] were modified by 1 November to 'consider matters relating to picketing referred to the Committee by the national officials or at the request of the areas'.[90] Similarly, the edict 'that no industrial dispensations be granted without the sanction of the National Union',[91] had become by 1 November that the NCC would 'consider requests for dispensations referred to the National office'.[92] The dilution of NCC objectives reflects its failure to subordinate the area strike organisation to a national strategy. The NCC had neither the power nor sufficient information to achieve these objectives.

3.3 The administration of the strike in Yorkshire

In contrast with the 1972 strike, in which the four Yorkshire panels developed their own strike organisation, in 1984 their role was confined to implementing the picketing strategy agreed at the SCC. In 1972 left influence was greatest in the Doncaster and Barnsley Panels which took over the organisation of the strike from the right-wing area leadership. With the left in power it was inevitable that area resources would be more effectively harnessed, but this restricted the opportunity of branches and pickets to influence strike strategy. Picketing targets were agreed at the SCC, whose members conveyed the decisions back to their panel strike centres. Although meetings were held in the strictest secrecy, several members of the SCC believed one of the committee regularly leaked details of the targets. Deficiencies in picketing strategy and tactics were evident in the local administration of the strike in Yorkshire.

Picketing strategy

The SCC allocated targets to particular panels on geographic criteria. Thus North Yorkshire usually covered Lancashire, while Doncaster and South Yorkshire picketed Nottinghamshire. The panels met regularly to organise the details of deployment. The North Yorkshire Panel, for example, met on Monday, Wednesday and Friday evenings to arrange the picketing of targets agreed by the SCC. On Tuesday and Thursday evenings target meetings were held at which picketing strategy was discussed prior to deciding targets at the SCC the following day. A sealed envelope with the time it was to be opened written on it was issued to an official from each branch. Pickets were then instructed by telephone or word of mouth to assemble at the deploying point (pit gates, club or panel strike centre) at the appointed time. Sometimes further information was supplied to ensure pickets were appropriately dressed: 'we'd be told "training shoes tomorrow" for a gallop or "boots" for a shove'. Such information, coupled with the departure time and amount of petrol money distributed, gave some indication of the possible targets to pickets and police alike (see Chapter five). At several branches the envelopes were opened in the evening (on occasions in a public house) and targets openly discussed.

At all collieries strike organisation was characterised as 'hierarchical': directives were issued by the area and panels with little opportunity for branches or pickets to influence strategy. At forty-seven per cent of collieries the character of strike organisation remained hierarchical throughout the stoppage. At only fifteen per cent of pits was strike organisation viewed as being more 'centralised', involving the National Union, and this was limited to the Orgreave era and, to a lesser extent, the early weeks of the strike. While several respondents expressed regret that the National Union did not exert more control, those involved at panel and SCC level were generally disappointed with the quality of national organisation. At thirty-eight per cent of collieries, strike organisation was described as being more 'autonomous', involving either information from pickets affecting SCC decisions or branches deciding their own strategy. Even in these cases, however, it was emphasised that decentralised strike organisation was a temporary aberration from the hierarchical norm. For eight per cent of collieries, decentralisation was associated with the Orgreave era (typically because the South Yorkshire pits had local information not readily available to the SCC), while in twenty-three per cent of pits greater autonomy developed towards the end of the strike.

Many activists believed that the major weakness of the strike organisation was the failure of the SCC, panels and branches to delegate activities. As Steuben argued: 'strike leadership should be much broader than the regular union leadership; for the greater the participation of the rank and file, the stronger the strike'.[93] Most authorities on strike organisation are agreed on the value of local strike committees involving active members. Community-based strike committees were established in Scotland.[94] In Yorkshire, however, only three collieries, Royston Drift, Highgate and Maltby, established strike committees separate from the regular branch committee. In every other colliery the branch acted as the strike committee. Where committee members had the necessary skills and enthusiasm, this arrangement had the advantage that they were familiar with the Union bureacracy. It also meant that they were reluctant to short-circuit it when direct contact with the SCC or NCC might have been necessary. Branch officials were generally (with some notable exceptions) committed to the strike and administered SCC decisions to the best of their ability. Other committee members, however, showed varying degrees of enthusiasm for the strike: one youth representative actually left the industry and applied to join the police force! Local officials, like SCC members, were often too preoccupied with strike administration to participate in picketing and thereby breached an acknowledged rule of strike organisation: 'ask others to do only what you are prepared to do'.[95] Pickets and their informal leaders were not represented as such on the committee so local strike organisation was often weaker than it need have been. Different skills are needed during times of peace and times of conflict: 'any intensely-fought dispute will throw up new leaders and the way must be kept open for them'.[96] The division between activists and bureaucrats, apparent at the level of the SCC, was reinforced at branch level.

At ninety-three per cent of collieries, pickets fed intelligence reports back to the branch or strike centre concerning routes to targets (avoiding police road blocks), and coal movements. New targets were also identified: collieries, workshops, oil terminals, coal depots, power stations and coke works. Information from pickets led the North Yorkshire Panel to formulate the strategy of blockading the Selby and Humber Bridges. The initiative for the successful Selby Bridge deployment came from a rare meeting of the North Yorkshire Left, at which members argued the need for 'a victory, even a hollow one, to boost flagging morale'. Information concerning routes and the best times for picketing was usually acted upon but twenty-four per cent of collieries reported that the panels and SCC ignored advice about targets. In some cases pickets became frustrated and organised autonomous action in the evenings.

Throughout the strike the SCC was faced with conflicting objectives of making picketing effective and maintaining control over branches and pickets. Branch autonomy was discouraged by the SCC and a number of branches expressed concern about the control that the SCC could exercise over their affairs. At the end of April: 'It was unanimously requested that either a Council meeting or Executive take place as some delegates were expressing resentment at the control exercised by the Coordinating Committee. This matter was to be put to the officials'.[97] Nevertheless, the SCC continued to exercise its authority over branches and continued to be defied by some of them. Problems with Grimethorpe Coalite plant prompted the SCC to issue a statment that 'branches should not be in a position to disregard the opinions of this Committee'.[98] Any aspect of branch activity which might involve members in representing the NUM was liable to be vetoed. For example, in May the SCC ruled that: 'All people speaking on behalf of the NUM should be arranged by the General Secretary. Branches should not have the authority to arrange speakers without reference to Mr Briscoe'.[99]

In early April the committee agreed that pickets should only go to designated targets and it was stressed that more discipline was needed amongst pickets. The SCC agreed that notices to this effect were to be placed at strike centres, and target meetings were held to encourage feedback from pickets. Nevertheless, those pickets who preferred to choose their own targets continued to do so: 'The men didn't like committees, didn't like decisions coming from panels and committees, including the target meetings. The rank and file began picking their own targets and going on from one to another, using their picketing money for more petrol'.[100] To enforce their control the SCC ruled that union support (financial and legal) would be withdrawn from those involved in 'independent picketing'. When Doncaster pickets went to London to raise funds without authorisation, the SCC agreed that 'no representation or assistance be given to these pickets'.[101] However, at the end of May the SCC was still struggling to enforce its picketing instructions. The SCC arranged for Doncaster Area to picket wharves but at 7 a.m. no-one had arrived at the targets. SCC orders had been countermanded and the majority of pickets had massed outside Coal House in Doncaster. SCC members were furious and recognised that unless their control of picketing was accepted by the rank-and-file then it would be impossible to implement any strategic resistance to NCB tactics: 'unless they start going where told to go then we might as well pack up'.[102] Indiscipline by pickets detracted on occasions from potentially successful deployments. In early June, for example, South Yorkshire pickets were sent to Cotgrave and got through in large enough numbers to be effective, but only thirty-five pickets were left when the 'afters' shift arrived because the majority had travelled on to Orgreave.

Picketing tactics

The Yorkshire Area SCC focused upon deploying pickets to particular targets rather than with ground tactics at the target. The first indication that the SCC was becoming aware of the shortcomings of picket organisation was two months into the strike in early May. After discussing the deficiencies of picket leadership, the SCC suggested that each *group* sending pickets out should have a team leader and that the men should stay in twos and threes rather than large groups, 'we have to get some discipline into our ranks'.[103] Steuben advised strike organisers to select and brief 'picket captains' before the strike where possible: 'the men in charge of the line are entrusted with major responsibility in any strike . . . To bring the strikers to the picket line, to keep them on the line, to maintain discipline in face of provocation, to observe all movement in and around the plant and to sense the mood of the strikers themselves'.[104] McIlroy likewise recommends 'one individual should be placed in charge of each group or shift of pickets'.[105] The SCC suggestion of team leaders was never universally adopted and the absence of effective picket leadership was frequently criticised by activists. In the 1980 steel strike, however, the Rotherham Strike Committee had modelled its picket cells led by 'picket marshalls' on the NUM, whose local officials advised the ISTC.[106] Within the first few weeks of the miners' strike, picketing arrangements had stabilised to the extent that groups of flying pickets repeatedly travelled with the same drivers. Frequently neighbours would picket together to minimise picking-up time at deployment. Drivers were expected to return with all their passengers and to report any arrests to the strike centre. Members of a group tended, therefore, to 'look out for each other' at the picket, even if they did not stay together as a group. Although this amounted to a *de facto* cell structure, it arose by accident rather than by design.

At forty-six per cent of Yorkshire collieries there was no formal leadership at the target. Branch officials or committee men took on these responsibilities at thirty-nine per cent of pits, while other picket leaders were appointed at fourteen per cent of pits. Informal leaders often emerged, especially for the duration of a tactical crisis, but such leadership encouraged autonomous action beyond the scope of SCC decisions and undermined the authority of delegated leaders attempting to direct pickets. All picket leaders were cautious and gave only a sporadic, temporary lead because police snatch squads arrested anyone consistently appearing to be an organiser or 'ring leader'.

Steuben emphasised the communication role of picket leaders:

> The policy-making group among the strike leaders can learn a great deal from the picket captains about the morale of the men, their thinking, the degree of stability of the strike and other vital information which has a direct bearing on shaping strike strategy. The picket captains, in turn need to be constantly briefed and imbued with confidence in victory. It is for them to transmit this confidence to the men on the line.[107]

The miners had no equivalent to the steel-workers' 'picket co-ordinators' who acted as a communication link between the strike committee and picket cells. The vacuum was filled by those 'prepared to have a go' at a particular time. Both delegated and informal leaders were able to exert most influence where their immediate followers constituted a major group at the target. When the leaderless majority saw concerted, organised action they invariably joined in.

The absence of a direct link between pickets and the SCC forced communication via the bureaucracy of Union branch and panel, which explains why few SCC decisions were influenced by information from the rank-and-file. Had co-ordinators been appointed, they would have enabled the SCC to direct operations more closely and could have provided feedback to the committee from the picket line. Both the control and authority of the SCC would have been strengthened.

During May the SCC suggested that each *panel* should instruct its members where to stand on a picket line and hence take the initiative away from the police who invariably confined pickets behind police lines. Steuben stressed the importance of 'constantly orienting, briefing and instructing the pickets' by distributing directives on 'what to do and what not to do on the picket line'.[108] McIlroy similarly argued that 'pickets should be briefed as to discipline and behaviour on the picket lines'.[109] Deployers and branch officials usually gave broad verbal instructions to pickets, although at twenty-nine per cent of collieries even this advice was lacking. Advice invariably included instructions to ensure that they reached the designated target because 'a few individuals wanted to be turned back by the police and pocket the petrol money'. At forty-three per cent of collieries, pickets were told 'not to commit any offence' and from about June 'not to throw stones' following the SCC-panel directive to 'refrain from throwing missiles'. Often the advice appeared contradictory: 'peaceful but effective picketing', 'get involved, don't stand back but don't let the police provoke you'. At eleven per cent of pits the advice given was ambiguous: 'look after yourself', 'don't get arrested', 'don't get caught bricking', 'don't throw stones from the rear' (since this endangered pickets at the front). At another eleven per cent of pits, pickets received more precise instructions, although these altered as the strike progressed. Once it became obvious that the police action prevented 'peaceful persuasion', pickets were advised to use verbal abuse and to be 'as vocal as possible'. Pickets were later told to stay together and advised of tactics to employ on the picket line.

Activists cited Orgreave and the Selby Bridge as targets where successful tactics were formulated beforehand. However, regardless of whether it was a sensible strategy to focus resources on Orgreave, the Union's tactics at the coking plant on 18 June were deficient in several respects. The secrecy of the tactics to be employed at the picket was so complete that activists were not briefed beforehand so did not know what was going on. The problem was highlighted when a member of Yorkshire Area Executive who was communicating with other organisers with a 'walkie-talkie' radio narrowly escaped attack by a Scottish miner who thought he was a plain-clothes policeman. The radio communication (which was almost certainly monitored by the police) was only used to pass information and not to direct operations. Unlike the police, the Union had no effective leadership or channel of command and no lieutenants to translate strategy into ground tactics. Although a small number of pickets succeeded in entering the plant, having killed a police dog which attacked them, they left to avoid arrest as they had received no instructions on how (or whether) to sabotage the plant. The strategy had been limited to getting through the police lines rather than to stopping the plant. Union officials, and the vast majority of rank-and-file members, were reluctant to cross the threshold of illegality and fearful of being accused of insurrection. Beynon highlighted the resultant problem of the 'plastic pickets': 'only a minority (perhaps ten per cent) would turn up regularly for picketing . . . Of those who did turn out, only a minority were prepared to "do the business".'[110]

In response to police violence, pickets at one South Yorkshire colliery organised 'catapult crews to retaliate when attacked'. When militants in Doncaster distributed details of how to deal with horses and dogs, however, they were reprimanded by the SCC for 'advocating violence'. Union officials wanted pickets to succeed but were reluctant to accept responsibility for the actions that were necessary to bring about success. Officials used euphemisms for violence such as 'hot it up a bit today lads', or simply let it be known that fines would be paid. Most activists were critical of the lack of formal instructions and some believed 'it was cowardly of the union branch to give a nod and a wink to acts of violence and then disclaim them'. The officials' dilemma was exemplified at Rossington on 9 July 1984 when pickets, acting on a tip-off from area headquarters, barricaded themselves in the colliery with a dozen management 'hostages'. Local and area officials were persuaded by the police to convince the miners to release their hostages, thereby dissociating themselves from the action. The inadequacy of guidelines for picketing tactics reduced the effectiveness of picketing and undermined the authority of the SCC because pickets decided their own tactics in the light of circumstances at particular targets. Because pickets reacted to police tactics, rather than having formulated their own beforehand, the police were able to have a determining influence on the character of picketing and the Union lost the initiative.

In Yorkshire, the role of panel strike centres and branches was limited to the day-to-day administration of strike operations. The SCC failed to develop mechanisms to encourage the panels and branches to influence strategic decisions, and provided no effective command structure for translating strategy into tactics. The Yorkshire SCC repeated the mistake of the NCC in attempting central control and discouraging branch autonomy. Even at branch level, officials attempted to control the activists but rarely incorporated them into the strike organisation. In picketing, the crucial business of the strike, formal organisation gave way to informal arrangements. Picketing tactics, essential to make the strategy effective, developed only where informal leaders emerged.

3.4 Patterns of activism in Yorkshire

Picketing was designed to secure a complete stoppage of work, prevent the movement of fuel, and maintain morale among the strikers. While mass-pickets were ineffective in relation to the first two objectives, the effect of picketing on morale was in direct proportion to the numbers involved:

> The lack of mass picketing creates a real danger that the mass of strikers may become rusty and dispersed and thus not prepared to meet sudden strike-breaking moves. Lack of mass picketing also prevents the workers involved from becoming active participants and turns them into passive observers . . . strikes in which workers are passive observers and not active participants are in constant danger of ending in defeat.[111]

Callinicos and Simons note that 50,000 miners were on strike in Yorkshire and criticise Taylor for asking SWP members who wanted to expand the Orgreave action, 'where am I going to get 15,000 men from?'[112] Yet this *was* the essential problem: there was only a small core of activists prepared to become involved in the strike.

As the strike progressed and it became clear that the role of picketing would fall on the shoulders of a limited number of pickets, the SCC became concerned with ensuring that pickets did a full shift's picketing. Picketing the day shift was considered to be tactically most effective, even though this put great stress on the pickets who often ended up doing over twelve hours on picket duty.[113] Although aware of the proportion of members actively involved in the strike, neither the Area nor National Union attempted at any time to analyse where support was strongest or weakest among the membership, except on the most rudimentary geographic basis.

Extent of activism

During the peak period of flying picketing (March – August 1984) the SCC deployed a daily average of about 4,500 pickets. The flying pickets represented about ten per cent of the Yorkshire strike force but were not evenly drawn from the four panels. Doncaster supplied the highest number of pickets, representing some eleven per cent of that area on an average deployment; the smallest number came from North Yorkshire and represented about six per cent of the area. Barnsley and South Yorkshire produced roughly the same average daily deployment, midway between the two other areas, representing about eight per cent of each area. Differences between the panels in picketing involvement therefore conformed very closely to the initial response to the strike call.

Over twice the average deployment, about twenty per cent of the total strike force, became involved in flying picketing at some stage. The numbers involved for at least part of the strike also varied between the panels, although to a lesser extent than did the average deployment. The highest involvement was in Barnsley, representing twenty per cent of that area. North Yorkshire had the lowest involvement, representing seventeen per cent; South Yorkshire was marginally higher with eighteen per cent involved. The *turnover* among flying pickets was therefore highest in Barnsley (two hundred and seventy per cent) and North Yorkshire (two hundred and sixty per cent). Doncaster experienced the lowest turnover (one hundred and eighty per cent), while South Yorkshire matched the average turnover for the whole Yorkshire area (two hundred and thirty per cent). Variations in turnover between the panels were associated with a variety of problems. The relatively high turnover in Barnsley and North Yorkshire meant that pickets sometimes lacked the experience to be effective. The stability of the Doncaster picket teams, however, enabled autonomous action which caused problems of control for the SCC.

There was significant variation in the numbers of flying pickets at different collieries within each panel. In Doncaster the percentage of flying pickets ranged from about eleven per cent at Brodsworth to ninety per cent at Highgate (a small branch). At all but three collieries over twenty per cent of the workforce became involved at some stage. In North Yorkshire the percentage of flying pickets ranged from about four per cent at Ricall to thirty per cent at Kellingley. At six North Yorkshire collieries over twenty per cent became flying pickets, while at five pits under ten per cent were involved. In Barnsley the proportion of flyers ranged from eight per cent at Woolley to forty-two per cent at Ferrymoor Riddings. At twelve Barnsley pits over twenty per cent were involved and Woolley was the only pit with under ten per cent of flyers. In South Yorkshire involvement in flying picketing ranged from eight per cent at Thurcroft and Brookhouse to forty per cent at Wath. At five collieries over twenty per cent were involved but only Thurcroft and Brookhouse had under ten per cent of flyers. The proportion of members involved as flying pickets at each branch showed no significant correlation with the status of the colliery, the ballot results in 1981 and 1982, nor with the initial reaction of the membership to the strike call.

When the emphasis moved on to local picketing, a slightly larger number of the area's strikers was involved, representing about twenty-three per cent of the total. Barnsley saw the largest mobilisation of local pickets (thirty per cent) and North Yorkshire the smallest (fourteen per cent), while Doncaster and South Yorkshire were close to the overall average. Within each panel there was again considerable variation between different collieries. The proportion involved in local picketing in Barnsley ranged from eighty-eight per cent at Dearne Valley colliery to thirteen per cent at Grimethorpe. At eight collieries over thirty per cent were involved while at three collieries less than twenty per cent participated as local pickets. In North Yorkshire the proportion of local pickets ranged from four per cent at Kellingley to thirty-three per cent at Ledston Luck. Ledston Luck was the only pit with over thirty per cent involved, while at eleven collieries fewer than twenty per cent became local pickets. In Doncaster, local picket involvement ranged from one hundred per cent at Highgate to eleven per cent at Brodsworth. At three collieries over thirty per cent were involved, while another three saw fewer than twenty per cent. In South Yorkshire involvement in local picketing ranged from seventy-seven per cent at Cadeby to seven per cent at Wath. At four collieries over thirty per cent were involved while at seven under twenty per cent were active as local pickets. The proportion of members involved as local pickets at each branch showed no significant correlation with the status of the colliery nor with the ballot results in 1981 and 1982. However, the percentage of members who picketed locally is *negatively* correlated with the initial branch reaction during mobilisation.[114] The initial branch reaction was probably a good indicator therefore of the future scale of strike-breaking, which event brought more local pickets out (see Chapter five).

Forms of activism

Activists among the membership undertook various duties associated with strike organisation. Picketing activities involved the most people and comprised deployment, local picketing and picketing other targets. Those involved in these acts are referred to respectively as *deployers*, *pickets* and *flyers*. Like all social action, industrial conflict is circumscribed by codes, but these may be violated and the strike involve 'forms of action no longer generally sanctioned by custom'.[115] During the miners' strike, accepted post-war conventions of industrial conflict were suspended and in several instances picketing gave way to acts of sabotage against NCB and police property. Sabotage is therefore considered as action distinct from picketing, but with the related object of preventing production and imposing costs on the employer. Those engaging in such action are described as *saboteurs*. In addition to picketing, actions ranging from gestures or social ostracism to actual physical violence were directed at dissuading people from returning to work. Those involved in systematic attempts to discourage potential or actual strike-breakers, whether through offering advice, issuing threats or actually committing acts of violence against them or their property, are referred to collectively as *persuaders*. By far the commonest form of persuasion entailed local officials visiting a potential strike-breaker and offering support. While actual physical attacks were rare, spontaneous threats were commonplace and 'in so far as threats arouse dread, they are effective, even if they are devoid of intent'.[116] In Darfield, for example, towards the end of the strike, a young miner confided in his friend that he was seriously thinking of returning to work. The next day he found a cross painted on his door, which proved sufficient warning for him to stay on strike to the end without a direct threat of violence. It is impossible to assess the extent of involvement in the milder forms of these activities since occasionally a whole community would ostracise or ridicule a strike-breaker. Neither is it possible to assess involvement in personal acts of intimidation.

Theories of activism

Before the strike, and during its early weeks, when Yorkshire activists discussed the problems of mobilisation and strike organisation there was a recurrent theme that certain individual miners would be less likely to vote for action, would do nothing to make it effective by way of picketing, or could even conceivably become blacklegs under certain circumstances, although scabbing was hardly contemplated as a serious threat in Yorkshire until much later in the strike. When the different theories concerning variations in the degree of strike support are disentangled, six sets of factors emerge: age, length of service, occupation, kinship, community and politics. Each will be examined in turn against the empirical evidence from the Yorkshire coalfield.

Theories concerning the effect of age upon strike support were both the most widely articulated and the most contradictory. It seemed that no age group would support a strike, as each had distinct reasons for not taking action. In ten years the average age of the mining workforce had *fallen* by ten years, effectively a shift of twenty years, and many branch officials feared that support would be weaker among the younger miners: 'these young lads, living at home, spend all their money down the pub, won't go a fortnight without being out on the piss'. Yet younger activists believed support would be weaker among older miners who were 'just after their redundancy money and won't do anything to jeopardise it'. One 58-year-old offered another explanation for his reluctance to become involved: 'I've done my stint in 1970, 1972 and 1974: I'm getting out; let the young 'uns fight for their own jobs.' Of course, those in between these age groups often had mortgages, young families and other commitments so were held to be 'in the Board's pocket . . . they can't strike or they'll lose everything'. All of these difficulties specific to particular age groups were real but if the major participants in the strike were drawn unevenly from the various age groups, this would suggest that the factors affecting a certain group might be more important determinants of strike involvement in general. The age distribution of participants is therefore worthy of further investigation.

At about twenty per cent of the collieries deployers were not characterised as any particular age group, but at forty per cent of collieries they were identified as 'middle aged' (thirty-six to forty-nine years), and at almost thirty per cent of pits as 'family men' (twenty-one to thirty-five years). Pickets were deployed from panel strike centres and at branch level. The strike centres were normally manned by panel members while branch deployment often involved the treasurer, who would control the picket money (£1 per day plus petrol money for drivers). Where the branch treasurer failed to undertake these duties for reasons of health, lack of enthusiasm, or whatever, the secretary or another committee member would substitute. The age distribution of deployers therefore illustrates the typical ages of branch officials and panel members. At thirty-nine per cent of the collieries (local) pickets were drawn evenly from all age groups but twenty-seven per cent of the pits reported a disproportionate involvement of family men and twenty-one per cent of middle-aged. About seven per cent reported 'old-timers' (50+ years) as overrepresented in both deployment and local picketing. At twenty-seven per cent of collieries there was no age bias among flyers but disproportionate numbers of family men were reported at forty-one per cent of pits and of 'youngsters' (sixteen to twenty years) at twenty-seven per cent of pits. Saboteurs were non-existent, unknown to the respondent or drawn from all age groups at twenty-three per cent of collieries, while at thirty-eight per cent they were characterised as family men and at thirty-six per cent as youngsters. Persuaders were either not reported or of all ages at eighteen per cent of collieries, but at forty-five per cent of pits family men, and at twenty-seven per cent youngsters, predominated. The view that any particular age group failed to become involved in strike organisation is unfounded but there is evidence of specialisation in the age pattern of involvement. Middle-aged and old-timers were more likely to be involved in deployment and local picketing, while youngsters were more likely to be flyers, saboteurs and persuaders. Those in the age range twenty-one to thirty-five years, who might be expected to have experienced the greatest financial and family pressures, were prominent in every activity and were most often the best-represented group of pickets, flyers, saboteurs and persuaders.

Length of service was, for many people, a variation of the age theme but it represents an analytically distinct hypothesis. Miners in 1984 were thought to be qualitatively different from those of 1974, a view not without foundation since fifty per cent of the 1984 workforce had not been in the industry during the strikes of 1972 and 1974. Crudely put, 'newcomers', those who joined the industry since the 1974 strike, were not experienced in the sort of tactics necessary to make a strike effective: 'they've no union experience, never been involved in a serious strike, know nothing about picketing'. One Nottinghamshire miner, who was to become a striker, argued 'most of these lads have just joined the industry, they're from the army or engineering, they've no idea'. A miner from Royston Drift commented: 'those that came to the pits after 1974 got the benefits of what we fought for in the big strikes of the 1970s; they were drawn in by the wages but played no part in the struggle to win them and if the going gets rough they'll move on to another industry, assuming there's any jobs to go to'. The consensus before the strike could be formulated as a hypothesis that strike involvement would be lower among newcomers than among the 'old hands' (those with at least ten years' service). Deployment was primarily undertaken by old hands at forty-eight per cent of Yorkshire's collieries and by newcomers at eighteen per cent, while thirty-four per cent reported deployers in both categories. Pickets were drawn evenly from the two groups at forty-one per cent of pits, while thirty-four per cent reported a disproportionate involvement of newcomers compared with twenty-five per cent where old hands played the major role. Flyers were predominantly newcomers at fifty-four per cent of pits, and old hands at seven per cent, while thirty-nine per cent reported them as coming from both groups. Sabotage was either unknown or involved both groups at forty-six per cent of pits, while newcomers were thought to be responsible at forty-one per cent of pits and old hands at thirteen per cent. Persuaders were similarly thought to come from both groups at forty-one per cent of collieries, to be predominantly newcomers at another forty-one per cent and to be old hands at the remaining eighteen per cent. Overall, therefore, both groups participated equally in each aspect of strike organisation at about forty per cent of pits. Where there was a disproportionate involvement, old hands tended to be deployers while newcomers were predominant as flyers, saboteurs, persuaders and, to a lesser extent, as local pickets. The evidence suggests that, far from being reluctant participants in the strike, newcomers were the backbone of much of its organisation in Yorkshire.

Before the strike, many miners believed the degree of support for action varied with occupation. Miners are often assumed to caricature the homogeneous 'isolated mass' of the Kerr-Siegel hypothesis,[117] yet the differing propensities to strike of various mining occupations has been well documented.[118] Hypotheses advanced before the strike were particularly confusing, however. 'Surface workers won't back the strike; they say they can't afford it but what they mean is they have more chance of getting the same money or better in another industry', claimed a face-worker. 'Craftsmen see themselves as above the rest of us and can get work outside the pit... they'll disappear once the strike is underway', said another. According to a craftsman, however, 'facemen will be hardest hit as they're on the big money . . . craftsmen have become more prepared by the overtime ban as we've always worked too much overtime'. In defence of the surface-workers, one suggested 'we're the only ones that can afford a strike . . . we don't have big mortgages like face workers as we earn bugger all'. Some miners identified face-workers as a key group, others included all underground workers. Four categories are therefore distinguished in the analysis: face-workers, craft-workers, elsewhere-below-ground, and surface-workers. At fifty-five per cent of collieries deployers were drawn from all occupational groups, while at twenty-five per cent face-workers were disproportionately involved and at eleven per cent craft-workers predominated. Since branch and panel positions are dominated by face-workers and, to a lesser degree, craft-workers, their preponderance as deployers at some collieries is less surprising than the occupational diversity of deployers at the majority of pits. Pickets showed even greater occupational diversity, with seventy-one per cent of collieries reporting that all job categories were proportionately represented. At twelve per cent of collieries face-workers were thought to be overrepresented among local pickets, while at an equivalent number of collieries, elsewhere-below-ground or craft-workers predominated. Similarly, flyers were evenly drawn from the different occupations at seventy-five per cent of pits, while at fourteen per cent craft workers were over-represented. Sabotage was either unknown or involved all occupations at fifty-nine per cent of collieries, while at twenty-eight per cent underground workers (divided equally between face and elsewhere) were involved. Persuaders were drawn from all occupations at fifty-four per cent of collieries, but at the remainder face, craft or elsewhere-below-ground workers were overrepresented. Overall, occupation was irrelevant as a determinant of strike participation at between one-half and three-quarters of Yorkshire collieries. In the minority of pits where there was occupational bias, surface-workers alone were consistently less involved.

The question of kinship was often confused with length of service, but discussion resolved its supposed role as a separate factor. The importance of oral history and socialisation in the forms of trade-union action appropriate to coal-mining are to be stressed here: 'it doesn't matter if a lad's only been at work a fortnight, if his father, his brothers, grandfathers and uncles are all miners, he'll learn from them'. Conversely, it was claimed, 'even blokes that haven't worked anywhere but the pit don't have the same commitment if they don't come from a mining family'. Although it was rarely acknowledged by miners before the strike, women play a very significant part in this socialisation process. The hypothesis reduces to an expectation that those from a 'traditional' mining family would show more support for the strike than would those from an 'alien' (non-mining) family. The importance of kinship as a factor influencing strike participation is difficult to investigate because the majority of Yorkshire miners come from traditional mining families: 'Relatively few men who are strangers to mining want to endure its risks voluntarily. Only those who grow up in the environment of mining, for whom the costs are an everyday feature, become immune to them. The mining family, therefore, serves to perpetuate the mining industry'.[119] The empirical evidence, however, produced an overwhelming consensus that those from traditional mining families were overrepresented in every activity from deployment and picketing (at eighty-two per cent of collieries) to sabotage and persuasion (at seventy-five per cent of collieries). At the majority of remaining collieries, respondents were either uncertain of kinship ties or thought these were unrelated to strike involvement. At the few collieries where those from an alien family were thought to have participated more, their major activities were flying (at nine per cent of pits) and local picketing and persuasion (at seven per cent of pits). While kinship appears to have influenced strike participation, its usefulness as a general explanation is limited by the rarity of miners drawn from alien families. While it may be said that those with traditional kinship ties were over represented among the activist *minority*, they still formed in absolute terms the majority of those who remained inactive. Moreover, those from non-mining families were less likely to live in a pit village, and the community effect may have been more important than kinship.

Relatively few miners anticipated the importance of area of residence as a factor before the strike. Those who did offered a version of the Kerr-Siegel community integration thesis.[120] Kerr and Siegel attributed the high 'propensity to strike' exhibited by coal-miners throughout the world to their closed communities, which reinforce collective consciousness and maintain isolation from outside social pressures. Although the thesis was originally tested on international inter-industry data, it is arguably better formulated at the level of communities and the individuals living within them, which is how the theory was articulated to us by the miners themselves. Miners living in a pit village, or at least a mining community, were expected to enjoy greater support and be encouraged to become involved: 'If you live near the pit, you're naturally expected to go picketing with everyone else. People help each other to overcome their inhibitions and get involved . . . but if you live away from the village, neighbours might be hostile and there can be transport problems'. Initially we examined the hypothesis that those living in a mining community, *insiders*, would be more likely to become involved in the strike than *outsiders* living in a more integrated community. The evidence from Yorkshire suggests that insiders predominated as local pickets at seventy-one per cent of pits and as persuaders at seventy per cent of pits, with only nine per cent of collieries reporting higher involvement of outsiders in each activity. Saboteurs were insiders at sixty-three per cent of collieries and outsiders at eleven per cent. Outsiders were more prominent in other activities. Deployers were outsiders at eighteen per cent of collieries and insiders at sixty-one per cent, while flyers were outsiders at twenty-five per cent and insiders at fifty per cent. Since deployers were union officials, they will have been accustomed to involvement irrespective of area of residence. It is possible that outsiders were more involved in flying because they owned transport in order to travel to work. For each activity about one-quarter of collieries reported equal involvement of insiders and outsiders. Many miners live away from the pit but in a mining community, whether associated with another colliery or simply a community within the travel-to-work area of several pits. We found little difference in strike involvement between those living in *the* pit village and those living in *a* mining community. Local pickets were more likely to come from the local village, while flyers were as often from another mining community. Miners typically picketed with those from the pit nearest their home, which was not necessarily their own pit. The important determinant is, therefore, the existence of a cohesive mining community, even where this is divorced from the colliery.

Finally, some miners expressed the view before the strike that an individual's predisposition to become involved was governed by their politics above any other factor. 'Conservatives' were characterised as those with a privatised lifestyle, whether or not actual Tory voters: 'They keep to themselves, don't go to the Club and have no loyalty to the Union. Even if they come from a mining family and live in the local community, they behave as though they don't'. Such conservatives were not expected to participate in the strike. There was consensus that 'radicals', broadly defined as members of the Labour Party or any other political grouping on the left, would be in the forefront of the forthcoming conflict. It was also clear that the vast majority of NUM members could be described as 'non-political', typically loyal Labour voters but lacking any wider political consciousness. From mobilisation, through the organisation and continued maintenance of the strike, radicals would have to struggle for the support of the non-political majority in the face of powerful institutional opposition (see Chapter five). The empirical evidence suggests that radicals were only overrepresented as saboteurs at forty-three per cent of collieries, and as persuaders at forty-five per cent. Non-politicals were overrepresented as local pickets at sixty-eight per cent of collieries and as flyers at sixty-three per cent. Deployers were reported as predominantly non-political at forty-eight per cent of collieries and as radicals at thirty per cent. At almost a quarter of pits deployers, saboteurs and persuaders represented a wide political cross-section.

Identifying the 'typical' activist is far from straightforward. Although in round terms only 10 per cent at any one time and 20 per cent over the whole strike were active, this minority generally reflected the diversity of the mining workforce in Yorkshire. At almost half the collieries, most groups of miners were involved in most activities. This diversity is exemplified by the profile of local pickets at the majority of pits: drawn from all age groups, with varying degrees of experience and reflecting all occupational groups, local pickets were generally non-political, from traditional mining families and living in mining communities. Deployers differed only in being primarily middle-aged old hands, while flyers tended to be family men who were newcomers to the industry. Saboteurs and persuaders were distinguished as family men, more often radicals than non-political. Age and politics therefore appeared to influence the type of involvement; younger and more radical miners were involved in flying, sabotage and persuasion, while older and non-political miners confined their activities to deployment and local picketing. Occupation proved largely irrelevant to strike involvement. Kinship and community showed the greatest consistency over all activities: activism was clearly associated with, although not exclusive to, traditional mining families and, especially, mining communities.

The miners' strike in Yorkshire was led by activists and sustained by them, yet they had little formal control over strike strategy because the Union failed to develop an adequate organisation to harness their skills and encourage a wider involvement. The patterns of activism demonstrate how only a small minority became active, and how the Area Union failed to recognise the importance of community organisation in the strike.

Chapter four

Strike maintenance

The 1984–85 miners' strike was the largest strike ever to have lasted for a year in Britain. It therefore entailed exceptional hardship and posed unprecedented problems for strike maintenance. This chapter examines the support miners received from the labour movement, the development of women's action groups, and strike maintenance within the Union. The reservations displayed by the leadership of the labour movement about the strike are shown to have prevented an adequate representation of the miners' case. Variations in the degree of trade-union solidarity and the failure to mobilise TUC support are analysed. The support networks which emerged demonstrate the contrast between the leadership and membership of the labour movement. The successful maintenance of the strike is attributed to the organisation of women's support groups. The Union's strike maintenance provision, in terms of both material support and propaganda, is analysed and found to have been deficient. After assessing the nature and extent of hardship experienced in the Yorkshire coalfield, we analyse how individuals coped with this and consider how the Union could have organised more effective maintenance.

4.1 The labour movement

Both wings of the labour movement were involved in raising money for the miners but each was uniquely qualified to support the NUM in other ways. The LP could have publicised the miners' case in three ways: party press officers could have assisted the NUM in countering the media offensive; information from the NUM could have been distributed to the CLP network from Walworth Road, and the PLP could have pressed for a parliamentary debate on the coal crisis. Likewise, the TUC could have taken the role of co-ordinating industrial action in support of the NUM. In fact, neither institution fulfilled these roles. Although rank-and-file LP members and trade-unionists demonstrated immense financial and moral support and occasionally industrial solidarity for the miners, these activities were organised despite the lack of official support.

The Labour Party

The LP leadership did little to publicise the miners' case and the strike brought into sharp relief divisions within the PLP. These divisions enabled the leadership to avoid giving the support expected by the NEC and LP Conference. The Campaign Group and individual left-wingers saw the miners' strike as a stand against unemployment and the attack on public ownership on which the LP could have built a campaign against the Thatcher Government: 'Most of the pressure we were exerting was clearly on the leadership to adopt a more supportive position on the dispute; to support Labour Party initiatives that were being taken . .. There was a difference between the NEC of the Party and the Parliamentary leadership . . . the Party nationally was very much supportive'.[1] The leadership and PLP right-wingers, however, believed the dispute, or more specifically its conduct, would damage support for Labour among the social democratic middle ground. Opposition leader, Neil Kinnock, struggled to maintain unity while distancing the LP from the miners' strike. The resultant equivocation satisfied neither left nor right and the media used the strike against the LP *irrespective* of Kinnock's position (see Chapter five). A Marplan poll on 15 November 1984 attributed the Tories' lead to the miners' strike and four months after the strike had ended Kinnock blamed Scargill and Benn for Labour's defeat in the Brecon and Radnor by-election.[2] The ambivalence of the LP leadership towards the miners' strike was symptomatic of the wider contradictions within the party; while dependent upon the support of working-class militants, it has consistently failed to satisfy their aspirations for a new socialist order.[3]

The PLP was even reluctant to provide the minimal publicity of parliamentary discussion. Dr David Owen for the SDP and Jonathan Aitken for the Conservatives accused Labour of avoiding a debate on the strike, while Skinner confirmed that Kinnock was not 'altogether keen'.[4] The Labour leadership first initiated a debate on the mining dispute on 7 June 1984 when the strike was in its thirteenth week;[5] later attempts to force a debate were less successful. When pressed by Kinnock on 3 September, Thatcher refused to recall parliament for an emergency debate on the coal strike. In January, when the strike was weakening, a dozen left-wing MPs refused to sit down in the Commons in an effort to force a debate, but Kinnock condemned them as 'undisciplined' and claimed they were 'not helping miners'.[6]

The miners' strike entailed both substantive issues concerning the causes of the crisis and procedural ones relating to how the dispute was handled by the parties. The leaders of the PLP were cautious when attacking the Government's handling of the dispute for fear of exposing NUM tactics, about which they had serious misgivings, to Tory criticism. However, in avoiding discussion of the procedural issues, the PLP also conceded the substantive argument by default and misssed the opportunity to demonstrate the logic of a socialist energy policy and promote Labour interests. Moreover, when Labour leaders joined the Tories in criticising the NUM, they were given maximum publicity which damaged support within the labour movement. Whatever position the PLP adopted would, of course, have been used by the media and the Conservatives to undermine either the strike or the party.

In the early weeks of the strike, the NUM Executive decided 'they wanted no political or TUC involvement'.[7] The decision reflected a justifiable distrust for the PLP and TUC leadership who had shown little enthusiasm for any of the key industrial disputes of the 1980s. The decision had the effect of excusing the PLP leadership from putting the case for coal but failed to prevent them from criticising NUM tactics, which were attributed to Scargill. The two major criticisms which PLP leaders made of NUM tactics concerned violence and the ballot, issues maintained on the agenda by the Government and media (see Chapter five). Kinnock repeatedly criticised picket violence, arguing that this obscured the miners' case.[8] During the key phase of the NCB back-to-work campaign, James Callaghan suggested that the NUM should ballot the membership on a return-to-work.[9] Stan Orme maintained a principled position on the ballot throughout the strike, replying to a call for a ballot from John Carlisle (Con., Luton N.): 'that is an internal matter for the NUM',[10] but admitted that the absence of a ballot influenced the PLP's avoidance of a major parliamentary debate before 7 June: 'the NUM did not want the matter discussed because they didn't want the issue of the ballot debating'.[11]

The Opposition motion in the first main debate on the miners' strike criticised the Government on both procedural and substantive grounds:

> this House strongly condemns the Government for its mishandling of the coalmining dispute and the way in which severe hardship has been inflicted on miners, mining families and communities and for its abandonment of the principles and objectives of *Plan for Coal*; calls for the withdrawal of the plan to lose 20,000 jobs and close over 20 pits in the coming year, which was a unilateral decision of the NCB; and believes that the Government should then call a tripartite meeting to project the *Plan for Coal* forward based on investment and an expansion of the coalmining industry which will be essential for Britain's needs.[12]

Opposition Energy Spokesman, Stan Orme, challenged Thatcher's claim of 11 May that the Government 'are not going to intervene in the coal dispute', citing the Ridley Report as evidence of their involvement but also criticising Walker for having failed to speak to 'both sides' after thirteen weeks.[13] In May Jack Dormand (Lab., Easington) had called upon the Home Secretary to establish 'an inquiry into police activities in the present coal mining dispute', while Kevin Barron (Lab., Rother Valley) criticised both the intercept policy and the 'proforma used in the Nottinghamshire coalfield'.[14] Several Labour members returned to this issue during the June debate in the light of their experiences. At the LP conference Kinnock was careful to condemn both 'the violence of the stone-throwers and battering-ram carriers' and 'the violence of cavalry charges, the truncheon groups and the shield bangers'.[15] Such even-handed condemnation of *all* violence had an uneven impact: it gave credibility to media accounts of picket violence but failed to counter the myth that police violence was isolated or a response to the miners' aggression. The following day Leslie Curtis, Police Federation Chairman, claimed that certain officers would be unable to work with a Labour administration[16] and Kinnock was moved to praise the police force as 'the finest in the world', explaining that they had been misused by the Government.[17]

By the time the substantive issues were discussed in the June debate both parties had rehearsed their positions. In introducing the debate, Orme condemned the Government for closing collieries before exhaustion contrary to the expansionist intention of *Plan for Coal*.[18] Walker reiterated that the Government was 'financing the biggest investment programme in history',[19] which Tony Marlow (Con., Northampton N.) claimed would reduce energy costs.[20] Orme argued that investment expenditure on modernisation, as opposed to 'new coalfields and major new workings', had fallen by thirty per cent since 1979 and that 'if there had been investment in what are described as uneconomic pits, they would not be uneconomic'.[21] The debate over investment was futile without a recognition that *any* measures to increase production would only exacerbate the *demand crisis* unless the market was expanded in proportion. When Benn reminded Walker that the production targets of *Plan for Coal* were to be cut from 130m tonnes to 100m tonnes,[22] Gerald Howarth (Con., Cannock and Burntwood) was able to demand 'what will happen to that extra 35m tonnes?'[23] Conservative policy therefore appeared rational: low-cost production was to be substituted for high-cost production in a contracting market. The central defect in this policy, the sterilisation of reserves which Ann Clwyd (Lab., Cynon Valley) denounced as 'industrial sabotage',[24] was given insufficient prominence by both the PLP and the NUM.

While parliamentary debates and questions were obvious ways in which the PLP could promote the miners' case, Parliament was an ineffective forum for broadening support. Even the limited discussion of the substantive issues which took place was not reflected in media coverage but was safely confined to the parliamentary vacuum. The PLP could have been more effectively used to promote the miners' case *outside* Parliament, but the strike had run for six months before the NUM formally sought such assistance by suggesting that 'the leadership' participate, and other front-bench spokesmen, in a series of rallies'.[25] Although Scargill and Kinnock had 'constructive talks' on 25 September, relations between the two leaders were poor and were not improved by the LP Conference. The situation deteriorated further when Kinnock publicly condemned Roger Windsor's visit to Gadaffi without first establishing the exact circumstances of the meeting.[26] After Scargill publicised the union's decision to invite Kinnock to appear on the NUM platform the Labour leader's refusal was widely reported.[27] National officials finally met LP leaders on 13 December: 'the Labour Party confirmed its total support for the Union and its campaign for the industry and had called upon the TUC/LP Liaison Committee to meet urgently and discuss further the stronger financial, industrial and political action'.[28] The LP also announced there would be a rally or demonstration in the New Year and in January the NEC were told of various

LP support campaigns including a 'party political broadcast dealing with unemployment and the arguments of the miners'.[29]

It was therefore ten months into the strike, and three months after NUM officials made overtures, before the PLP leadership gave the miners public support. This delay advertised the leadership's reservations about the miners' strike and inhibited rank-and-file support. Despite the lack of leadership, Labour 'Members responded as individuals to the calls for campaigning'.[30] The Campaign Group consistently publicised the miners' case and joined NUM picket lines, but moderate Labour MPs from mining constituencies, like Roy Mason, Allen McKay and Alec Woodhall, also showed their support by visiting strike centres, kitchens and picket lines, and by taking up individual cases involving police harrassment or denial of social security benefits. It was not until 16 November that Kinnock, son of a Tredegar miner, promised his constituents that he would visit a picket line, which he did on 3 January 1985. Kinnock's action prompted other MPs to assess the support they had given: 'There was a point after Christmas when the Yorkshire Group became sensitive to criticism that MPs had not done enough . . . we went to picket lines as a group then rather than as individuals . . . that was good in publicity terms. We had a good turnout of MPs'.[31] A Campaign Group member stressed the *limited* nature of this gesture: 'I regret I wasn't able to get the Yorkshire Group, as a group, to support picketing on more than one occasion'.[32] Martin Redmond argued 'it was difficult trying to decide what the Union wanted . . . we were for ever lobbying the Yorkshire officials asking . . . them to give us information on which we could act'.[33] In contrast, Benn considered: 'I was very well briefed and informed and I didn't do anything of substance without checking out with them [NUM] because I felt that this was the right thing to do. I was in much closer contact with the NUM than the parliamentary leadership was throughout the strike'.[34]

The LP was an important source of funding for the strikers. Many Labour MPs who actively supported the strike saw their major role as fund-raising and accepted that the basic material needs of miners and their families had to be met in order to sustain the strike. From May onwards Labour MPs contributed a minimum of £5 per week to the NUM strike fund; members of the Campaign Group also donated £600 directly to NUM regional food centres.[35] The CLPs were the major focus of party fund-raising for the miners, organising collections, raffles, benefits and social events for mining families either independently or as part of regional support groups.

The Secretary of the PLP believed that the LP had never: 'engaged in such active fund raising for any cause . . . I've been in the Party 28 years and I can't remember activity at local level to such an intensity . . . in parts of the country that couldn't be more remote from mining areas.'[36] In Bradford an estimated £25,000 was raised during the strike, although a local MP argued that the level of support 'on a per head basis is miserable; we should have done very much more out of a basic catchment area of three to four hundred thousand people'.[37]

LP and Communist Party members in Shipley miners' support groups raised about £6,000 over the year, while in Leeds supporters collected as much on a weekly basis. In Chesterfield, the Women's Support Group largely consisted of LP members who raised on average £7,000 per week, a dramatic contrast to the normal fund-raising of the local LP which 'had probably never raised and spent more than £5,000 a year'.[38]

Miners' support groups sprang up all over the country, and were particularly concentrated in the large urban conurbations and centres of militancy. On Merseyside fourteen support groups collected £1m,[39] while, at the other extreme, Waveney Town (Suffolk) collected an average of £25 per week.[40] Support groups, whether independent or within LP constituencies or branches, were mostly formed by the end of May 1984. Although the groups varied considerably, there were thirty people in the average group which raised £119 per week according to the LRD survey.[41] Most of the money was raised in weekly street collections, so representing a small return for the effort involved. Some CLPs, Hull for example, raised regular sums through a membership levy; had the levy been organised on a national scale, the miners would have received a substantial regular income throughout the strike.

The absence of national co-ordination or even encouragement within the LP increased the tendency for fund-raising to become fragmented. Leeds Food Force was established by local LP and Communist Party activists to channel resources centrally to the Yorkshire Area NUM. Although street-collecting licences were obtained, there were too few activists available for street collections, so miners were invited to collect. Miners from Sharlston collected in Otley, and from Ledston Luck in Leeds, on the understanding that the money would go directly to the branch concerned, not the Area NUM. The collections, like the numerous twinning and adoption arrangements involving individual collieries and support groups, were a crucial source of income, but such fragmented activities undermined the Area Union's efforts to ensure equity between the branches (see below).

The trade unions

The contrast between the activities of the PLP leadership and individual Labour MPs was mirrored in the difference between the TUC General Council and individual union officials. Left-wing union leaders consistently supported the miners and right-wingers opposed them, while those with the balance of power prevaricated. The leaders of the two giant general unions both urged the TUC to support the miners; whereas Moss Evans (TGWU) argued the miners must not be starved into submission, David Basnett (GMBATU) hoped the NUM would go to the TUC.[42] The national officials of the NUM failed to persuade ISTC leaders to encourage supportive industrial action among steel-workers. As a result, the steel problem distorted NUM strike organisation throughout the crucial first six months (see Chapter three). There was rather more support among GMBATU and TGWU members in electricity supply but this was patchy and contrasted with the hostility of EETPU and EMA members towards the miners. Solidarity action therefore focused upon the transport unions in an effort to cut off the fuel supply to steel-works and power stations.

While the leaders of all the major transport unions called upon their members to support the NUM, there was enormous variation in the response of the rank-and-file. The TGWU had the least success, although Alex Kitson explained 'if any of our members come into difficulty they will have the full support of the TGWU for observing any picket line'.[43] The fragmented ownership of road haulage and the prevalence of small operators and self-employed drivers made it difficult to monitor breaches of Union policy. Many drivers were either non-union or held token membership whereby union dues were paid by the employer, a practice accepted by the TGWU on pragmatic grounds. Therefore, when the TGWU expelled members who were driving coal past pickets at Llanwern, haulage operations were unaffected.[44] The South Yorkshire Panel compiled a list of thirty unionised haulage firms crossing NUM picket lines, which the Grimsby TGWU office asked to refrain from carrying coal.[45] The most effective solidaritv action was taken by the NUS in preventing coal shipment from the North-East to Thames power stations. The British collier fleet was brought to a standstill and Sam McCluskie was confident that the ITF would ensure that 'no coal is loaded on British ships'.[46] In addition, the NUS gave the first floor of their offices to the NUM to co-ordinate support of other unions.[47] A month later Scargill met union leaders from forty-nine countries at CGT headquarters in Paris and received assurances from Australian, French and Polish unions that coal shipments to Britain would be stopped.[48]

Since most coal had been moved by rail before the strike, the attempt to restrict coal movement pivoted on supportive action by rail workers. On 3 April the Federation of Rail Unions called upon members not to move coal or cross NUM picket lines. Rail-workers at depots like Tinsley nerar Rotherham, where the NUM strike was solid, had already approached local NUM officials to make such action more effective. Where the majority of NUM members were still working, however, BR management suspended rail-workers who refused to work normally in an effort to maintain vital supplies to the Trent power stations. At Coalville in Leicestershire, the NUR branch reached a compromise with management to operate trains from opencast sites. In May, Scargill met leaders of the TGWU, NUS, NUR and ASLEF to discuss ways of tightening control over coal movement.[49] By June these four unions had agreed to blockade *all* fuel for power stations but the blockade proved impossible to maintain owing to weaknesses in road haulage and the fact that some power stations are fed by pipeline. When management pressed for normal working at Coalville the ASLEF and NUR branches voted to stop moving any coal and members were again suspended. At Shirebrook, centre of NCB strike-breaking operations (see Chapter five), rail-workers voted to black all coal movement and thirty-two drivers were suspended but BR claimed coal deliveries were unaffected.[50] The following day rail unions instructed members not to cross picket lines or transport any fuel into power stations.[51] Even after the NUR annual conference increased strike pay from £1 to £11 per day and Jimmy Knapp promised all coal movement would be stopped, coal trains were still running from Shirebrook depot.[52] Derby branches of the NUR and ASLEF threatened action over Coalville in August and the national leadership spoke of action over victimisation at Shirebrook in September. Nevertheless, with the exception of a temporary reversal following a meeting addressed by Knapp and Ray Buckton, support steadily weakened at Shirebrook and BR used substitute labour at Coalville.

Printworkers also took direct action in support of the miners. In April SOGAT82 threatened to prevent newspapers being printed where the strike was reported unfairly or where the right of reply to a particular report was refused.[53] When Lord Matthews, owner of the *Daily Express*, was pressurised by SOGAT82 into allowing the NUM to reply to a feature 'the truth that Scargill dare not tell', the Editor threatened to resign and journalists initially refused to handle the NUM reply,[54] although it was finally published on 15 May. Print-workers also prevented the *Sun* from printing a picture of Scargill waving in what the caption implied was a Nazi salute on 14 May. The following month newspaper proprietors refused the print unions' request for a quarter-page statement of support for the miners to coincide with a regional day of action and no Fleet Street papers appeared on 27 June.

The supportive action taken by the NUS was the most complete and consistent, but was insufficient without the continual support of dockers, which only developed on two occasions (see Chapter three). Nevertheless, when dockers and print-workers took action this was generally effective. The support of rail-workers was more difficult to maintain but much more effective than attempts to halt road haulage. Support among power-workers was extremely uneven and virtually non-existent among steel-workers. These variations in solidarity action may be explained by the industries' relative economic security and differences in the unions' leadership and workplace organisation. While the fear of unemployment was a factor militating against support in general, the economic security of the newspaper houses and shipping companies was unlikely to be affected in the short term by action workers took to support the miners. British Rail was more vulnerable but rail-workers appreciated that jobs in freight would be lost in proportion to the contraction of coal-mining capacity. The road haulage industry had experienced surplus capacity for some time, competition was fierce and drivers underemployed. The opportunity to haul scab coal represented a bonanza which saved some small operators from bankruptcy. Steel-workers undoubtedly felt most insecure and believed solidarity action could immediately determine the location if not the scale of job losses. Power-workers, however, were aware of their long-term security, so the fear of unemployment was relatively unimportant in their case.

The quality of union leadership clearly had an important influence upon solidarity action. The leaders of all the transport unions gave tremendous support to the miners, as did the leaders of the print unions. The leaders of the

GMBATU, representing the largest proportion of power-workers, were initially indifferent to the miners, although they appeared supportive later. The leadership of the EETPU and EMA, however, representing key power-workers, along with the steel-workers' leader, were openly hostile to the miners. While union leaders are constrained by membership aspirations, the diversity of opinion among the rank-and-file gave most leaders considerable scope to promote a particular policy on the miners' strike. It required some effort on the part of leaders to encourage support for the miners if they wanted to prevent solidarity from being undermined by the media. The effectiveness of such intervention was largely dependent upon the nature of workplace organisation and the degree of control exercised by officials. Highly developed workplace organisation in newspapers and on the docks enabled union policy to be translated into action. Although workplace organisation is underdeveloped in shipping, the high degree of centralised control which the NUS leadership has over the membership was wielded in the miners' interests. The rail unions had neither strong workplace organisation nor centralised control which made the solidarity of rail-workers more problematic. Although the TGWU generally promotes workplace organisation, none exists in most road haulage companies and the leadership was unable to exert control. Plant-level union organisation in steel was severely damaged after the 1980 strike when a generation of activists was removed through redundancies. The right-wing leadership therefore had little difficulty in maintaining opposition to the miners' calls for solidarity. Similarly, despite fairly extensive workplace organisation, the power industry is characterised by centralised collective bargaining and a significant degree of corporatism at all levels so management and union leaders together were able to discourage membership support for the miners.

Unions supporting the miners hoped to extend action through the trades councils and regional councils, despite the attempts of the General Council to distance the TUC from a strike which they believed was damaging to the labour movement and would be difficult to win. When the Yorkshire and Humberside Regional Council of the TUC organised a day of action in support of the miners on 21 May, Len Murray claimed that the Regional Council was 'exceeding its authority' and acting in direct contravention of a General Council decision not to intervene in the miners' strike.[55] Ronnie Fieldhouse, Secretary of the Bradford branch of SOGAT82 declared 'Mr Murray cannot supersede any individual decision of an individual trade union'. Murray's opposition contributed to the uneven response to the day strike; civil servants voted not to take part while printers, bus-workers, rail-workers and other public-sector workers showed overwhelming support.[56] The TUC was similarly unenthusiastic about the days of action organised in Wales on 12 June and in the South-East Region on 27 June when transport workers brought the London underground, buses and commuter rail services to a standstill.[57]

A month later at a Tolpuddle Martyrs rally, Buckton warned trade-unionists not to wait for the TUC to organise action in support of the miners but to 'come into the fight now because their fight is your fight'.[58] In the week before Congress, however, Scargill told a rally in Ayrshire that the full support of the TUC would be *demanded*.[59]

The NUM prepared a motion for Congress calling for 'total support', while other unions considered amendments for a 10p per week levy of trade-unionists (NUR), no crossing of NUM picket lines (ASLEF) and no use of fuel supplied by non-union labour (NUS). The NUM motion was expected to gain enormous support at Congress but there was speculation over whether the NUM would leave the amendments separate and risk their being lost or incorporate them into their substantive motion to commit the TUC to physical support.[60] To avoid such commitment at Congress, the General Council invited the NUM leaders to a meeting beforehand in Brighton on 30 August and convinced them to request the NUR, ASLEF and NUS to withdraw their amendments in return for a compromise statement of verbal support from the General Council:

> The General Council condemn the NCB's efforts abetted by the Government to run down the coal industry and affirm total support for the following:
> (i) support for the NUM's objectives of saving pits, jobs and mining communities;
> (ii) concerted campaign to raise money to alleviate hardship in the coalfields and to maintain the union financially;
> (iii) to make the dispute more effective by:
> (a) not moving coal or coke, or oil substituted for coal or coke, across NUM official picket lines, or using such materials taken across NUM official picket lines;
> (b) not using oil which is substituted for coal.
>
> The NUM acknowledge that the practical implementation of these points will need detailed discussions with the General Council and agreement with unions who would be directly concerned.
>
> The General Council call for a fresh commitment of all to an expanding coal industry.
>
> The General Council call on the NCB to resume negotiations immediately with the NUM to resolve this damaging and costly dispute in line with the *Plan for Coal*.[61]

Right-wing union leaders expressed dismay at the General Council statement. Sirs accused Murray of sacrificing industrial peace for one week's peace at the TUC,[62] while John Lyons (EMA) questioned the ability of the TUC to organise effective action.[63] In fact, the General Council's statement did not materially advance the strike at all. The phrase relating to 'practical implementation' implied that other unions would have to agree before supportive action would be forthcoming, effectively the position that already existed, and, most significantly, gave an opportunity for increased TUC involvement.

On the first afternoon of Congress, Murray asked for endorsement of the General Council statement, reminding delegates of their 'special debt to Britain's miners'.[64] Scargill was given a tumultuous welcome when he moved the NUM's Motion 25 and supported the statement of General Council. Speaking of the 'determination and courage' of his members who had been on strike for six months, Scargill asked delegates to give the same 'unstinting support that has already been given by our colleagues in the NUS, the NUR and in the TGWU because without their support we would not have survived'.[65] Knapp called for '*commitment* – effective action', while Slater argued that the cost of solidarity was no excuse for failing to support the miners.[66] Like Scargill, Slater implored the ISTC to support the NUM and promised NUS support if the steel-workers faced difficulties as a result. Buckton had difficulty in maintaining order when Eric Hammond (EETPU) attacked the General Council's statement as 'dishonest and deficient',[67] and warned that the EETPU would not be stopped by this statement. Hammond accused Council of voting for point (iii) 'because they thought it would be ineffective: they rely upon the power workers to disagree'. Lyons also objected to the EMA being expected to 'solve the miners' dispute for them' since this would make power workers 'responsible for hitting the elderly and sick . . . We will not do it . . . for our enormous power to be used in this way would threaten democracy itself'.[68] Both the General Council statement and the NUM motion were overwhelmingly carried, with the EETPU, EMA and ISTC representing the only serious opposition. However, the euphoria of Congress was not translated into direct action; the gap between rhetoric and reality had simply grown wider.

Following Congress, the NUM national officials met leaders of the ISTC and the power-workers' unions at Congress House but failed to win their support.[69] A month later the TUC agreed to co-ordinate further meetings to make the strike more effective.[70] Fears that the TUC could not deliver the support promised were vindicated. At the Trent power stations TUC guidelines were ignored as TGWU members worked overtime.[71] In Yorkshire, however, no coal was accepted at power stations and power-workers adopted the oil embargo. Like the NUM, the TUC only succeeded in gaining the support of certain sections of the movement, and in November the NUM Executive was informed that: 'the General Council of the TUC had recently decided to enlarge the Liaison Committee with the NUM in order to determine further means by which the Congress decision . . . could be more fully implemented. Furthermore meetings were being convened by the TUC between representatives of the transport and electricity power industries in order to try and strengthen TUC policy.'[72] Scargill told a rally in South Wales that the strike would have been won already had the TUC given the NUM the same level of support which the CBI had given the NCB.[73] The same day Norman Willis, who had succeeded Murray as TUC General Secretary, described violence as 'alien to our trade union tradition' and abandoned his speech to South Wales miners when a noose was lowered from the gallery above the platform.

The threat of sequestration of NUM funds (see Chapter five) brought another crisis in the Union's relations with the TUC and, at a lengthy meeting at Congress House on 2 December, the parties failed to reach agreement on action in the event of sequestration. The next day the SDC voted not to pay the fine, not to purge contempt, not to assist the receiver to obtain Union funds, and called upon the General Council to organise supportive action. After further discussions on 6 December with national officials, the TUC Liaison Committee reiterated support for the NUM and agreed to discuss ways of enabling the Union to continue to function.[74] Sequestration represented a qualitative change in the conflict which should have prompted the TUC to mobilise whatever industrial action it could in defence of the NUM. However, the majority of the Geneal Council were not in favour of industrial action and instead offered the NUM an alternative office and financial assistance to keep the national office operational. Whether unwilling or simply unable, the General Council failed to mobilise industrial action in support of the NUM and became increasingly involved in negotiations which attempted to negate NUM policy (see Chapter six). At the SDC which finally terminated the strike, Scargill told delegates 'the fact that this Union today is facing the present dilemma . . . is because we have been isolated by the movement that should have come to our aid'.[75]

The response of the trade-union movement was not uniform, but all levels were involved in raising money for the miners' strike. Yorkshire and Humberside Regional TUC raised over £400,000 and the trades councils within it as much again, as well as organising collections of food, clothing and toys.[76] In Barnsley the Trades Council established a support cabin in the town centre in June 1984 to provide an advice centre for strikers and act as a focal point for the collection of money, food and clothing. By March 1985 collections at the cabin, from an area almost totally dependent upon coal-mining, amounted to £10,673.[77] Of the twenty-four trades councils which responded to the LRD survey, 'some had set up special miners' support sub-committees, while others were very clearly linked closely with local miners' support committees'.[78] According to the LRD survey, the average trades council collected £230 per week, but there was an enormous variation, with Edinburgh collecting £1350 per week and Harringay £1500 per week.

Trade-unions donated large sums centrally to support the NUM. The GMBATU, for example, provided over £1m. Some unions stipulated how the money should be used: 'the ISTC, made responsible by the International Metalworkers Federation for distributing its contributions of £450,000 in kind and cash, made sure it all went in the form of food parcels and toys'.[79] The 127 individual workplaces and trade-union branches responding to the LRD survey collected an average £110 per week, ranging from £5 per week collected by a member of BFAWU in Oldham to £525 per week raised by the JSSC of Rolls Royce, East Kilbride from 1,300 members.[80]

At the start of the strike the NUM made an international appeal to mining unions concerning both the question of imports and financial donations. Agreements were made with these unions regarding the organisation of collections among their own members and means of generating support from other unions in their respective countries. The appeal was extremely successful and it was reported to the NCC in July that 'the International Appeal . . . is currently yielding twenty per cent of the total income received into the Miners' Solidarity Fund'.[81] The Swedish miners' union introduced a membership levy, while the French made an initial donation from national funds and then organised regular colliery collections, as well as selling a 'postcard of solidarity' addressed to the British Ambassador in Paris, which had raised £4,000 by July. The Australian Coal and Shale Employees Federation paid for two NUM representatives to visit Australia on a fund-raising tour. In Germany the IGBE made an initial donation of 20,000DM and organised regular collections at colliery level. The CGT and ICTU also made regular donations to the Miners Solidarity Fund, while children's holidays were provided in France and the Soviet Union by supporting trade-unionists.

Christmas was a critical period for strike maintenance when miners were under most pressure to abandon the strike. Children's toys poured into the pit villages from European trade-unionists, while support groups financed Christmas parties. Where the parties involved the whole community, as in Upton, they were an enormous success. Miners at Kellingley declared that 1984 was their 'best Christmas ever', but once the celebrations were over it was difficult to maintain unity within a workforce drawn from several communities (see Chapter five).

4.2 Women's action groups

The financial and material needs of the coalfield communities were met by women's action groups, which were formed in a variety of ways in Yorkshire. One of the earliest, Barnsley Women Against Pit Closures (BWAPC) began when five women wrote to the *Barnsley Chronicle* in support of the strike 'with the invitation to like-minded women to join us'.[82] Founder members of the group toured local villages encouraging women to establish similar organisations to support mining families.[83] In Sheffield, Kath Mackey and Pat Berry, who had been closely involved with the steel-workers' strike, began collecting food for miners. After making contact with Brookhouse and Treeton NUM branches, these two arranged a meeting of about forty women, mostly from outside mining communities, and formed Sheffield WAPC.[84] In North Yorkshire the early groups were begun primarily to help single miners without family benefits. Eight groups operating in North Yorkshire, largely autonomously, first met at the Barnsley rally on 12 May 1984 and agreed to organise an area meeting, which was attended by thirty-five women. By the end of the strike there were thirty-four groups in North Yorkshire,[85] while in Barnsley two separate women's groups each co-ordinated about sixteen village groups.[86]

Among our respondents in Yorkshire who completed the women's section of the questionnaire, thirty-two per cent were involved in at least one activity to sustain the strike. Of the women involved, seventy-six per cent were wives or girlfriends of activists who themselves spent at least five days per week on activities related to the strike, while only ten per cent were associated with miners who spent fewer than three days per week on strike business. Therefore, while there were women who became active despite the lack of commitment to the strike shown by their partners, overall there was a close correspondence between the activism of women and that of their husbands or boyfriends. More women were involved in kitchens than collecting and more collected than went picketing. Moreover, women associated with activist miners predominated among those who went picketing more than among those who ran the kitchens. While the women's groups struggled to operate autonomously, the pattern of women's involvement was nevertheless conditioned by traditional gender roles within coalfield communities.

It is possible to estimate the level of women's involvement in Yorkshire overall using the results of our survey in conjunction with the level of involvement among miners analysed in Chapter three. On this basis, 4·7% of Yorkshire coalfield women had an involvement of a collective kind at some time during the strike. Naturally all women and men were involved in maintenance within individual striker households. Overall, 3·7% of women were involved in manning kitchens, collecting or picketing, while three per cent were members of a women's action or support group. Therefore, over half of the women involved in maintenance activities were members of a formal group. Women from communities wholly or partly dependent upon coal-mining were twice as likely to be involved as miners' wives living outside traditional mining communities. There may have been a higher proportion of women who would have liked to have become involved but were discouraged by their husbands:

> I went to Park Lane College a few times with my husband. The woman there was pleased to see me and said I ought to encourage more women to come. We would have liked to come, but the men didn't want it; they liked to go to all the talk and then socialise in the pub afterwards and have drinks bought for them, but they didn't want us there. She didn't understand: even if we wanted to come, we have to do as we are told and our place is behind the kitchen sink.[87]

Another reason advanced for the small proportion of women involved in the strike is that some support groups, once established, did not always welcome newcomers who might disrupt working relationships:

> They just wanted to do it themselves. I felt that I wanted to help but could find no real way of doing this. Just a small clique controlled the support. They should have had meetings and let more women come along and get involved. More could have been helped then. There must have been lots of people needing help that they didn't know about and more involvement would have helped to get to those people.[88]

Some women's groups, however, found it difficult to recruit supporters: 'Many people seemed frightened to get involved; quite eager to have that meal and any other service we could offer, but when we asked if they could spare an hour it was always "well . . . " This attitude could have really disheartened the best of us but there was always a minority of us who would not give in.'[89]

Feeding strikers

Feeding, clothing and keeping the community warm became the primary responsibilities of the women's groups. The kitchens and food parcels organised by the women's groups probably sustained the strike more than any other factor. Strike kitchens involved about two per cent of Yorkshire coalfield women; seventy-seven per cent of those involved were associated with activists spending at least five days per week on strike duties. Women living within communities wholly or partly dependent upon coal-mining were four times as likely to be involved in running kitchens as women living outside traditional mining communities. Most kitchens were started in the first four months of the strike, although several South Yorkshire and Barnsley pits were late starters. There was enormous variation between kitchens in food provision: a 'meal' in one kitchen might represent a 'snack' in another. The number of meals served in the kitchens depended upon the volunteers available to run it, the size of demand from the local community, the resources the group was able to raise and the facilities available for the mass-production of meals. Many kitchens were open only three days a week, as was the case at Lundwood (Dodworth), Darfield and Royston kitchens.

The highest average meal provision per head was in South Yorkshire and the lowest in North Yorkshire, with the Barnsley and Doncaster provision about mid-way between them. Within each area there was also significant inter-colliery variation. Kellingley support kitchen produced on average 600 meals per day for a workforce of 2,300, while Manton's kitchen, serving a workforce of 1,165, produced 70 to 100 meals per day. Some miners, particularly in the Castleford area, had a choice of kitchen, whereas others had none near their home. South Kirkby Women's Group began their kitchen in a portakabin for the first two or three weeks of the strike then moved into a tent, having prepared meals throughout without proper kitchen facilities, or even mains water and drainage. Despite the inadequate facilities, eight to ten women provided up to 570 meals per day, five days per week.

The Union did make some provision for food, establishing a kitchen in Barnsley Civic Hall which served between 500 and 600 meals per day, five days a week. In some pits, such as Wath, Shireoaks, Rossington and Prince of Wales, the branch and the women's group co-operated to provide food. In Maltby, one of the three Yorkshire collieries which established a local strike committee, six women participated in the discussion and decisions of the committee. Some kitchens organised by branches were only for the use of pickets, as at Bullcliffe Wood and Woolley. The co-operation between a few Union branches and women's groups in South Yorkshire contributed to the area's high average meal provision. North Yorkshire had the lowest average meal provision because kitchens could not be organised where, as at Selby, the community was dispersed and divorced from the colliery.

Women's groups occupied the enormous vacuum left by *ad hoc* organisation by the NUM, and often did so with such success that the branch relinquished responsibility for providing food entirely to the women, as at Brodsworth. Nevertheless, the women's groups occasionally clashed with branch committees over the arrangements for operating kitchens. A kitchen run by women to provide meals for miners and their families from Woolley and neighbouring North Gawber was supported by the smaller branch but the Woolley branch gave minimal financial support and when the group protested that they were feeding more people from Woolley than from North Gawber, the Woolley branch advised them to turn Woolley people away.[90] Some kitchens were sited close to the pit, as in Mapplewell, the village which surrounds North Gawber colliery, while others were established in town centres like Rotherham or Barnsley. The kitchens contributed both to physical and spiritual maintenance of the strike: 'kitchens often became centres of strike organisation . . . with less than half of the active pickets attending union meetings, the kitchen became an important place for the pickets to meet and discuss'.[91] Brian Beeston, Thurcroft delegate, endorsed this view: 'whenever you want to get something going you've got to get people together and talk, that's why the soup kitchen was a success, because it drew people together'.[92] Those who felt humiliated at eating a meal in the kitchens were denied both the physical and moral support that the experience brought to the people who made most use of the provision. In North Yorkshire, all eleven kitchens were in established communities with high concentrations of miners: four in Castleford, and one each in Knottingley, Kellington, Allerton Bywater, Pontefract, Sharlston, Ackworth and Featherstone.[93] The location of kitchens made logistic sense in that more people would have access to them, but it meant that those who lived outside the community and therefore had least moral support, were also denied material support.

Food parcels offered an alternative which avoided the 'soup kitchen' stigma, but which also did little to break the isolation of those most reluctant to use the kitchens. Where miners lived in a dispersed community rather than around the pit, as at Selby, it was impossible to establish kitchens and, in some cases, it was even difficult to distribute food parcels or vouchers: 'I only got two food parcels during the whole strike: one was delivered to me and the branch wrote to me to collect the other. This was one of the factors that made them go back to work.'[94] Some branches used the distribution of food parcels to get men to the branch in order to maintain their involvement and support and to provide an opportunity to monitor morale. At Edlington (Yorkshire Main) food parcels were only distributed at branch meetings, while at Sharlston miners had to collect them from branch officials at the local club. Some branches organised food parcels as an alternative to kitchens and some to supplement the kitchens, and the contents of food parcels varied greatly. At Wath, Kiveton Park and Park Mill, for example, it was claimed that every miner regularly received a food parcel, while at others, such as Cadeby and Barnborough, none was organised. In North Yorkshire twenty-four groups produced food parcels but these were restricted to £4 in value for people receiving DHSS benefits. There were occasional disputes over disparities between individual's food parcels,[95] and complaints:

we had these big tins from the CGT with a picture of a chicken on the outside. Everyone thought 'great – a tinned chicken'. When they opened them it was chickens' feet in water. They said 'we aren't eating them' but we had to try it; it was soup, broth, it was nice. The best thing we had, though, was the beans from the Soviet Union; everyone wanted more of them.[96]

In the main most people were pleased to receive anything although odd individuals felt it was just as humiliating to receive 'charity' in the form of food parcels as to eat in the strike kitchens.[97]

Even the most well-organised groups found it difficult to identify those in need and to deliver parcels to them on a regular basis: 'most groups aim to give out a parcel every week, but in practice many just work down their list and start again when everyone has had a parcel . . . It all depends on how much money they have managed to raise that week'.[98] The provision of food parcels could have been far more effective with better communication between the women's groups and branches, and with co-ordination at area level.

Financing the operation

Funding was a major concern of the women's groups and involved almost as many women as the kitchens. Overall about 1·4% of coalfield women were involved, eighty-two per cent of whom were associated with activists spending at least five days per week on strike duties. Women living in traditional mining communities were twice as likely to be involved as outsiders. The women's groups' collections were crucial because the money coming into the Union was used to finance the picketing operation: 'None of the money went directly to the women's groups, but was given to the branches to decide how to use . . . If the women had not got themselves organised there would have been no systematic help for families in terms of food and money'.[99] The Yorkshire Area provided each group with £120 to help it become established but thereafter they were expected to raise their own finance. Women organised street collections, raffles and socials and made contacts with trade-union branches, trades councils, LP constituencies and other miners' support groups to arrange regular donations. Their fund-raising activities took women beyond the parochialism of the coalfield communities to the large metropolitan areas and even abroad, bringing them into contact with radicals, blacks, gays, famous entertainers, and intellectuals.

When the Yorkshire Area distributed money from the regional TUC between the four panels, none was designated to BWAPC by the Barnsley Panel. Similarly, NUM officials argued at a meeting organised by Barnsley Trades Council that all money should come into the NUM as the branches were using the funds to deal with hardship, yet, according to Miller, the NUM 'wasn't able to organise any kind of systematic back-up for the families because it didn't make that a priority'.[100] Thus when the Barnsley Panel asked the BWAPC to accept responsibility for financing their kitchen, the women refused because they believed the panel would still expect to control it. In North Yorkshire the women's groups similarly had to resist attempts by the Union to control them through their funds. When a women's appeal in North Yorkshire raised £16,000, area officials demanded that the money be administered by the NUM; the women's representatives were summoned before the North Yorkshire Panel but refused to relinquish their funds.[101]

Picketing and rallies

About 1·5% of coalfield women became involved in picketing at some time, the largest number being during November when whole communities turned out to deter local strike-breakers; eight-four per cent of these women were associated with activists spending at least five days per week on strike duties. Women living in pit communities were twice as likely to be involved in picketing as those from communities not dependent or only partly dependent upon coal-mining. In some cases, picketing by women prompted the formation of a women's group:

> It started because I couldn't stand the TV making out that the wives weren't behind their men. I was so angry and frustrated for a week that in the end Brian . . . took me round the wives of some other militants in the pit. Ten of us sat up half the night talking about what to do and five of us decided to go and picket Thoresby in Nottinghamshire that night.[102]

The administration of picketing was a major preoccupation of the Union at all levels (see Chapter three) but no attempt was made to co-ordinate the women's picketing.[103] At the Barnsley rally Scargill encouraged women to join the picket line but many NUM branches refused to entertain the idea and even went so far as to refuse legal cover for women who were arrested while picketing.[104] Bentley women's group refused to accept the branch ruling: 'when they wouldn't let us go picketing we all got together and played hell with the Union. The treasurer of the Union locally wasn't keen at first and he said we weren't on the books any more (that means you are not covered by the NUM if you are arrested) but that were just a threat'.[105] While some rank-and-file miners felt that women had no right to picket unless they were members of the Union (and would raise the same objection about men who were not miners), often there was a genuine underlying fear that women would be at risk. Fears for the safety of the women were vindicated when a group was arrested after confronting police who were protecting working miners: 'I always thought, and the men said to us the day we were arrested, "you'll be alright, they won't do it to you, you're women". Its just not true: they treat women just the same. Sometimes I think they treat you worse.'[106] Some men saw the women as 'morale-boosters', demonstrating that they were completely behind the strike and emphasising the unity of the communities against the employer, the Government and the police. Moreover, they were notoriously successful in turning men back when there was an opportunity to communicate with prospective scabs.[107] However, the women's skills on the picket lines only increased fears for their safety: 'I don't like my mam going on the picket line and I don't like going with her; I keep thinking that she's going to be arrested, she never shuts up'.[108] Many women (and men) were afraid to go picketing because of the violence and preferred to support the strike in other ways. Those with children were particularly concerned about what would happen to their family if they were arrested.

The women's groups also played a major role in maintaining strike morale through organising rallies, which involved about 3·5% of coalfield women in Yorkshire. Women associated with activists even predominated in rallies; some eight-two per cent who took part in these activities lived with miners who spent at least five days per week on strike duties. Among our women respondents who took part in rallies, those from pit communities were over twice as likely to attend as outsiders, while those from communities partly dependent on coal-mining fell between the two. When the Women's Support Groups organised a rally in London for 11 August without formally consulting the National or Area Unions, the NCC both welcomed the rally and expressed the hope 'that details of any further rallies be notified to the National Union'.[109] The NCC was concerned that an organisation was emerging which was not subject to the control of either the areas or the National Union and agreed there was a need for 'effective liaison and co-ordination between local groups and appropriate area secretaries especially with regard to the organisation of rallies and other events'.[110] The NCC asked national officials to initiate discussions between area secretaries and women's groups but in many cases neither the Area Executive nor the support groups appeared eager to work together. The support groups feared losing control over the organisations they had built, while Area Executives were reluctant to accept responsibility for organisations over which they could not maintain authority. Barnsley WAPC, which had from the outset made clear they were 'a campaigning group, not just a support group', resented being instructed by a member of the Barnsley Panel: 'I've found you a place to run a kitchen, go tomorrow and start serving meals', and refused to accept this directive.[111]

Inevitably there were tensions within the women's groups, as there were within the NUM. Women activists whose husbands were branch officials were sometimes assumed to be speaking on their behalf, 'expressing the "do this, do that" line of the NUM'.[112] Where the relationship between the group and the branch committee was strained, the wives of officials were occasionally excluded: 'I wanted to get more involved in the strike but found it difficult to do so. I took some food that had been donated to my husband to the kitchen. They said that they were often short-handed and I offered my services. They took my telephone number but never contacted me. They didn't want Union wives involved.'[113] Such conflicts of personality caused the Barnsley WAPC group to split in November 1984, with Anne Scargill establishing an independent

Barnsley Miners' Wives support group. The split added to the financial difficulties of BWAPC and one week the shortage of funds was so acute that there was no money at all for the kitchens.[114]

As local support became more organised, women's groups began to meet within each area, so that 'informal structures grew within the women's movement and regional co-ordinating groups emerged'.[115] Eventually a Women's National Office was established in Sheffield in early July, staffed by volunteers including Jean McCrindle of Northern College, who had been a member of BWAPC before the split. The location of the office enabled national officials to intervene in the organisation of funding for women's groups. Funds raised for the national women's rally were collected at national office and the NCC were concerned that 'there may be confusion among those donating funds and there must be some financial accountability'.[116] Following further meetings between national officials and the women's groups, Scargill told the NCC that the groups had agreed to consult the Union when organising future events.[117] In some instances, women's groups did seek the authority of the National Union, as when the Abertillery and Mynach Women's Support Groups arranged to collect donations from Holland.[118] For the most part, however, the women's groups continued to operate autonomously and any contact with the Union was usually at national level rather than with the areas. Within the groups there was some concern over national co-ordination of the women's action from inside NUM head office: 'the offer of help – a room, secretary, and, it seems, someone to look after the money for national events – seemed to have been set up by others who were party to some kind of discussion, but no discussion with the women's groups'.[119] The decision to hold a national women's rally in London in August 1984 was also taken without discussion by the women's groups. Some women would have preferred to have held a demonstration in Nottinghamshire, but when a conference was held at Northern College to consider the detailed arrangements of the national event 'it was all cut and dried before we got there'.[120]

The maintenance of the strike owed more to the emergence of the women's groups than to any other single factor. The women's groups destroyed the stereotype of families undermining a strike through their demands on strikers; they clothed and fed their communities and developed elaborate support networks. Moreover, they were a manifestation of the mobilisation of communities behind a strike that was about the preservation of those communities. National officials encouraged the formation of women's groups, welcomed women joining picket lines, and attempted to establish liaison between the groups and area officials. The National Union was powerless to do more, given the autonomy of the area unions. Yorkshire Area leaders were less enthusiastic about the women's groups and more concerned to control them, fearing both decentralisation of decision-making and the possibility of the groups acting as a Trojan horse from national office. The organisation of kitchens was viewed by both area and branch officials as a legitimate activity for women's groups and was therefore delegated to them without providing adequate support or involving them in decisions about the strike. Ambiguous and sometimes acrimonious relationships between the women's groups and union branches were the inevitable result of failing to integrate the two local maintenance networks. The women's groups were organised on a community basis, whereas the branches were pit-based and therefore structurally inappropriate for maintenance activities. Had community-based strike committees been established, as we argued in Chapter three, maintenance activities could have been integrated, empowering the women's groups while simultaneously providing a structure for their co-ordination and control.

4.3 Internal strike maintenance

It is impossible to assess the amount of money, food and other items collected on behalf of the miners. Nevertheless, from the evidence in Yorkshire, Benn was correct in describing the donations of money, food, clothing and toys as 'on a scale of which there is no parallel in any industrial dispute in living memory'.[121] One estimate put the total collected at £60m,[122] but such a sum only amounts to £9 per striker per week, so would not have been sufficient to ameliorate the appalling hardship experienced. Although the National Union established a special fund, large sums of money came directly to pit branches and support groups, many of which kept no record of donations.

Union maintenance provision

The National Union established a Miners' Solidarity Fund (MSF) to avoid the money donated to alleviate hardship being sequestrated if other Union funds were lost. The three Yorkshire politicians who acted as trustees, Richard Caborn, Bill Michie and David Blunkett, recognised the difficulties in attempting to assign funds equitably: 'we tried with a very rough rule of thumb to equal the monies out which was very difficult . . . there were all sorts of discrepancies . . . nevertheless we did what we could'.[123] The trustees were aware that 'twinning' arrangements between pits and support groups were common but had no details of branch income. Naturally a small area like Kent which could draw more easily upon the large and relatively prosperous London region for support was in a better position than a massive coalfield like Yorkshire where unemployment was higher, incomes were lower and coal-mining dominated particular localities. Nevertheless, existing income to branches was largely ignored in the allocation of funds from the MSF on the basis of the numbers on strike. The trustees also attempted to identify where the strike was weakening and whether miners were returning because of hardship: 'we had to monitor not just the hardship cases but how many were out, where they were out and where they were going back in and where there were particular problems so far as morale was concerned'.[124] Once the MSF was underway donations increased, particularly during the Christmas period. However, at certain periods of the strike the flow of funds fell off dramatically: 'After a major act, as portrayed on television, an act of violence . . . everyone used to assume by the way it was presented in the media that it was the miners at fault, and then it fell off'.[125] The Orgreave fiasco and Libyan affair reduced the numbers subscribing to the fund. The TUC and LP Conferences, however, dramatically increased the money received by the fund. Since the level of donations was unpredictable, it was impossible to assess likely income for the next week. Although the money raised was nowhere near the normal income of a mining family, it nevertheless allowed miners to continue the strike, so tremendous effort was put into fund-raising. Support came from a variety of organisations, abroad as well as in the UK.

The autonomous activities of colliery branches and local support groups to raise funds from international sources threatened to disrupt the arrangements made through the Appeal. The NCC expressed concern on two grounds: firstly, they feared that further financial demands by individual pits and areas could result in the foreign unions ceasing to donate to the central fund, and, secondly, they believed that there was a risk of the central fund being fragmented, which would increase inequities between areas and pits. The NCC therefore resolved: 'Fraternal unions should be instructed not to provide any assistance to any individual or group unless this has been discussed and agreed by the two Unions and the individuals had a letter of authorization signed by one of the officials'.[126] Shortly afterwards, in an effort to prevent unauthorised collections, the NCC agreed: 'That national guidelines be established that assist areas with arrangements for building up support and collecting cash contributions on the understanding that all monies raised should be sent to the Miners' Solidarity Fund for onward distribution to areas on an equitable basis. Arrangements for adoption of pits and food supplies should be, where appropriate, co-ordinated centrally to avoid duplication or omission.'[127]

The NCC found it necessary to insist that Union collectors be issued with credentials and circularised the areas with details.[128] Areas which persisted in undertaking unauthorised collections abroad were reminded of the procedure by the NCC, but with little effect.[129] In December the ICTU contacted the NCC concerning the authorisation of collectors from Armthorpe branch and the Sheffield Womens' Support Group, who had contacted the ICTU independently.[130]

Although the National Union was responsible for collecting large sums of money during the strike, areas, pits, support groups and individuals continued to raise money wherever and whenever they could without consulting the NCC or even, in many cases, the area union. The majority of this money never passed through the MSF but was administered by the group which had organised the collection. Hence Heathfield's comment that ninety per cent of support to the MSF came from abroad;[131] the National Union was able to maintain more effective control over international fund-raising, whereas UK funds were directed to individual pits and women's groups. While the NCC sought to centralise and control collections in the interests of equitable treatment, they had insufficient knowledge of specific circumstances at collieries to do this successfully. Moreover, collectors from branches and support groups usually took a parochial view of their activities and were not prepared to collect on behalf of others who were not making these efforts. The LP also agreed to an official collection by the NUM at the 1984 Conference and ten collectors were appointed from the North-Western Area. The LP was asked to ensure that all collections at the Conference or fringe meetings were sent to the MSF.[132] However, the collection did not take place because the North-Western Area refused to accept responsibility on behalf of the National Union.[133] The NCC was determined to avoid similar problems in the future by making its own arrangements, rather than relying upon areas to collect on their behalf.

Within Yorkshire, financial considerations were of immediate concern to the SCC; plainly money was vital to facilitate the picketing operation. The pool of money which the SCC had to draw on was limited and the SCC was concerned that the best use be made of this limited resource. On occasions these objectives brought the SCC into direct confrontation with branches. On 8 May it was noted that there had been a complaint regarding payment for pickets in Nottinghamshire; 'thirty Royston Drift pickets . . . were complaining that the amount of cash they had been given was insufficient for their needs'. The SCC responded that 'if we were having to pay for lodgings . . . we . . . bring our pickets back'.[134] In the event, pickets found free 'lodgings' in Ollerton Working Men's Club, Trent Polytechnic, and so on. But pickets were still dissatisfied with payment: 'Armthorpe pickets had complained most of all and it was suggested that they should contact their branch secretary and he could write into the union with any complaints.'[135] Occasional dissatisfaction with picketing payments surfaced later; for example, on 4 July Glasshoughton pickets were requesting double payment at weekends: 'They say they are covering power stations 24 hours a day with some difficulty on a weekend.'[136]

The SCC were also involved in using strike funds to maintain the strike, having been extremely critical of the lack of financial assistance from central funds for striking Nottinghamshire miners. In June reports to the SCC from Welbeck suggested that 'lads on strike can no longer go picketing because they have no petrol'. The DHSS had stopped all payment, and so had the Nottinghamshire Area Union; there were 'no payments for picketing any pit so lads on strike were staying in bed'.[137] The SCC agreed to help these Nottinghamshire miners with finance and to meet the NCC and national officials 'because money is not getting through to the lads'.[138] After meetings with representatives from Welbeck, Manvers, Shireoaks, and Beighton, South Yorkshire agreed to give financial assistance to establish a food kitchen in Welbeck. Fears had been expressed at the meeting that 'because they had no finance at all then they were likely to start drifting back to work'.[139]

Branches were late to become involved in maintenance activities and received little guidance from either the national or area unions on the need to establish kitchens, organise food parcels, raise finance, monitor hardship cases and disseminate propaganda. Food needs were met first, generally because of the action of women's groups, which were also involved in fund-raising. Some branches deployed collectors early on but most concentrated on picketing and relied heavily upon twinning arrangements. The SCC attempted to co-ordinate financial support by acting as a clearing system for collections and donations. Although the system was intended to redistribute finance according to need, it acted as a disincentive to branches to organise collections. Some branches initially co-operated, particularly in the Barnsley Area, but within six months most were organising support independently. Autonomous collections and twinning arrangements vastly increased branch income, but increased inequalities between pits in proportion. Fragmented fund-raising also introduced competition between pits to secure the support of key unions and even led to disputes over collecting 'pitches' in the cities.

Strikers' income

In the absence of adequate financial provision by the Union, it is necessary to examine the sources of income to striker households among our 1,830 respondents. There were four categories of income available to strikers, each of which played a part in the strike maintenance: state, personal, commercial and union. As part of the Ridley formula the Government introduced a number of measures designed to minimise state payments to strikers and their dependants (see Chapter five). These changes vastly reduced strikers' entitlements, but it has been estimated that 42,000 miners' families were in receipt of supplementary benefit amounting to a total of £26m.[140] These figures suggest that nationally thirty-two per cent of miners received an average weekly payment of under £12, while the remainder received nothing from this source. Our survey of 1,830 miners in Yorkshire found a higher proportion of miners, forty-seven per cent, received social security payments which represented the major source of income for thirty-seven per cent of respondents. There was no significant difference between activists and non-activists.

Supplementary benefit was the major source of income for fifty per cent of married men whose wives were not working, while eighty-five per cent of single miners received nothing. There was no significant difference between those living in mining communities and outsiders. Although forty per cent of owner-occupiers ranked this as their major source of income (compared with thirty-four per cent in rented accommodation), fifty per cent of them received no social security. A Kellingley miner whose family comprised two pre-school children and a wife who was not in paid employment, explained how they received only £11.75 in supplementary benefit. Their basic needs were assessed at £21.65 for the wife and £9.15 for each child; £39.95 in total. From this total was deducted £6.50 family allowance for each child and the £15 notional strike pay.

Child allowance became an important source of income for all striker households which included children, since it was not subject to any deduction.

Overall, child allowance was the most important source of income for only thirteen per cent of respondents, but among the fifty per cent of our respondents with young children this was a major source of income for eighty per cent of respondents and the most important source of income for twenty-two per cent. Dependance upon child allowance was higher among those in NCB housing (twenty-one per cent) than among owner-occupiers (fourteen per cent), and higher among face-workers (eighteen per cent) than among craft-workers (nine per cent). Tax rebates were largely irrelevant because the stoppage began towards the end of the financial year. Only ten per cent of respondents received any refunds from the Inland Revenue and none ranked this their major source of income.

Since most authorities in coalfield areas are Labour-controlled, the local state provided strikers with substantial support, which some observers regarded as crucial to the continuance of the strike.[141] However, there were variations in the degree of support provided by local authorities; of the two metropolitan county councils, for example, South Yorkshire found more ways of helping strikers than West Yorkshire. This variation created anomalies at collieries like Royston Drift, where miners living in Royston (South Yorkshire) fared better than those living in Ryhill (West Yorkshire). Selby miners received no assistance from the local authorities, whereas Barnsley, Sheffield and Doncaster authorities provided free meals, food and clothing vouchers, and lodging allowances for single miners as well as deferring or waiving rents and rates. Some areas gave the miners help more willingly than others. While Rotherham Council provided £268,095 of assistance to mining families in the first seven months of the strike,[142] this was only after pressure from activists: 'From the beginning of the strike, Barnsley Council both reclassified miners as dependants for rent and rates purposes (which reduced deferred debt considerably) and were paying single miners living at home with their parents a lodging allowance of £8.20 per week under the Housing Benefit regulations. In Rotherham this loophole was not discovered until about November.'[143] Rent or rate rebates were the most common form of local authority support; in Yorkshire fifty-two per cent of our respondents who were householders obtained such help. The variations in support given by local authorities in Yorkshire conformed with political differences and with the degree of influence exerted by NUM members in local politics. Radical authorities provided support irrespective of whether the miners were politically influential, as in South Yorkshire, or not, as in Sheffield and Leeds. Traditional Labourist authorities, such as Barnsley, Doncaster, Rotherham and Wakefield provided support because of the miners' political influence, but where this was less powerful, as in West Yorkshire, less support was forthcoming. The Conservative authorities of North Yorkshire and Selby denied the miners any support.

With restrictions on state benefits, personal sources of income became more significant. Miners and their families who had savings were initially cushioned from the deprivation which the strike entailed, but savings were soon exhausted by a strike of such duration, and hardship intensified. Families with members employed outside the coal industry were naturally in a better position than families where all those employed were on strike; unfortunately the traditions of mining made the latter situation more common than it would be in any other industry. Wives' earnings were the major source of income for sixty-nine per cent of those respondents with working wives, representing thirty-five per cent of the total. There was no significant difference between activists and non-activists. Naturally, single miners and those whose wives were not in paid employment, representing fifty-six per cent of respondents, received nothing from this source. Among those living in pit communities, forty per cent received income from wives' earnings and thirty-one per cent ranked it their major source of funds, whereas among outsiders fifty-two per cent received income from this source of funds, whereas among outsiders fifty-two per cent received income from this source and forty-one per cent regarded it as their major source of finance. In communities partly dependent upon coal-mining, forty-four per cent received earnings from wives and thirty-five per cent ranked this their major source of strike income. Similarly, while thirty-one per cent of those in NCB housing had wives' earnings (and twenty-seven per cent ranked it most important), fifty-one per cent of owner-occupiers received income from this source (and thirty-nine per cent ranked it their major income). The distance from the centre of the coalfield community and home-ownership (also more prevalent away from the pit village) was directly related to female economic activity rates and hence availability of wives' earnings for strike income. Relatively few respondents overall (eleven per cent) received financial support from sons or daughters employed outside the industry, although twenty-seven per cent of those who did (three per cent of the total) ranked it their most important source of support. The only significant departures from the eleven per cent average were among those in NCB housing, where seventeen per cent received income from children (and ten per cent ranked it their major source), and outsiders, of whom nine per cent received such income (and two per cent ranked it their major source). The difference is not explained by the age structure of respondents in the two categories, so must reflect differences in the rates of non-mining employment among children of respondents in the two groups.

In addition to income from wives and children whose employment was unaffected by the strike, striker households attempted to increase their income from a variety of sources. Casual work undertaken by strikers themselves represented the second most important source of personal income. Among activists, twenty-seven per cent obtained casual work during the strike and for nine per cent this was their major source of income, whereas thirty-six per cent of non-activists obtained work and seventeen per cent rated it most important. These figures suggest that overall thirty-four per cent of Yorkshire strikers obtained casual work, which was the major source of finance for fifteen per cent. Whether or not they were activists, only twenty-two per cent of those in pit communities obtained casual work and eleven per cent rated it as their major income. Among activists not living in pit communities, twenty eight per cent obtained casual work and nine per cent ranked it their most important means of support. Among non-activists, casual work was obtained by forty per cent of those from communities partly dependent upon coal-mining (and eighteen per cent ranked this their main income), and by forty-seven per cent of outsiders (among whom twenty-two per cent ranked this their major source). The availability of casual earnings, as well as wives' earnings, was therefore greater away from the traditional coal-mining centres, and community had more influence upon casual earnings than did the degree of activism. More craft-workers (thirty-four per cent) than others (twenty-four to twenty-six per cent) obtained casual work, but surprisingly fewer craft-workers (ten per cent) than underground-workers (fifteen per cent) ranked it their major source of finance. While striking craft-workers who obtained casual work were able to maintain their consumption better than others, in undercutting builders, bricklayers and electrical contractors, they damaged any solidarity that might otherwise have been expressed by these workers and added to the impact of overall demand deficiency upon the local economy.

Among our 1,830 respondents in Yorkshire, forty-two per cent received loans or gifts from relatives and friends; the proportion was highest among single miners (fifty-three per cent) and lowest in the case of those with working wives (thirty-six per cent). There was no significant difference between activists and non-activists. In Thurcroft, it was found that 'Help from relatives was almost universal and came in many different forms. It was distinguished from other help of this kind by the fact it occurred on a regular basis. From the summer onwards it became the most important source of support in the wider community'.[144] Help from friends was less regular, though equally important. Finally, some households were forced to sell possessions in order to live during the strike. Among our respondents, eighteen per cent, regardless of whether they were active, resorted to such measures. Single miners were an exception, with only six per cent having sold possessions, presumably because the group mostly comprised single men with little accumulated wealth. Rented goods such as televisions, videos and telephones were the first to be returned: 'one local firm had so many returned that they ran out of storage space and told their customers to hang on to them'.[145] Cars were sold when road tax or MOT tests were due and some miners' wives were even forced to sell their wedding rings.

The support which strikers received from commercial sources naturally tended to take the form of loans rather than gifts. Special offers from local traders were taken up by seven per cent of respondents in Yorkshire overall, ranging from nine per cent among those living in a pit community to two per cent among outsiders, and from twelve per cent among single miners to four per cent of those with working wives. Banks and building societies generally did not press for payments but made individual arrangements according to family circumstances. Bank overdrafts were obtained by twenty-eight per cent of all respondents; the highest proportion was among outsiders and the lowest among those in NCB housing. Among our respondents, fifty-five per cent of owner-occupiers suspended mortgage payments, but the accumulation of debts forced three per cent to lose their houses. Mortgage debts were blamed for the large return-to-work at Selby; a group of Wistow miners had agreed to return when their mortgage arrears reached £3,000. Those in council housing faced fewer problems, but miners in private rented accommodation were often evicted as landlords refused to postpone or waive payment of rent. Insurance, finance and mail order companies were unsympathetic and inflexible. People were forced to cash insurance policies for their surrender value and because they were unable to maintain payments. Finance companies harassed those in arrears with hire purchase payments and repossessed goods, while mail order firms blacklisted defaulters.[146] Among our respondents, twenty-nine per cent attempted to defer hire purchase payments or were forced to return goods; the highest proportion was among those in NCB housing and the lowest among outsiders. There was little point in creditors pressurising striking miners who were without the resources to repay debts, but as the strike dragged on, it was often said: 'it's when we go back to work and everybody wants paying that worries me'.[147]

The Union could not provide strike pay but picketing payments gave some income to activists. The major forms of subsistence in which the Union had an involvement (in some branches a *minimal* involvement) were strike kitchens and donations of money or food. From the usage of strike kitchens among our respondents who were active in the strike (seventy-four per cent) and by those who were not (thirty-five per cent), it is estimated that thirty-nine per cent of miners overall made use of these facilities in Yorkshire. A higher proportion of single miners (sixty-eight per cent) and miners whose wives were not working (sixty-seven per cent) used the kitchens than those with working wives (fifty-nine per cent). Community played an important part in patterns of usage but within the distinction between activists and others. Thus, for activists, kitchen usage ranged from eighty-three per cent in pit communities, through seventy-four per cent in communities partly dependent upon coal-mining, to forty-nine per cent among outsiders. For those who were not active, kitchen usage ranged from forty-two per cent in pit communities, through thirty-one per cent in communities partly dependent upon coal-mining, to twenty-nine per cent among outsiders. Since strike kitchens were located within coal communities, it was naturally more difficult for outsiders to use the facilities. However, those who were not active in the strike were significantly less likely to use the kitchens than activists living in the same type of community. Far from acting as a meeting place and morale-booster for the non-activists, the kitchens served to *maintain* their isolation from the activists.

Overall, donations of food and money from support groups were received by fifty-five per cent of our 1,830 Yorkshire respondents. Among those living in pit communities forty-seven per cent received such support, whether active or not. In communities partly dependent upon coal-mining, however, fifty-nine per cent of activists received donations, compared with forty-seven per cent of those who were not active, while among outsiders the difference between the proportion of activists (seventy-two per cent) and others (fifty-nine per cent) receiving such support was even greater. While most branches adopted a policy of using donations and food parcels to compensate for those who had less access to strike kitchens, the policy was rather more successful in relation to activists than others. Unaccountably, only fifteen per cent of single miners reported receiving donations, compared with sixty-nine per cent of those with working wives and forty-five per cent of those whose wives were not working.

The NUM was largely dependent upon outside expertise for advice on strikers' entitlements to state benefits and ways of dealing with creditors. A guide produced by Leeds TUCRIC proved particularly useful; over 1,000 copies were sold during the strike.[148] A leaflet outlining miners' entitlements was also prepared by TUCRIC and Chapeltown CAB, which WYMCC published for distribution to strike centres, miners' welfares and union branches. During the strike the Claimants' Union and Merseyside Welfare Rights Resource Centre each prepared material aimed at striking miners,[149] although neither was circulated as widely as the TUCRIC guide which had been available from before the dispute. A more comprehensive hand-book was unfortunately not available until August 1985, although it drew extensively upon experience in the miners' strike.[150]

Citizens' Advice Bureaux in the coalfields were inundated with inquiries, seventy bureaux were involved directly and others indirectly. In May a mobile CAB was funded by WYMCC to visit villages away from existing bureaux and near to centres used by support groups: 'although each site was visited only once a week the worker was quickly accepted as a source of help'.[151] In the first six weeks the West Yorkshire mobile CAB dealt with 400 enquiries involving 207 callers. A similar service was introduced in the West Midlands by WMMCC, while another mobile operated in the North-East. In the Doncaster Area the three bureaux in pit villages were particularly in demand and Stainforth CAB ran additional sessions in the Miners' Welfare. The Citizens' Advice Bureaux were extensively involved in producing advice leaflets themselves and in co-operation with others. In July some sixty welfare rights workers met in Sheffield with NUM representatives to discuss more effective means of providing strikers with advice. Yorkshire Area officials proposed the establishment of contact points in strike centres to refer complicated inquiries and maintain updated information.[152] The NACAB recognised the need to co-ordinate the work of Citizens' Advice Bureaux with other organisations to gain the maximum benefit from limited resources:

> it was particularly important to work with the NUM in order to target those miners who were in most need. Without contact with the NUM it was difficult to reach miners . . . The demand for CAB involvement varied with the extent to which the local NUM was able to give advice. In some local unions the provision of advice was quite extensive and

the CAB was not encouraged to become involved. In others, the NUM welcomed CAB and worked closely with local bureaux.[153]

A number of women's groups organised advice sessions for those wishing to claim supplementary benefit. Barnsley WAPC co-operated with welfare rights workers to establish advice centres in miners' welfares and eventually obtained an office in the Centre Against Unemployment.[154]

A minority of branches were extremely well-organised to provide general advice and access to specialists. Manton, for example, established a centre in Worksop which was regularly staffed by a branch official and volunteers: 'we knew hardship would be a problem as soon as the Notts miners didn't come out. Lots of scabs from Notts collieries live in Worksop so Manton strikers felt *relatively* worse off, especially the young lads who could see the others going out of a night. It's not so bad if everyone's in the same boat'.[155] The branch committee also made regular visits to some members' homes to assess their difficulties and anticipate their needs. Although Manton was exceptional in respect of strike maintenance, the pressures from Nottinghamshire were so great that the strike was eventually broken there (see Chapters five and six).

Strike maintenance had to be devolved since the National and Area Unions could do little more than monitor local activities, operate a clearing house for donations and facilitate contacts with outside support groups. However, in Yorkshire, the Union failed to establish appropriate structures to enable the effective maintenance of the strike at local level. Strike maintenance was left to the branches in the same way as the administration of picketing, but maintenance necessitated community-based organisation which the majority of Yorkshire branches could not provide. By default, therefore, strike maintenance became the responsibility of women's groups, families and individuals. Individuals struggling to cope with hardship who were isolated from community action became increasingly likely to return to work. The fact that the strike lasted so long and remained so solid in Yorkshire demonstrates the success of the *ad hoc* strike maintenance, but the Area Union could have made it even more effective. Had the Union explicitly endorsed collective community maintenance through establishing local strike committees, it might have been possible to counter the individualistic response to hardship on which the back-to-work movement was grafted.

Hardship

To prevent miners from being starved back to work, it was necessary to monitor hardship and identify those whose need was most acute. Although every branch was concerned to alleviate special hardship, none in Yorkshire attempted systematic monitoring of the entire membership. Hardship cases were only identified when individuals presented themselves to the branch and asked for assistance. Even where branches delegated responsibility for acute hardship cases to a sympathetic individual, some members were reluctant to seek help:

'we could only help those that came forward, we had no way of knowing otherwise'.[156] A few branch officials took the initiative towards the end of the strike and visited people's homes to see whether they were coping but this was invariably in response to suggestions from neighbours or friends that a particular individual was 'having it rough' or, more often, 'thinking of going back'. Not all branches handled hardship cases with sympathy and sensitivity:

> I wouldn't have gone to the branch however bad things had got. It was like a DHSS tribunal; they treated you like it was their own money and you were scrounging. They told one lad who was in real trouble, with debt collectors and such like; 'it's the same for all of us, if you can't manage there's always the meshed bus'. The man was in tears when he told me what had gone off. He didn't scab but he'll never forget how he was treated.[157]

Doubtless some problems arose because it was often those who had received most help from the branch who went back to work. Officials were very bitter about the help that was 'thrown back in our faces' but it confirms that monitoring was inadequate and that the branches could only make a small contribution to what had become chronic hardship.

The Union was unprepared for the scale of the hardship involved in a strike of such duration and saw maintenance primarily in terms of picketing. Jack Taylor acknowledged the problem after the strike: 'The sad thing about the dispute was the tremendous hardship people suffered in Yorkshire. The Union could not provide for that. You would never have dreamt that the hardship people suffered would take place. We ought to have had more people collecting money, but the lads saw it as their duty to picket, perhaps the Union was not organised enough.'[158] Hardship took the form of shortages of food, clothing and coal, as well as financial difficulties of a general kind or specific to housing, electricity or gas. Overall, more Yorkshire collieries (thiry-six per cent) rated general financial difficulties as the most widespread. Financial problems were reported to be particularly prevalent at Selby and in the Western Calderdale corner of the Barnsley Area. Financial problems were therefore the major problem at the *periphery* of the Yorkshire coalfield, where the strike cracked without the influence of Nottinghamshire (see Chapters five and six).

Electricity and gas bills were the most extensive specific form of debt, reported to be the major form of hardship at thirty-two per cent of Yorkshire collieries. Although electricity and gas bills were a problem throughout the coalfield, pits in the Dearne area, near South Elmsall and around Wath, representing the *core* of the coalfield, particularly stressed hardship of this form. Barnsley WAPC demonstrated outside the electricity showrooms in Barnsley to dissuade the YEB from disconnecting supplies in miners' homes.[159] Pontefract Women's Group told both the gas and electricity boards that they would mount a picket at any house threatened with disconnection.[160] When the YEB installed coin meters the miners suffered the added disadvantage that 'the rate was bumped up so that the debt would be paid for as well as the actual units consumed'.[161] When the gas board installed coin meters, mining families found they had insufficient money to use fuel: 'so we used to sit with a rug round us all afternoon. It was really, really cold. We used to put the gas burners on the stove on a couple of times to keep the kitchen warm for the kids because I used to stand them in the sink to give them a bath'.[162] Some branches managed to negotiate for families to pay the YEB £1 per week to avoid disconnection.

Food shortages were ranked the most extensive form of hardship at eleven per cent of Yorkshire collieries, and all those we interviewed accepted that there was a serious *qualitative* deterioration in the food consumed throughout the strike. Food shortages were reported as most serious at Doncaster collieries. In the more integrated communities of North Yorkshire, South Yorkshire and Barnsley, non-mining residents were able to make casual donations of food. In Doncaster, however, the pit communities are all isolated and relatively few inhabitants are employed outside coal-mining so most food had to be collected away from the local communities.

Housing was cited as the major problem at nine per cent of collieries, particularly those at Selby and near Sheffield, where a high proportion of miners were owner-occupiers. Single miners without dependants were particularly affected if living with parents since they were refused assistance with housing costs. Some months into the strike a loophole was discovered whereby single miners could claim housing benefit, but the regulations were amended from 18 December 1984 to prevent such claims.

Lack of coal was the major problem at seven per cent of Yorkshire collieries, particularly near Sheffield, where many miners lived away from their pits, and Doncaster, where management took a hard line against coal-picking. With the majority of miners dependent upon solid-fuel heating, men, women and children were involved in riddling coal from pit stacks. When the NCB attempted to prevent coal-picking at Grimethorpe there were twenty-nine arrests, which sparked off violent clashes in the village.[163] Shallow drifts were also dug into the sides of stacks and along railway embankments, which led to many injuries and a few deaths, including some children. Less labour-intensive means of obtaining coal were also employed: 'the coal train passed near our home . . . The 16 tonne wagons had to stop to cross the road, so we'd go behind one, knock the catch and drop 16 tonnes of coal on the railway line. We were never short of coal'.[164]

Clothing was rated the major form of hardship at only four per cent of collieries, all within the Doncaster Area, probably for the same reasons that food shortages were more common in Doncaster. It was easier to 'make do' with clothes than with shoes, which had to be replaced or repaired. Babies' and childrens' clothes were also in demand because they grew out of clothes over the duration of the strike. Clothes collections were almost as extensive as food collections but required more volunteers to sort the materials donated. At several collieries a section of the women's support group washed and ironed clothes before sorting them and running a 'shop' in the local welfare or club.

The major forms of hardship altered over the duration of the strike: while food shortage was chronic, shortages of heat, light and clothes became particularly serious in the winter. One of the trustees of the MSF commented:

as the strike went on different patterns of hardship developed . . . at the end of it was a total package of hardship, of massive bills that the miners and their families had got and also the difficulty of sustaining people who had no reserves at all, who sold their cars, who had got shut of stuff out of the house. It was *real* hardship, something that I've never seen before.[165]

Propaganda

The dissemination of propaganda by the Union was the area where strike maintenance could have been most dramatically improved. In the first few months the National Union had produced a leaflet on the strike for public distribution which was modified and reprinted by the Yorkshire and Humberside Regional TUC for the one-day stoppage on 21 May. A second leaflet was produced during the summer but was not as widely circulated in Yorkshire. In November the NCC proposed the Union produce an up-dated fact sheet, but there is no evidence that this suggestion was acted upon.[166] Although special strike editions of the *Miner* and *Yorkshire Miner* were printed, distribution was problematic. Having relied upon NCB co-operation in the past and distributed the papers at the pits, the Union had failed to develop an independent communication network with its own membership. Without lists of members' names and addresses, even at branch level, only the activists who were regularly picketing and passing through strike centres, welfares and kitchens were likely to see the Union's papers. Because the majority of strikers received no communications from the Union, those who were isolated geographically or through their inactivity were unaware of the strength of the Union's case, of how solid the strike had remained and of the extent of support available from branches in cases of acute hardship. NCB communication was altogether more efficient and effective; publications and letters were regularly mailed directly to every miner's home (see Chapter five). The NCB took and retained the propaganda initiative, so strikers were more aware of the Board's case than the Union's. Moreover, the direct communication made it appear that NCB management were more concerned than NUM officials about strikers' welfare. While management were aware that the strike could be eroded once individual strikers were won over, Union officials appeared to take membership solidarity for granted.

There were several *ad hoc* attempts to meet the propaganda needs which the Union had failed to fulfill. In Yorkshire 'rank-and-file' information sheets were produced by *Socialist Organiser* and *Militant* in particular, but these were not directed at local issues and were treated with some suspicion by the majority of miners. At branch level there was a desperate need for locally-produced information sheets or newsletters to maintain support for the strike and break the isolation of the non-activist majority. Fewer than ten branches produced news sheets during the strike and, in all but one case, these were a response to the strike-breaking in November. The single exception was Ledston Luck where the *Tannoy* was produced from May on a regular basis and was widely circulated around the neighbourhood of the pit, as well as being distributed among support groups in Leeds. Since the *Tannoy* was largely the work of a single activist in the pit, it is clear that other branches *could* have developed news sheets earlier without detracting from other activities.

In Yorkshire the major propaganda initiative from the Union side arose out of a day school organised for Lancashire miners by a former Kellingley miner who was a student at Northern College when the strike began. About thirty activists, mostly from Yorkshire, and two members of the Area Executive attended the day school on 9 January. Mick Clapham and Dave Feickert of the NUM Industrial Relations Department introduced a discussion on the relevance of *Plan for Coal* and we presented the group with an up-date of our projections on the scale of job losses which the NCB restructuring would entail. In the afternoon session it became clear that the activists were seeking more tangible practical assistance to keep the strike solid. A Wheldale miner feared that the Union leadership would be 'willing to settle for the status quo . . . leaving the NCB to take the initiative on pit closures'. Several echoed the need for the Union to regain the initiative and thought that this would best be attained by getting the issues across at branch meetings and in the wider community. An official from Savile was concerned that 'the strike isn't solid any more . . . we need to get information to the rank-and-file', while an Ackton Hall miner warned 'even the activists aren't getting information because we're spending all our time picketing'.

The group resolved to prepare an information leaflet which was drafted and printed at the expense of local supporting trade-unions such as ASTMS and SOGAT82. Every Wednesday for the remainder of the strike, a growing number of activists, about 200 in all over the period, attended the meetings to draft leaflets, arrange their distribution and discuss tactics. The first leaflets were concerned with the issues of the strike: *Plan for Coal*, new technology, broken agreements, and the NUM alternatives for promoting coal use. One broadsheet, published with the help of Leeds District Miners' Support Group, put forward the arguments for politically marketing coal through the local authorities. From February, however, all the leaflets were concerned with stemming the return-to-work, and included details of the latest negotiations which were often telephoned through at the last minute. Leaflets described the preconditions imposed by the NCB and advised members to 'stay on strike to force the Board to talk'.[167] To counter the media back-to-work barrage, the group publicised cases of miners rejoining the strike and reminded members that with eighty-six per cent of Yorkshire still on strike, 'at the current rate it will take another 263 weeks to get Yorkshire miners back'.[168] The group attached particular importance to countering disinformation, so provided details of the TUC document and the decisions of the NEC and the February SDC.[169]

Since the first meeting had been postponed from before Christmas, the initiative came too late materially to affect the outcome of the strike:

> Lots of men have only heard what's happening via the media. There is a rumour going round that after twelve months on strike you can be sacked. I'm not making any excuses for scabs but there's probably some men that have scabbed because they've believed the rumours. This is one area where the Union can leaflet and let people know what their rights are. They have actually started now with leaflets, day-schools and meetings but they should have been doing this some time since.[170]

Despite the success of the leaflets and the enormous potential of the group meetings, the area leadership were indifferent or hostile to what they regarded as a dangerous rank-and-file movement (see Chapter seven).

As with strike organisation, the major deficiency with strike maintenance was the Union's failure, particularly within Yorkshire, to recognise the importance of community mobilisation. The strike only survived because the women's groups held the communities together. Instead of developing the potential of the women's groups and autonomous propaganda activities through establishing strike committees, the Union concentrated upon maintaining control through its formal structures.

Chapter five

Strike-breaking strategy

Royden Harrison contrasted labour's 'inspirational' view of history with capital's 'analytical' approach to history.[1] In other words, capital learns rather more from experience than does labour. This contrast is nowhere more apparent than in the case of the NUM and Conservative governments. The NUM was largely unprepared for the conflict and attempted to employ strategies and tactics developed during the 1970s. The Conservatives analysed their defeats of 1972 and 1974 and developed a complex strategy to cope with any future conflict. While claims that the 1984 strike was deliberately provoked[2] are exaggerated (see Chapter two), the Government was clearly prepared to resist a strike in order to implement its policy for the coal-mining industry. Five interrelated strands of strike-breaking strategy can be identified: firstly, contingency planning sought to avoid disruption by maintaining coal supplies and alternative fuels; secondly, trade-unions were weakened by the creation of a legal context of industrial relations which emphasised the rights of individuals; thirdly, the police force was militarised in pursuit of 'law and order' policies; fourthly, the strategy was underpinned by ideological support through the media which determined the agenda for debate, and finally, in an economic context of recession in which organised labour was already weakened, American 'union-busting' tactics were imported. None of these developments was specific to the miners' strike but the separate strands were entwined around this event. Each will be considered in turn.

5.1 Contingency planning

In contrast with systematic denials of government intervention in public-sector negotiations, from the onset of the miners' strike the Thatcher Government directed contingency operations. The machinery for this intervention was the Civil Contingencies Unit (CCU), successor to a series of strike-breaking organisations developed by the British state since 1919.[3] In 1947 the Attlee Government established an Emergencies Committee to cope with the docks strikes and further refined the Emergencies Organisation in response to more serious docks strikes in the two following years when states of emergency were declared. Regional committees to supply volunteer labour in the event of a national emergency were organised by the Emergencies Co-ordinating Committee (ECC) of F3 Division in the Home Office. This period demonstrated the problems associated with substituting troops for skilled workers and the need to involve representatives of the industry affected in any emergency planning. A fuel and power sub-committee of the ECC was subsequently established on which the NCB was represented:

Its report, which was completed in December 1948, laid the basis for emergency fuel and power planning ever since. Many of the recommendations were tactical. If a coal or transport strike was at all likely, coal stocks should be built up discreetly, for example by the railways, gas and electricity industries reducing services, 'if possible without assigning the reason for such action'. Even if an emergency occurred at the beginning of winter when coal stocks were at their peak, 'general restrictions of consumption (by Order) would be the only prudent policy'.[4]

In 1964 the Conservative Government introduced another Emergency Powers Act making Defence Regulation 6 permanent. States of emergency were declared during the 1966 seamens' strike, and the 1970 strikes of dockers and electricity workers. It was the 1972 miners' strike, however, which caused a re-evaluation of contingency planning. Reginald Maudling, Home Secretary and head of ECC at the time, defended his decision not to deploy troops to Saltley coke depot: 'if they had been sent in, should they have gone in with their rifles loaded or unloaded?'.[5] In the wake of Saltley, Heath arranged a review of civil emergency planning machinery under Lord Jellicoe, the Lord Privy Seal. The ECC of the Home Office was replaced by the CCU, a standing committee of the Cabinet Office. The Cabinet Office Briefing Room (COBRA) was created as the operations centre for the CCU, enabling direct communication with Chief Constables and regional military commanders. Within nine months civil servants working for the CCU had analysed, with the assistance of the relevant management, the strike capacity of sixteen major industries, including coal-mining. The 1974 miners' strike was a more low-key affair (see Chapter one) but indirectly led to a further revision of contingency planning. The Tory moderates' decision to call a general election was opposed by Thatcher and Sir Keith Joseph, who advocated resisting the miners. Labour's election victory appeared to vindicate the Thatcher-Joseph stance and gave credibility to the right wing within the Conservative Party.

Once elected as party leader, Thatcher established a committee under Lord Carrington, former Secretary of State for Energy in the Heath Government, to analyse the lessons of the 1973–74 dispute. From the secret evidence of businessmen, civil servants and senior management in the nationalised industries, the Carrington Report concluded:

Strong unions and the advanced technology operated by their members, especially in fuel and power industries, had irrevocably shifted the balance of industrial power in their favour . . . the armed forces could not be used on any large scale to break strikes as they had insufficient men for the task . . . higher priority should be given to contingency planning in periods of political and industrial quiet and . . . the best brains in Whitehall should be devoted to it . . . Skilled use of the 'publicity weapon' was deemed crucial to the public's will to resist.[6]

The Ridley strategy

The questions raised by the Carrington Report profoundly influenced the strategy developed by the Conservatives during opposition. The final report prepared by Nicholas Ridley for the Conservative Party policy group on the nationalised industries recognised the likelihood of a major public-sector confrontation in response to new financial objectives and the programme of privatisation which they proposed to implement. The Ridley Report classified industries according to their strike vulnerability, identifying water and electricity among the industries where strikes should be particularly avoided, and steel among the industries where the unions could be defeated. Coal-mining was regarded as an intermediate industry in terms of strike vulnerability but, in an annexe to the report, its authors anticipated that there would be: 'a major challenge from a trade union either over a wage claim or over redundancies . . . in a 'vulnerable industry' such as coal, electricity or the docks and have the support of the 'full force of communist disrupters'.[7] The Ridley formula to counter this threat became central to contingency planning after the election of the first Thatcher Government and was widely quoted during the strike:

*Return on capital figures should be rigged so that an above-average wage claim can be paid to 'vulnerable' industries.
*The eventual battle should be on ground chosen by the Tories, in a field they think could be won (railways, British Leyland, the civil service or steel).
*Every precaution should be taken against a challenge in electricity or gas. Anyway, redundancies in those industries are unlikely to be required. The group believes the most likely battleground will be the coal industry. They would like a Thatcher government to: (a) build up maximum coal stocks, particularly at the power stations; (b) make contingency plans for the import of coal; (c) encourage the recruitment of non-union lorry drivers by haulage companies to help move coal where necessary; (d) introduce dual coal/oil firing in all power stations as quickly as possible.
*The group believes that the greatest deterrent to any strike would be 'to cut off the money supply to the strikers, and make the union finance them'. But strikers in nationalised industries should not be treated differently from strikers in other industries.
*There should be a large, mobile squad of police equipped and prepared to uphold the law against violent picketing. 'Good non-union drivers' should be recruited to cross picket lines with police protection.[8]

The first principle of above-average pay settlements in 'vulnerable industries' was applied to the electricity workers. The funds to buy off the power-workers came from the miners as CEGB profits were maintained at the cost of the NCB (see Chapter one). The tactic helped preserve the power of the right-wing leaders of the EETPU and EMA, Hammond and Lyons, to resist any moves to support the miners. The isolation of the miners from other workers in the power industries complemented the use of Carrington's 'publicity weapon', a principle also applied to key workers *during* the 1984 miners' strike. On 2 April 1984 the Under-Secretary responsible for the Railways Directorate at the Department of Transport wrote to Ridley as Secretary of State for Transport seeking confirmation of the stance the Government expected BR Chairman Bob Reid to adopt in negotiations with the rail unions. The negotiations brought the threat of an overtime ban and on 13 April Ridley wrote to Chancellor Nigel Lawson suggesting the pay offer be increased in return for productivity improvements. The Government's acceptance of the need to buy off the railway workers was conveyed to the Department of Transport on 16 April by a Private Secretary to the Prime Minister. The following day Party Chairman John Selwyn Gummer wrote to Ridley: 'It seems to me to be critical at this juncture to avoid the risk of militants being strengthened in their attempts to block the movement of coal by rail, and to make wider common cause with the miners'.[9]

The second part of the Ridley strategy was manifest from the moment the Conservatives were elected, as the Government selected suitable battles to fight before the decisive coal war. In 1979 Michael Edwardes's 'survival plan' for British Leyland, which entailed the loss of another 25,000 jobs, provided the focus of the first confrontation. When Longbridge convenor Derek Robinson and leading members of the Joint Shop Stewards' Committee were sacked, action supported by the TGWU but not Robinson's own union, the AUEW, failed to win their reinstatement. The following year the defeat of the steelworkers' pay claim in a three-month strike effectively removed opposition to BS plans for a 'return to financial viability'.[10] In July 1982 after BR under Sir Peter Parker had made considerable progress with productivity negotiations, proposals to introduce 'flexible rostering' prompted a strike by ASLEF drivers. David Howell, Secretary of State for Transport, made the defeat of ASLEF a prerequisite for financial support even though the strike caused irreparable damage to the railway system.[11] The effect on the coal industry of the financial obligations imposed on the public sector made a conflict with the miners inevitable (see Chapter one).

The third part of the Ridley formula was designed to ensure the Government's success should such a coal strike materialise. Increasing coal stocks was a central element of the strategy, as Fig. 5 shows. During the first decade of nationalisation, total stocks hovered around 20m tonnes, but were allowed to rise to 30m tonnes in 1958 and approached 50m tonnes in 1959. While the official reason for stockpiling was to avoid hasty decisions over colliery closures, it also served to deter the Union from concerted industrial action against the run-down of the industry. Similarly, although stocks fell to 30m tonnes in the mid-1960s, they were allowed to rise to 40m tonnes in 1968 prior to the wave of closures which affected Yorkshire. Coal stocks had been allowed to fall to 20m tonnes by 1972, which partly accounts for the success of the strike in that year, and despite restocking to 30m tonnes in 1973 this proved insufficient insulation against the 1974 strike. Stocks returned to 30m tonnes in 1976 and remained at about that level until 1980. At the time of the 1979 general election, total stocks amounted to 27·7m tonnes, of which 12·4m tonnes were distributed to power stations. When the 1981 strike against closures began, stocks had increased to 37.2 tonnes, of which 16·2m tonnes were held at power stations. After the Government had postponed the major confrontation David Howell explained: 'Neither the Government nor I think society as a whole was in a position to get locked into a coal strike . . . In those days the stocks weren't so high. I don't think the country was prepared, and the whole NUM and the trade union movement tended to be united all on one side'.[12] Howell was replaced as Secretary of State for Energy by Lawson who 'strongly approved the tactics outlined in the Ridley analysis, and . . . systematically set about the means to make them work'.[13] When the overtime ban began at the end of October 1983, coal stocks in the UK had reached a post-war record of 59·9m tonnes, of which 33·7m tonnes were held at power stations. Since the first Thatcher Government had come to office, total coal stocks had increased by one hundred and sixteen per cent, while power-station stocks had grown by one hundred and seventy-one per cent.

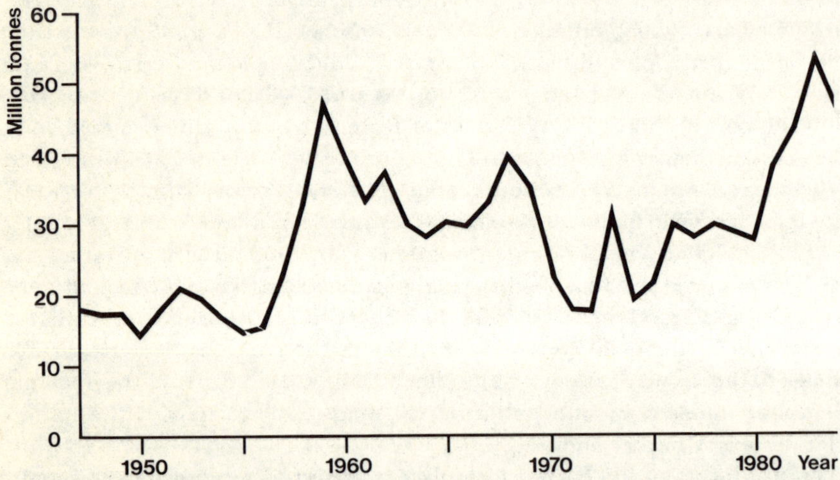

Fig 5 Coal stocks, 1947–84

In addition to building coal stocks, the Ridley formula advocated contingency plans for importing coal. The 1981 crisis was precipitated by increased imports and, in order to settle the strike, the CEGB stockpiled a further 3m tonnes of American and Australian coal at Rotterdam. The enormous increase in coal imports advocated by Turner[14] would require improved port facilities and coastal location of new power stations, but Tony Lane described how smaller ports were used to bring coal into the country during the strike.[15] Whether the arrangements for coal imports were the subject of detailed contingency planning before the strike is impossible to establish on existing evidence, but such planning was recommended in the Ridley Report. Similarly, while the report recommended the recruitment of non-union drivers, it is difficult to see how the CCU could have such a direct influence except through directing the NCB and BS to contract the services of particular hauliers. The evidence of scab drivers was plain enough during the strike but in many cases these were nominally TGWU drivers. Although the Royal Corps of Transport is known to have intensified its HGV training programme and drivers in the Orgreave convoy were seen to salute a senior police officer, there is scant evidence that troops were used in this capacity.

The Ridley Report raised the question of vulnerability of fuel supplies in terms of the attraction of dual coal/oil firing in power stations. A Cabinet Meeting on 23 October 1979 proposed a nuclear power programme as a more long-term solution (see Chapter one). The replacement in 1982 of Glyn England as CEGB chairman by Sir Walter Marshall, formerly head of UKAEA, has been interpreted as evidence of the Government's determination to implement its nuclear programme.[16] The nuclear strategy was of no immediate relevance to the miners' strike, however, given the lead time for building nuclear power plants. After the 1981 crisis, the CCU appointed Committee MISC57 to assess the energy industry's vulnerability to strike action. England reputedly regarded instructions to build coal stocks and substitute oil wherever possible as 'hysterical', whereas Marshall's opposition to coal is well-known.[17] Whatever the influence of MISC57, the vulnerability of electricity supply to action by miners was significantly reduced between 1981 and 1984. Dual firing was probably of equal importance to coal stocks and the three new oil-fired stations Grain, Ince and Littlebrook supplied eighteen per cent of the electricity used during the strike. It was oil rather than coal which kept the lights on (see Chapter six). In addition to increased flexibility in electricity generation, new computerised control of the system allowed sophisticated load-switching around the national grid. Throughout the strike the CEGB was able to reverse the power flow on the grid system from oil-fired power stations in the south to compensate for reduced coal-fired generation in the north.[18] Fuel flexibility and computer control enabled the CEGB to implement the coal stocks conservation strategy recommended in the 1948 ECC guidelines, without any reduction in service, from the outset of the overtime ban. When the strike began in March 1984 winter consumption and the overtime ban had only reduced stocks to 51·5m tonnes, of which 27·2m tonnes were distributed to power stations.

The two remaining parts of the Ridley strategy represented a frontal assault on strike activity: one part attacked strike maintenance, the other strike organisation. Before 1980 the state provided income which critics claimed 'subsidised', and thereby encouraged, strike action. The *Economist* argued in relation to the 1972 miners' strike: 'Not all men are willing strikers; in the miners' strike 45% of the men initially voted against it . . . they might go along with the mob a little less complacently if there was a penalty'.[19] Workers involved in a trade dispute, whether on strike, locked out or laid off as a result of such action, have never been entitled to unemployment benefit or supplementary benefit in their own right, but in the past could claim for dependants. Single strikers were also able to obtain Urgent Needs Payments from the DHSS in certain circumstances. In practice, the take-up rate among strikers entitled to benefits for themselves or dependants was under fifty per cent from 1954 to 1972 and under fourteen per cent of strike households actually received any benefits during this period.[20] In the 1972 and 1974 miners' strikes, however, the NUM advised members on social security entitlements; strikers' dependants received £8m and £5m in these years compared with an annual average of £2m in the 1970s and £52,000 Urgent Needs Payments were made in 1972 compared with £5,000 in the previous decade.[21]

After the 1974 miners' strike, the Conservative Political Centre produced a paper which argued that the state should not provide benefits to strikers' dependants.[22] Although systematic studies failed to establish any link between state benefits and strike activity,[23] when returned to office the Conservatives acted upon their assumption. The Social Security Act 1980 removed absolutely any entitlement of single strikers and ruled that all strikers were to be assumed to be in receipt of a notional sum of £15 strike pay per week. Ostensibly this move was designed to impose the costs of striking upon trade-unions but in practice few unions have the resources to provide strike pay in a protracted dispute, especially where the entire membership is involved. Orme described the fictional strike pay as 'one of the Government's most vicious moves against working people'.[24]

In November, Benn and Michael Meacher questioned Norman Fowler, Secretary of State for Health, on the Government's decision to take a further £1 from strikers' families by increasing the notional figure to £16.[25] Roy Hattersley described the move as 'callous and vindictive', and a number of MPs prevented Fowler from speaking.[26] As a result of the miners' strike, two addenda were issued to the 'S-manual' which DHSS staff use to interpret the 'Yellow Book' of statutes and regulations. The first, 'Mineworkers' Strike: Supplementary Guidance', issued on 16 March 1984 by the Chief Supplementary Benefit Officer, instructed DHSS staff to deduct all 'notional resources' received such as food parcels, vouchers or anything to which a value could be attributed, from the benefits to be paid to strikers' dependants.[27] The second, 'Guidance Additional to that in the S-Manual: Miners' Strike 1984', issued in May, directed that a whole range of support such as local authority meal vouchers for children, should be counted as income for the purposes of calculating benefit entitlement. In addition Specialist Claim Control teams were sent into mining areas to make random checks on claimants.[28]

The element of the Ridley strategy designed to confront strike organisation had been the subject of more commentary than any other part of the formula, perhaps because of its greater visibility. In response to inner-city disturbances in the summer of 1981,[29] new methods of policing were introduced to mainland Britain from Northern Ireland.[30] While the police gained experience in the 'riots', the miners had no such preparation. Urban unrest rather than industrial conflict provided the opportunity to militarise the police force, but the changes only became apparent to the miners once the strike had begun. Throughout the strike the overall policy of deployment of police was overseen by Cabinet Sub-Committee MISC101, serviced by the CCU. Committee MISC101 was chaired by Thatcher, and attended by Leon Brittan (Home Secretary), Lawson (Chancellor), Walker (Energy), Tebbit (Industry), Tom King (Employment), Ridley (Transport), Sir Michael Havers (Attorney General), Brigadier Budd (CCU) and David Goodhall (CCU).[31] The operations carried out by the police in accordance with the strategies developed by MISC101 are outlined below, after considering the legal changes which served to criminalise industrial conflict.

By definition, the NUM could only have prepared for the Government's contingency planning well before the strike. The first requirement was for the Union to have anlaysed the state's plans and formulated a strategy in response to these. The details of the Ridley Report were publicised six years before the strike, so NUM strategists must have been aware that any conflict would be heavily weighted against the Union if the policies were effected. In retrospect, it is clear that after the 1981 conflict the NUM should have taken action to reduce coal stocks, while campaigning within the NUM and the wider labour movement. The 1983 overtime ban and associated campaign were implemented too late to counteract strike preparations. The moves tc reduce social security entitlements for strikers' dependants should have demonstrated the need to make effective provision for strike maintenance. Similarly, recent changes in labour law and the policing of industrial disputes should have indicated the need to develop strike organisation which was not dependent upon 1970s' tactics.

5.2 The criminalisation of industrial conflict

The 1984 miners' strike took place in a more hostile legal and political climate than any other post-war stoppage. For 100 years the legal context of industrial relations in Britain was largely one of voluntarism, characterised by Flanders as a preference for collective bargaining, an insistence on the participants' autonomy and a desire to keep industrial disputes out of the courts.[32] This third element is important, for the courts are not merely inexpert, the legal system is designed to regulate relations between individuals not collectivities like trade-unions. Moreover, observers from Lord Justice Scrutton in 1920 to Michael Foot in 1978 have identified the class bias of the judiciary.[33] The class background of the judiciary makes it difficult for them to adjudicate independently between management and workers, particularly during times of intense social conflict.[34]

Laws governing the conduct of industrial conflict developed within the voluntarist framework. Hence, in place of positive rights to combine and to strike, trade-unions and industrial action remained inherently unlawful at

Common Law, but these judicial interpretations were legislated away by negatively expressed immunities in statute. Apart from restrictions on 'political strikes' under the Trade Disputes Act 1927, the first attempted departure from voluntarism in *statute* was Labour's 1969 Industrial Relations Bill. The Donovan Commission had prescribed voluntary reform of industrial relations rather than any extension of legal sanctions,[35] but the 1969 White Paper proposed legal enforceability of collective agreements and criminal sanctions against unofficial strikers.[36] Although the Government withdrew the Bill, and the TUC General Council gave a 'unanimous solemn and binding undertaking' to restrain unofficial action by members of constituent unions, the contradiction between the ideology of the Labour leadership and the Party's working-class foundation was illuminated.[37] Conservative thinking on labour law had been outlined earlier,[38] but their 1970 manifesto contained a firm commitment to curb industrial militancy through legislation. The Industrial Relations Act 1971 failed when unions co-ordinated their opposition, employers failed to use its provisions,[39] and the courts were unable to enforce their decisions.[40] That the defeat of the 1971 Act was a triumph for voluntarism was reaffirmed in the package of legislation introduced under the Social Contract by the 1974 Labour Government. The PLP appeared to have learnt from the struggle over the Tories' Act; the Conservatives learnt a different lesson.

A new legal framework

Between August 1978 and May 1979 the Social Contract disintegrated and 13·5m days were lost in strikes. The 'winter of discontent' became an election issue and the media launched a campaign against strikes in general and picketing in particular. During the 1979 road haulage strike, Conservative opposition was so successful that Callaghan added his criticism of 'secondary industrial action', even though this was the subject of much legal confusion.[41] Within two months of the Conservatives' election victory, Prior, Secretary of State for Employment, published consultative papers on the closed shop, picketing and secret ballots, outlining elements of the Employment Bill prouduced in December 1979. The Employment Act 1980 restricted the traditional immunity of individuals from tortious liability to pickets attending at or near their normal place of work (s.16) and to action not defined as secondary (s.17). Tebbit, who replaced Prior, introduced the Employment Act 1982 which made it lawful for an employer *selectively* to re-engage after three months workers dismissed in connection with industrial action (s.9) and which removed the immunity of unions from civil liabilities in the case of secondary action (s.15). King, Tebbit's successor, introduced the Trade Union Act 1984 which removed the immunities of unions over any industrial action authorised or endorsed by a union without reference to a secret ballot (s.10).

After the winter of discontent the Labour Government pressurised the TUC into issuing voluntary guidelines on picketing.[42] However, to complement the 1980 Act, Prior unilaterally imposed codes of practice on picketing and the closed shop. The Government's picketing code contained many of the restrictive provisions of the TUC Guide, but voluntary guidance was considered inadequate.[43] Whereas ACAS codes were used evidentially by tribunals, the 1980 Act (s.3) obliged *courts*, including the Criminal and High Courts, to take Prior's codes into account. While the picketing code is not *law*, it introduces restrictions on picketing just as effectively, suggesting a maximum of six pickets at each entrance of a workplace, and imposing the responsibility for controlling pickets upon the union.

The Employment Acts seemed to have been tailor-made for the miners' strike. During the first week it appeared that the NCB would use the courts to counter the domino mobilisation strategy. On 14 March 1984 Mr Justice Nolan granted an injunction against Yorkshire Area NUM to prevent the deployment of flying pickets. The area ignored the injunction (see Chapter three) and on 16 March the NCB was granted leave to bring action for contempt. Three days later, however, Justice Caulfield agreed to the NCB's request for an indefinite postponement of the action. It is probable that the Government feared such civil action could unite the NUM, and possibly sections of the wider labour movement, more than the original issue of the dispute: 'when we passed that legislation it was not in contemplation that a union would be split', Sir Michael Havers later admitted.[44] The Government was criticised for its apparent failure to direct the use of its own legislation: 'it weakens the authority of the courts to see its orders so flagrantly ignored' claimed Walter Goldsmith of the Institute of Directors, who feared: 'the policy of inaction may be encouraging the spread of secondary action in other industries'.[45] Lord Denning similarly argued: 'the law must be enforced; it must not be allowed to become a dead letter'.[46] The CEGB and BS also decided against using the civil law for fear of alienating their own workers who had been persuaded not to support the miners.

The new legislation was used by smaller employers to great effect. Richard and George Read, owners of a haulage firm prevented from carrying coal to Port Talbot by South Wales pickets, were granted an injunction by the High Court on 17 April against the Area Union.[47] On 19 July the Court awarded another injunction against the South Wales Area on behalf of Wight Contractors of Llanelli, who were being prevented from extracting coal from a colliery tip.[48] On 30 July Justice Park fined the Area Union £50,000 for their continued picketing of Read Transport, although the Union had written to the company to offer exemption from picketing.[49] South Wales miners attempted to prevent bailiffs from entering Union headquarters in Pontypridd but the sequestrators were able to obtain £707,000 which was frozen until the Union purged its contempt.

Actions by dissident miners

In contrast to the Government's reticence to encourage civil actions by public-sector employers, Ministers were reported to attach the 'highest importance to the actions brought by miners in Derbyshire and Nottinghamshire against the

Union'.[50] During the strike the NUM and almost every constituent area faced actions brought by 'working miners'.[51] These were actions brought under Common Law by individuals suing the union on the grounds that its rules had been breached. Although the Employment Acts appeared to play no part in these actions, they did influence the stipulations made by the courts in injunctions and orders. The courts generally found in favour of individuals and ignored the collective interests of the wider NUM membership. Strikers were angered that the 'right to work' was interpreted narrowly by the law as an immediate right of individuals to scab rather than in terms of opposition of pit closures.

The major actions derived from the weakness of the domino mobilisation strategy: area autonomy enabled the courts to use the results of ballots in areas opposed to the strike to undermine the constitutional authority of the National Union. Early action was 'limited and defensive'.[52] Injunctions were obtained restraining several areas from disciplining working miners. Where area ballots had rejected a strike, the courts held the action to be unofficial, even in the case of Lancashire where an Area Delegate Conference in May had voted for an official strike.[53] National Disciplinary Rule 51 was proposed to counter the problems of area autonomy, but instead provided an opportunity for further legal action culminating in Sir Robert Megarry's ruling on 18 July 1984 that the new disciplinary rule was unlawful (see Chapter two).

After these initial civil actions, a legal offensive was launched on behalf of the working miners and timed to coincide with other strike-breaking initiatives. For area strikes to be sanctioned by the National Union under Rule 41, the courts had to be satisfied that each area strike was called in accordance with the rules relating to that area. If it could be shown that just one area was in breach of its rules, the National Union could be joined as a defendant, and the legal process would gradually emasculate the NUM. The diversity and complexity of area rule-books provided ample opportunity for the courts to find prima facie cases of breach of rules sufficient for interlocutory injunctions to be granted. As a result, the Union was subjected to a bewildering array of legal attacks involving some twenty separate actions by dissident miners.[54] The two most damaging actions were ostensibly independent although both enjoyed financial support channelled through Chris Butcher, a key member of the working miners' movement (see below). The first major civil action was begun by Colin Clarke (Pye Hill) and John Liptrott (Sherwood) with £10,000 collected through Butcher's network. David Negus of Derbyshire solicitors, Ellis Fermor, engaged Michael Burton QC to pursue Clarke's case, in which it was claimed that the NUM had called the strike in breach of Union rules.[55] On 25 May 1984 Vice Chancellor Megarry granted an injunction to prevent the NUM from instructing Notts miners to join the strike, on the grounds that there was no basis for calling the strike 'official'. Although the Clarke case gave only an interim ruling, it fulfilled an important role in appearing to legitimise the police operations to prevent 'unlawful' picketing in Nottinghamshire.

The action which did most damage to the NUM was brought by Robert Taylor and Ken Foulstone, miners at Manton colliery, who sought interlocutory orders against both the Yorkshire Area NUM and the National Union, claiming that each was in breach of its rules through failing to hold a strike ballot. Butcher provided the funds for Taylor and Foulstone to begin the action, and introduced them to David Payne of the Newark solicitors, Hodgkinson and Tallents. Their case was supported by union documents from Yorkshire Area Office. Ottey believes Scargill suspected him of furnishing the solicitors with these materials,[56] but Ottey would not have had access to the Yorkshire SCC minutes. Payne contacted Sarah Hogg, Economics Editor of *The Times*, who advised on media coverage. On 9 August the case was adjourned until September, apparently to allow the NUM to prepare its defence, but also enabling the plaintifs to obtain financial backing. The same day a *Daily Express* feature called for support and within a month £34,000 was raised to cover legal costs.[57] Payne served a writ on Heathfield at the TUC, but the NUM failed to attend court proceedings. On 25 September, Taylor and Foulstone, accompanied to the High Court by Butcher, claimed that the NUM had 'circumvented democracy'. On 28 September 1984, Justice Nicholls granted injunctions restraining both the Yorkshire Area and the National Union from describing the strike as 'official', and from instructing members to respect picket lines or to refrain from working.[58] The strike in Yorkshire was declared unofficial because there had not been a valid area ballot, the 1981 vote being 'too remote' from the present dispute, despite the power vested in Council by Yorkshire Area Rule 53(a) 'to say when notices to cease work shall be tended'.

Several other cases had considered the legality of the strike in the so-called 'rule-book' actions brought by NUM members.[59] Derbyshire miners Roland Taylor and David Roberts pursued an action like the Yorkshire case and had the strike ruled unlawful in Derbyshire.[60] In two areas, however, the legality of the strike was upheld. On 24 September 1984, a Scottish judge refused to rule that the strike was illegal without a national ballot, a decision which was reaffirmed on 6 November.[61] Similarly, in South Wales, dissident miners claimed that the NUM, or NCC, had encouraged secondary picketing, thereby inducing breaches of contract and breach of the area rules. The judge found that there was no 'arguable case' that the South Wales Area rules had been breached,[62] although other aspects of the ruling were not beneficial to the defendants (see below).

The Taylor and Foulstone case had a significant impact because of the Union's refusal to abide by an order rather than because of the injunction itself. The NUM had not attended the proceedings and Scargill claimed that the High Court decision would 'not be accepted'.[63] Contempt proceedings were immediately initiated and, on 1 October 1984, Scargill was served with a writ at the LP conference. The NUM was fined £200,00 and ordered to purge its contempt by accepting that the strike was unconstitutional. Scargill was fined

£1,000 individually, but this was paid anonymously by persons who felt that jailing the President of the NUM might galvanise other workers into action and inflame the dispute. The NUM continued to refuse to co-operate and, on 25 October 1984, Justice Nicholls ordered sequestration of the Union's assets, estimated to be £8·4m. Price Waterhouse, the accountants appointed to seize the Union's funds, told the court on 9 November that they had only recovered £8,000 in the UK and found £3·25m in Dublin, which was frozen. By 18 November the accountants had traced the funds which the NUM had secretly moved abroad in March, because of 'a most staggering breach of confidentiality on the part of a banker'.[64] A further £4·75m was frozen in Luxembourg and £500,000 located in Zurich, but in accordance with European law a Luxembourg court returned £4·5m to the NUM and kept only £250,000 to cover the contempt fine and legal fees.

A second wave of legal actions was already underway to prevent the leaders from controlling Union finances and to obtain damages from them. On 1 November Clarke and fifteen other working miners began action to make the twenty-five NEC members personally liable for the £200,000 fine. As a result, on 30 November the court removed Scargill, McGahey and Heathfield as trustees and put the NUM's assets under the receivership of Herbert Brewer, a Derbyshire solicitor who was also a Tory councillor, and an offical of both the Conservative Party and the Institute of Directors. The NUM challenged the ruling in the Court of Appeal, but on 1 December Lord Justice Stephenson upheld the decision. This was the first time ever that a receiver had been appointed to manage the affairs of a trade-union. Accountant Michael Arnold was appointed receiver after Brewer's fitness to undertake his duties was challenged in court, and on 10 December the Government agreed to meet the cost if Price Waterhouse failed to obtain NUM funds. On 21 December the High Court instructed the TUC to provide details of all its assistance to the NUM since the strike began. Following the seizure of assets in Luxembourg, the fine was paid. In a second action brought by Derbyshire miners Taylor and Roberts, Justice Vinelott refused on pragmatic grounds to order two Derbyshire Area officials to repay £1.7m of Union funds expended in the course of the unlawful strike.[65] The National Union was unable to use its funds from 1 December, but financial support from the areas and other unions was maintained through clandestine cash operations.[66]

Impact of the law

The Clarke case constrained the strike in Nottinghamshire, while the action by Taylor and Foulstone eventually removed the NUM's resources. Each contributed significantly to the Union's difficulties, but in retrospect neither could be said to have made it impossible for the strike to continue. It is unlikely that Notts miners would have joined the strike, irrespective of the Clarke case (see Chapter two), and the maintenance of the strike had never depended upon the resources of the NUM (see Chapter four). Although Ottey resigned over it,[67] the Union's defiance of the law was itself unremarkable and probably did more to unify the NEC.

Two other actions, however, were decisive in bringing strike organisation to an end. In the Thomas case, it was accepted that miners picketing pits other than those at which they were normally employed, were deprived of traditional immunities, and that the South Wales Area NUM was vicariously liable for any damages arising out of this.[68] The judge ruled that secondary picketing was not necessarily an offence under s.7 of the Conspiracy and Protection of Property Act 1875, but that mass-picketing did amount to 'unreasonable harrassment'.[69] An injunction was issued restricting the number of pickets to six at the pits where the plaintiffs worked. A parallel action was brought by nineteen working miners in Yorkshire against twenty-eight members of the National and Yorkshire Area Executives, and the Yorkshire Area NUM. The day after the South Wales judgement, Justice Nicholls granted an injunction ordering the Yorkshire Area to cease organising picketing of the eleven pits at which the plaintiffs were employed, except for peaceful picketing at the pickets' own place of work by not more than six pickets at each entrance.[70] In these judgements the Employment Acts and the Code of Practice on Picketing became embodied in Common Law. The South Wales Area announced its intention to appeal against the ban on mass-picketing, although support was growing for returning to work without agreement (see Chapter six). The Yorkshire Area agreed to comply with the ruling on the eleven named collieries, the SCC was disbanded and members were advised to destroy documents in their possession.

Ultimately the law had a decisive impact upon the strike. Although the new employment legislation was not used by the major employers, it played a key role in exposing pickets to criminal prosecution. Courts took the picketing code into account when considering breaches of the law which commonly arise during picketing and were more likely to grant injunctions or impose related bail conditions. Police were more likely to attempt to restrict picketing, leading to further charges of obstruction. Thatcher's claim that it was *unnecessary* for BS to use the civil law because supplies were getting through[71] failed to explain why, in Lord Denning's words, it was 'high policy at Cabinet level not to use it but to call out, instead, hundreds and hundreds of police'.[72] Changes in civil law were an integral part of Conservative strategy, regardless of whether employers chose to take civil action, because the apparent 'civil illegality' of secondary action and flying pickets in the public mind *legitimised* the increased use of criminal law and police operations against pickets. In this way the strikers were *criminalised*, making it more difficult to generate solidarity among other workers. The Common Law actions taken by working miners reinforced this process, giving the appearance that the strike was sustained by intimidation, providing the courts with opportunities to rule on the legality of the strike, and publicising divisions within the miners' ranks.

If the NUM had held a national ballot, the actions which provided the courts with an opportunity to rule the strike unofficial could not have taken place.

There were compelling reasons for not holding a ballot (see Chapter two), but in any case a ballot would not have protected the Union from the crucial actions which brought an end to strike organisation. The mass-picketing outlawed in these cases was necessitated by the decay of the strike which, in Yorkshire at least, stemmed from the chronic hardship strikers suffered. The best protection against the legal attack on strike organisation would therefore have been more effective strike maintenance.

5.3 Police operations

When the new employment legislation was proposed, several Chief Constables and the Attorney General argued that it was unnecessary. The civil immunities afforded trade-unionists involved in industrial action were strictly limited in relation to picketing and gave no protection from criminal liabilities. The immunity restored by the Trade Union and Labour Relations Act 1974 (s.15) extended only to the purpose of 'peacefully obtaining or communicating information, or peacefully persuading any person to work or abstain from working' and excluded picketing a person's place of residence. Under the Conspiracy and Protection of Property Act 1875 (s.7) it is a criminal offence to compel any person to abstain from doing any lawful act through various means, including violence, intimidation, injury to property and following in a disorderly manner in the street. The Public Order Act 1936 (s.5) makes it an offence to use 'threatening, abusive or insulting words or behaviour . . . with intent to provoke a breach of the peace or whereby a breach of the peace is likely to be occasioned'. The Highways Act 1959 (s.21) makes it unlawful to 'obstruct free passage along a highway'. The courts have been able to build on Common Law as well as statute in developing crimes associated with picketing, but the Common Law charge of riot had not been used against post-war strikers until the 1984 miners' strike.

These extensive restrictions on industrial action support the view expressed in a judgement of 1896, which sought to make picketing inherently unlawful, that 'you cannot make a strike effective without doing more than is lawful'.[73] Bercusson explains that the legal system 'attempts to establish an equilibrium based on the maintenance of capitalist interests as owners and employers and sees the picket line as disrupting this equilibrium and thus illegal'.[74] Although the police have always had a wide range of offences with which to charge pickets, regardless of developments in civil law, they rarely enforced the law because this would invariably inflame industrial relations and employers usually expect strikers to return to work. As a result there developed a widespread myth among trade-unionists of a 'right to picket' which obscured the extent of criminal offences likely to be committed in the course of normal picketing. Any remnants of that myth were effectively destroyed in the first few weeks of the miners' strike. The decision to enforce the law and prevent effective picketing required a massive police operation which itself provided more opportunities for prosecution.

Police strategy

The CCU established the overall strategy for policing the strike but the deployment of such large numbers of police was only possible because of the National Reporting Centre (NRC). The NRC was established after the 1972 miners' strike to co-ordinate the deployment of Police Support Units (PSUs) between forces under the mutual aid system. Based at New Scotland Yard under the control of the President of the Association of Chief Police Officers (ACPO), the NRC recorded on computer details of police manpower and local conditions which were summarised as symbols on an operations map covering one wall.[75] The NRC had only been fully activated on three previous occasions, all since 1980, although it did monitor the situation during the 1974 miners' strike.[76] The NRC began monitoring the 1984 miners' strike on 13 March and was fully activated within that week.

In order to deploy PSUs effectively, the NRC depended upon high-quality intelligence concerning picketing targets. The responsibility for surveillance in an industrial dispute rests with MI5, who are known to have bugged a meeting between NUM officials and a leading Communist Party member during the 1972 strike.[77] In 1980, when Yorkshire Area NUM President, Scargill had complained that his home and office telephones were tapped.[78] Michael Bettany, a former member of MI5 now serving a twenty-three-year prison sentence, confirmed that Scargill's telephones were tapped twenty-four hours a day 'both during periods of industrial action and at all other times'.[79] During the 1984 strike there was evidence of extensive phone-tapping in South Wales,[80] Kent,[81] and Yorkshire.[82] Aware of this problem, strike centres communicated as little as possible by telephone, used previously agreed codes and laid false trails.

Police intelligence involved more than phone-tapping. When arrested, miners were questioned about their political beliefs and asked who was responsible for picket deployment. Plain clothes officers were discovered impersonating pickets to gain information. Frank Slater, South Yorkshire NUM official, was tailed by plain clothes officers on 22 March. The demonstration outside the Nottinghamshire delegates' meeting in Mansfield on 6 April was infiltrated by security personnel, who were also among pickets at Thoresby during the fifth week of the strike. On 9 April David Owen, Chief Constable of North Wales, admitted using plain clothes officers at Point of Ayr colliery, and on 10 April Brittan defended such tactics. The police obtained further information about numbers and directions of travel of pickets from strategically-placed 'spotter cars' on major routes and possibly from computer-controlled cameras on motorway bridges.[83]

Such information facilitated the 'intercept policy' which had been planned in November 1973 for use in the 1974 miners' strike.[84] Although the policy was not implemented in 1974 because the NUM used fewer pickets, it was employed during the Stockport Messenger dispute in 1983. In the 1984 strike Yorkshire pickets heading for Lancashire were intercepted on the M62, while Kent miners were stopped at the Dartford tunnel.[85] Police at roadblocks demanded to know the destination and the purpose of the journey; unless satisfied that the occupants were not striking miners likely to commit a breach of the peace at some unspecified place, the driver was ordered to turn the vehicle around or face arrest for obstructing a police officer in the execution of his duty under the Police Act 1964 (s.51). Nevertheless, a police officer must have 'reasonable grounds' and 'reasonably anticipate' a breach of the peace before an individual can be charged with obstruction. Although the law only gives police the right to stop a vehicle and ask a driver's name and address, the legality of the intercept policy was upheld by the courts.[86] Roadblocks were used most effectively to prevent pickets reaching Notts pits, the CCU having identified that coalfield's strategic importance. Police statistics suggest that during the major campaign of picketing against Nottinghamshire, between the end of April and the beginning of August on average 2,000 pickets were stopped each day,[87] while only 1,200 reached their targets.[88] On police evidence sixty-two per cent of pickets deployed were stopped, while the Union's figures for deployment give the proportion as thirty-seven per cent; the mean of the two estimates suggests fifty per cent of flying pickets were stopped. The intercept policy was an integral part of surveillance as the registration numbers of pickets' cars encountered at roadblocks were entered on the 'Stolen and Suspect Vehicles Index' of the Police National Computer as 'seen in noteworthy circumstances'.[89]

Those pickets who managed to reach their targets encountered some of the 8,000 police officers deployed on a daily basis at the peak of the strike. The police on picket duty were organised along military lines into PSUs (twenty men, two sergeants and one inspector, trained in riot-control techniques and equipped with riot shields, helmets and flame-proof boiler suits) from as many as fourteen different forces, under the direction of a 'ground commander'. These methods prompted widespread rumours that troops were being deployed in police uniforms to the picket lines. Individuals claimed to have recognised soldiers (sometimes relatives) and clothing workers reported an increased volume of production of police uniforms. Benn wrote to Thatcher asking 'whether any members of the armed forces have been working in police uniforms, or have been involved in any other way in the police operations'.[90] Thatcher's reply gave her assurance 'that no authorisation whatever has been given to service personnel to get involved in police operations in connection with the miners' strike or to work in police uniforms'.[91] Much of the circumstantial evidence, such as absence of identification numbers, and some officers' apparent ignorance of the police complaints procedure, demonstrates how the policing of industrial disputes had altered.

Police tactics

At the picket line, police would allow a maximum of six 'official pickets' to stand near a colliery entrance while the remainder were kept at a distance and termed 'demonstrators'. In a celebrated case in 1974 it was held that 'pickets who assemble in unreasonably large numbers do have the purpose of preventing free passage',[92] so the mass-pickets from Saltley (1972) to Warrington (1983) were probably unlawful, although during the miners' strike one authority thought mass-picketing was not a criminal offence.[93] In another important case,[94] the counsel for the defendant argued:

> if pickets have become demonstrators as they do in such a case, and seek to make their views known from behind a cordon of police officers at a certain distance, then picketing becomes a shouting match, and in my submission there is nothing more likely to be productive of conflict than if pickets are to be restrained behind cordons when there is no suggestion that they are other than peaceful pickets.[95]

The official pickets were not guaranteed any opportunity of communicating with persons crossing the picket line. While pickets are probably entitled to signal to drivers requesting them to stop, they have no right to force vehicles to stop,[96] and police action to prevent pickets even from signalling to a bus bringing strike-breakers has been supported where a breach of the peace was reasonably anticipated.[97] Police also attempted to restrain pickets from using offensive language: one inspector helpfully advised pickets that they could call scabs 'bounders'.[98] Miners complained of verbal taunts from the police. The otherwise innocuous question, 'what's your kid getting for Christmas?', was particularly provocative where police officers, and scabs in the areas hardened against the strike, waved money at the pickets. Arbitrary orders were used to provoke obstruction or assault, and women were deliberately insulted and verbally abused 'in order to incite violence from . . . the men'.[99]

As well as verbal abuse, pickets were subject to physical violence, not confined to the isolated cases of individuals losing their temper but as part of a systematic operation. Many miners described as typical the pattern of relations which we witnessed ourselves on several occasions. Initially, pickets would assemble in a disorganised fashion to face a relatively small line of policemen in conventional uniform. Police numbers would increase as transit vans carrying PSUs arrived and stood by. Before the strike-breakers arrived, the police cordon would be strengthened and the pickets confined in one area. Tension would increase as the scabs arrived and pickets would push against police lines; tempers could flare on either side at this stage. Police would then drive a wedge into the pickets to allow snatch squads to make arrests, having identified picket leaders with the help of infiltrators[100] or video cameras.[101] Pickets would then be advised to disperse or would simply be attacked by 'short-shield men' or mounted police.

Although local picketing became routinised into a half-hearted ritualistic push when the scab van passed, there was little evidence of the stabilisation of police-picket relations described in the 1980 steel strike,[102] let alone the 'amicable' relations observed in the 1970 Pilkington's strike.[103] The limited stability of local picketing was fragile and liable to be destroyed by the arrival of flying pickets or 'flying police'. As Field noted, 'the South Yorkshire Police were not interested in an accommodative approach; their strategy was to impose tight control over the placing of pickets'.[104]

The majority of activists whom we interviewed felt that the aggressiveness of policing was determined by the ratio of police to pickets more than by picket behaviour or the police forces involved (although the Metropolitan Police had the worst reputation). Police brutality was alleged during the first week of the strike when pickets were kicked after being knocked to the ground at Thoresby colliery. On 21 March three Welsh miners were roped to railings for three hours and the following day police used crowbars to smash the windscreens of pickets' cars. In the fifth week truncheons were used on pickets at Cresswell and thereafter riot police, horses and dogs were regularly used on the picket lines.

Orgreave was viewed by many miners as the epitome of organised police violence: pickets were *allowed* to mass at the coking plant,[105] to enable the police to inflict maximum damage,[106] and to keep the pickets away from Nottinghamshire.[107] The Orgreave trial brought to light a confidential police manual 'Public order tactical operations', which instructed short-shield officers 'to disperse and/or incapacitate' members of a crowd, recommended police horses 'advance . . . towards the crowd at a canter', and advised rhythmic banging of truncheons on shields and other techniques to 'create fear'.[108]

South Yorkshire was an important battleground, both because of Orgreave and because the Doncaster and South Yorkshire pickets had led the picketing of Nottinghamshire. In May the police besieged the Notts village of Blidworth in order to drive out Yorkshire pickets who were staying there.[109] After June, when Orgreave was no longer a distraction, the police mounted similar operations to Blidworth in Yorkshire pit villages. Large numbers of police occupied villages, imposed a curfew and restricted the movement or residents.[110] When communities resisted, the police on occasions sealed off the entire area and refused to admit journalists or TV crews. The most serious incidents involved police reprisals after major confrontations with pickets. Thus the events described by Waddington at Maltby on 8 and 15 June were connected with Orgreave.[111] On the evening of 9 July after the incidents at Rossington described in Chapter three, the police retaliated with a series of attacks and arrests in the villages of Hemsworth and Fitzwilliam.[112] After pickets barricaded the entrance to Markham Main following the appearance of the first scabs on 21 August, the police sealed off Armthorpe village and caused extensive damage in pursuing pickets through private houses.[113] After mass-picketing in Maltby, the village was attacked a second time on 24 September. The following month Grimethorpe received the same treatment when strikers assembled after arrests had been made for coal-picking on the pit stacks.[114] Although the community siege was most widespread in Yorkshire, it occasionally occurred in other areas, as at Easington, Durham, and Shirebrook, Derbyshire.

The impact of the police

By the end of the strike there had been 11,312 arrests and 5,653 miners had been tried for alleged offences.[115] Most of those arrested were photographed and fingerprinted,[116] a process which made even the miners themselves feel like criminals. According to Home Office figures, forty per cent of those arrested were charged under s.5. of the Public Order Act 1936 and a further sixteen per cent with obstructing a police officer. From their statistical analysis of the charges, Percy-Smith and Hillyard concluded:

> Throughout the strike, the police, media and the Government colluded to suggest that violence on picket lines was widespread and occurring daily. Yet only 878, or 8·4% of all charges, were for offences of assaulting a police officer, actual bodily harm, murder, wounding, or possession of an offensive weapon. The proportion of persons charged with these offences was in fact less than the proportion of miners who were injured or roughly handled during arrest.[117]

Moreover, one third of those arrested were either not charged or were acquitted.[118]

Once arrested, miners were either cautioned and released or detained to appear before the local magistrates. In Nottinghamshire, pickets were invaribly brought before the magistrates because the police could not impose conditions on bail. 'To prevent further offences', magistrates usually imposed the following bail conditions: 'Not to visit any premises or place for any purpose in connection with the current trade dispute between the NUM and the NCB other than peacefully to picket at his own place of employment'. In the first six months of the strike ninety-five per cent of those charged before Mansfield Magistrates Courts had such conditions attached to their bail.[119] Benn told the Commons that magistrates in his constituency had expressed 'anxiety because the magistrates' clerk has the bail conditions stapled to a duplicated form before the case even comes up'.[120] The significance of this strategy should not be underestimated. As Christian has argued: 'Criminalisation is at its most effective . . . when a final determination of guilt or innocence can be avoided. Because of this the criminal courts' importance to the police has been less as judges of individuals' guilt or innocence, than as agents for outlawing strikes and legitimizing police control of their movements and blanket prohibitions on picketing.'[121] The courts' strategy depleted the numbers who could picket and removed those with most picketing experience. *Breach* of bail conditions brought the risk of imprisonment, even in the case of relatively minor offences which in themselves would not carry such a sentence. Like any deterrent, the effectiveness of the threat depended upon its being carried out: when activists like Malcolm Pitt, Kent Area President and Keith Hammond from Prince of Wales colliery were jailed for alleged breaches, they served as examples to other miners on bail.

In total, between 150 and 200 people were imprisoned during the strike; of those remaining in jail six months after the strike ended, forty-two were serving sentences of eighteen months or more.[122] The success of the criminalisation strategy was demonstrated by the failure of the labour movement to mount an effective campaign for the release of these prisoners (see Chapter six).

The style of policing and the whole process of criminalisation which went with it provoked violence by pickets. Picket violence was used to legitimise the policing that had brought it about, helped stigmatise the miners and undermined their case: like the ballot, violence provided an excuse for those who were reluctant to support the NUM. In reality there were a few serious convictions for violent offences, but the acquittal rate among those charged with the worst common-law offences was spectacular. On 30 November 1984 David Wilkie was killed when pickets dropped a concrete block from a motorway bridge onto the taxi in which he was driving a scab to Merthyr Vale. Merthyr miners Dean Hancock and Russel Shankland received life sentences for his murder.[123] Similarly, two Wakefield miners were jailed for causing grievous bodily harm to Fryston strike-breaker Michael Fletcher whom they attacked in his home, although four others charged with riot and assault in the same case were acquitted.[124] Three miners were also cleared of threatening to kill back-to-work leader Robert Taylor in Manton. On 15 July 1985, thirteen miners were cleared of riot and unlawful assembly at Doncaster Coal House by Sheffield Crown Court. Two days later another fourteen pickets were also acquitted of charges of riot and unlawful assembly in connection with the Orgreave mass-pickets.[125] The seven miners who were charged with riot and affray, and a further eighteen charged with unlawful assembly, were also all acquitted by Nottingham Crown Court.[126] The riot charges brought against thirty-eight miners after the Mansfield rally were also dismissed by the courts.[127]

Even NUM leaders were forced to criticise the killing of the taxi driver, yet Yorkshire activists had long feared a lorry driver or policeman would be killed because police operations prevented normal picketing. It must also be remembered that two NUM strikers were killed on the picket line, while others died through accidents related to the strike. Police methods did not go unchallenged. A report by the South Yorkshire Police Committee expressed concern about 'the volume and nature of the complaints about the police and criminal justice', the bulk of which it supported.[128] The Home Secretary was criticised and asked to establish a public inquiry into the policing issues raised by the dispute.[129] These criticisms prompted Chief Constable Wright to publish a major report on policing the strike,[130] and to submit two minor reports to the Police Committee.[131] The conflict between the Police Committee and the Chief Constable demonstrated how little control the police authorities had in practice.[132]

Public concern over the controversial police methods led to the formation of 'Operation Policewatch' following a public meeting at Sheffield Town Hall on

6 April 1984. Policewatch observers monitored police activities, submitted reports to the NCCL and published a report on their findings.[133] The NCCL report into the strike adopted the law's narrow interpretation of individual liberties by accepting 'that freedom not to take part in a strike is as much a fundamental right as the right to strike'.[134] Their call for 'all those involved in the strike and the police operations [to] . . . refrain from violence',[135] succeeded in alienating police and strikers alike. The debate over civil liberties obliterated the main issues of the strike to such an extent that policing was discussed in Parliament before the major issues of the strike (see Chapter four). The police operation did not *make* the strike important; rather it reflected the importance which the state attached to this conflict.

Police operations illustrate the contrast between the lessons which the NUM and the state drew from the 1972 miners' strike. The NUM leadership remembered the Saltley victory and developed a strike strategy which would replicate the mass-picketing tactics. What appears to have been forgotten are the events at Longannet power station, about which it was written shortly afterwards:

> If Saltley was the greatest victory for the pickets, Longannet power station held the greatest warning for the future . . . The Electricity Board concentrated on getting police reinforcements in order to smash the picket lines. The special 'hard' police squads used for dealing with demonstrations were brought in from Glasgow and Edinburgh . . . The police literally smashed into the pickets, arresting the picket leaders who had been trying to co-ordinate the activity through loudhailers. Mick McGahey, President of the Scottish Area of the NUM, was assaulted by police and when thirteen pickets were arrested they were handcuffed and kept in jail without bail. It was only the calling off of the strike that prevented further tough action by the authorities.[136]

The police had refined these tactics since 1972, as the more politicised Yorkshire activists knew from their experiences at Hadfields, Grunwicks and, in particular, the Stockport Messenger. The intercept policy, snatch squads and riot police had all appeared in strikes since 1972, so their use in the 1984 strike was predictable. Orgreave demonstrated the failure of mass-picketing itself in these circumstances,[137] rather than the failure of area leadership.[138] Irrespective of which were logistically the most important targets, by concentrating pickets at a few sites, the NUM made it easier for the police to exert control. On the occasions when more dispersed activities were undertaken, such as the motorway and bridge blockades (see Chapter three), police resources were stretched. These actions, like autonomous acts of sabotage, created publicity with more economic use of activists, although it is doubtful whether they advanced the strike at all. The mass-picketing played into the hands of the Government by distracting attention away from the strike issues which *might*, had they been sufficiently promoted, have generated sufficient support within the NUM and other trade-unions, to have rendered mass-picketing unnecessary. The adoption of mass-picketing as a substitute for campaigning could only result in arrests and injuries in the face of such state power. The strikers were never prepared to mount an armed insurrection, but the Government had no qualms about acting as though they were.

5.4 The mass media

The mass media should not be seen as autonomous from the activities of the Government, the NCB and the police, since so much of their strategy for undermining the strike entailed the use of the media. Media hostility to strikes and trade-unionism is usually explained by the pattern of ownership and control, which encourages the reproduction of the social values of the dominant class as 'fact'.[139] The media played a crucial role in breaking the strike, which was reported predominantly in relation to three themes: Scargill, violence and democracy.

Scargill

Journalists bring complex social issues down to the common denominator of 'human interest' because it is easier to write trivia about people and because this avoids having to challenge the dominant social values. Personalisation of the miners' strike was also inevitable because of a kind of symbiosis which had developed between Scargill and journalists: each loathed the methods and respected the power of the other. As Yorkshire Area President, Scargill used the media to promote left policies within the Union and journalists used him as a rich source of statements guaranteed to frighten Tory voters. Scargill represented the radical policies which rejected Union co-operation in the pit closures of the 1960s. The enormous support he received for the national presidency was support for those policies, and reflected the belief among ordinary miners that Scargill would not compromise their interests. Many union leaders clearly believed that the national presidency would cause a moderation of Scargill's behaviour, if not his beliefs, but Yorkshire activists viewed him as the acid test of the 'iron law of oligarchy'.[140] Like A.J. Cook, Scargill is one of the few trade-union leaders to have defied the iron law.

Scargill's commitment to his members made it more difficult and more imperative for the media to discredit him. The 'gunfighter' principle came into operation with television interviewers; as each failed to get the better of Scargill the prize increased and the next interviewer's questions were more antagonistic. The television news repeatedly referred to 'Scargill's strike', which suggested a division between the leadership and the membership. The NUM's federal government and the origins of the strike (see Chapter two) made nonsense of such a view but, as Douglas notes: 'The pickets, deployed by area strike committees . . . went to locations known only to the committees but were known throughout the media as "Scargill's pickets". Even though he didn't deploy them, didn't know where they were going, or indeed where they had been until after they arrived back and the story was in the news.'[141] Even journalists with a reputation for being more analytical succeeded in convincing themselves that the strike owed more to Scargill's personality than to the crisis of the coal industry.[142]

Personal attacks on Scargill were most effective when appearing to come from within the labour movement. In 1982 ex-Labour Minister Lord George Brown had claimed 'Scargill's increasingly wild activities . . . are mischievous, mad or deliberately aimed at both wrecking his Union and the coal industry and politically destabilising the nation.'[143] Ex-Communist Jimmy Reid followed Brown's themes in his character assassinations during the strike, claiming that Scargill was 'obsessed by getting the miners out on strike' and warning of the dangers of 'Scargillism', which he described as 'the politics of seizing power not winning it through the democratic process'.[144] Vic Allen pointed out that Reid was exaggerating Scargill's influence within the Executive and offered a more accurate portrait of the miners' leader, but the damage had been done and Reid's analysis was widely accepted as authoritative.[145]

Violence

In addition to personalising the conflict, the media focused upon the means of the strike and ignored its ends. Violence was portrayed as the primary mechanism for sustaining the strike, while acts of violence by those opposed to it were ignored. During the 1972 miners' strike, accusations of violence against white-collar workers crossing NUM picket lines at Yorkshire NCB headquarters outweighed reports of the death of picket Fred Matthews. In the building strike later that year the media reported violence as flying pickets visited sites, but not the threatening behaviour of contractors, and the only serious injury was that of a strike organiser assaulted in his own home.[146] Limited picketing in the 1974 miners' strike and the consensus period of the social contract gave the media few opportunities to portray violence in industrial disputes. Mass-picketing in the Grunwick strike provided more newsworthy events in 1977, including a policeman injured by a bottle allegedly thrown from among the pickets. The media resumed its campaign for legal restrictions upon mass-pickets and secondary action. In the steel-workers' strike in 1980 the media gave a high profile to confrontation between pickets and scabs outside Hadfields.[147] In the 1977 fire-fighters' strike,[148] and the 1982 health service dispute,[149] the media called for restrictions on strikes in essential services while inner-city disturbances in 1981 fuelled the 'law and order' debate.[150]

Against this backdrop, it was inevitable that any attempt to use mass-picketing in 1984, combined with the confrontational tactics of the police, would provide the media with opportunities to divert attention away from the coal crisis onto law and order issues. When violence arose, the tabloids and the television gave such graphic portrayals that these images dominated public perceptions of the strike. For the majority of people, television news bulletins were the main, often the only, source of information about the strike. The brevity of news topics and the definition of 'newsworthiness' precluded analysis and polemic. The mass audiences of the evening news programmes were presented with a brief statement of 'fact', accompanied by whatever film had the most dramatic images. Camera crews only filmed when there was violence, so the majority of peaceful picketing went unreported and violent confrontation was projected as the norm. News programmes failed to report that police prevented film crews from entering villages under siege, a marked contrast with the way in which the same programmes dealt with reports from South Africa compiled under the security regulations. Since the dominant value system denied the legitimacy of the strike itself, violence became a major issue and the police were presented as protectors of law and order, acting in the national interest.

Violence became the paramount news issue when the NUM attempted to prevent the movement of coking coal: 'the disturbances at Ravenscraig and Hunterston now amount to significant breakdown of public order' claimed *The Times*.[151] During the Orgreave campaign in June the media implied that pickets were insurrectionaries, thereby legitimising police action. In the BBC News report of Orgreave on 18 June 1984, pickets appeared to launch an unprovoked attack on the police, whereas unedited ITN footage included the police violence which had caused the pickets to retaliate. In the Commons, Benn described how:

> when BBC television covered the Orgreave picket, it showed stones being thrown, and then cavalry charges being made by the police. I know from BBC editors who took part in that bulletin that there were three cavalry charges by the police before a single stone was thrown. But the BBC, pretending to be impartial, put out a bulletin designed to mislead the British people about the sources of violence.[152]

The tabloids covered the 'riot' in detail but generally ignored the Orgreave trials a year later when miners were acquitted of this charge;[153] other newspapers concluded that the law was defective when the Orgreave defendants were not convicted.[154] The Home Office White Paper on public order, published two months after the strike ended, drew upon the experience of the miners' dispute to support its proposals.[155]

Violence perpetrated *against* strikers was seldom reported, although there were several incidents in Nottinghamshire.[156] The picket-line deaths of David Jones and Joe Green were reported dispassionately while the miners (and their childen) who died digging coal from spoil heaps were largely ignored. The media orchestrated outrage when David Wilkie was killed driving scabs to work.

Democracy

From the outset the media campaigned for a national ballot and declared the strike undemocratic. The Union's failure to ballot the membership helped the media to present an image of a militant minority, following Scargill's political lead, sustaining the strike through the intimidation of the moderate majority. The ballot issue damaged the moral legitimacy of the strike, undermining support and giving credibility to the working miners' case. The Nottinghamshire vote against the strike provided further support for the media theme that violent pickets were interfering with a legitimate right to work. The newspapers eulogised the working miners' leaders: Butcher was 'the bravest man in Britain',[157] and made the ridiculous claim that 'in Yorkshire 80% of striking miners wanted to return to work for a national ballot to be held'.[158] Threats of violence made against scabs were given maximum coverage: 'they want to kill me', claimed Robert Taylor,[159] and there were calls for the 'vicious criminals' who attacked Fletcher to be 'hunted down'.[160] Media coverage of strikebreakers emphasised divisions within the NUM and portrayed the Government and the NCB as neutral.

Coverage of the back-to-work movement was a logical extension of the democracy theme which Hart and associates built upon: 'the whole "back-to-work" movement was of course a child of the media. It was a classical *self-fulfilling* prophesy worked on night and day to wear down the miners with feelings of isolation and futility'.[161] Occasionally media exaggeration of the numbers returning to work was blatant, as when the *Sun* showed an additional 15,000 ancillary workers in the 'flood back to work'.[162] Television news employed a more subtle form of dramatisation, whereby an insignificant number returning to work nevertheless established a 'record'. As the back-to-work movement was sensationalised, television news bulletins took on the character of a sports commentary or a game show. Attention was focused upon the working areas and, towards the end of the strike, on collieries where substantial numbers had returned. The return-to-work *was* 'news', since it was a departure from the norm of the strike which had remained solid for so long. However, television news reports portrayed the catastrophic as the norm and concealed how solid the strike was in the heart of the Yorkshire coalfield. The media portrayed those returning as opponents of the strike, rather than victims of the process of attrition. Having demonstrated how each breakdown in negotiations caused more miners to give up, the *Daily Mail* still described the return as 'a record snub for Scargill'.[163] The same democracy themes were used after the strike to promote the emergent UDM (see Chapter seven).

Effect of the media

The television and newspapers were not uniformly hostile to the miners' case. As Scargill himself acknowledged, programmes like *Newsnight* and *Channel Four News* were less biased, but the combined audience of these programmes was only about one-fifth of the number viewing each major ITN or BBC News broadcast. The left press, which devoted much space to sectarian disputes, represented an even less significant proportion of total newspaper readership.

It is difficult to assess the extent of media bias against the miners, but even the debate over media treatment of the strike was subject to distortion. Gus Macdonald, of Channel Four's *Union World*, argued that the NUM was receiving fair media coverage because Scargill's total exposure time was greater than MacGregor's.[164] Professor Hetherington (Stirling), whose chair is partly funded by the BBC and Independent Television, similarly based his conclusion that the NUM case was fairly reported by the television on quantitative rather than qualitative assessment.[165] However, Michael Tracey, from the Broadcasting Research Unit of the British Film Institute, revealed how Scargill was more often treated as an adversary in interviews than MacGregor was, and other commentators have noted how the terminology of news reports was biased against the Union.[166]

Media treatment of the miners' strike, particularly television news reports of picket-line violence, weakened support with the TUC and PLP leadership, and, more importantly, with the rank-and-file of the labour movement. Miners invited to address labour movement gatherings were frequently astonished that activists accepted so much of the media's image. The support of *strikers* was similarly affected, particularly among the great majority who failed to get involved with picketing and therefore received most of their information from the media and the NCB. Activists were able to compare media presentations with their own experience, but still depended upon media reports of the strike in other areas or at targets they had not attended. From our survey of 1,830 strikers in Yorkshire, there was a marked difference in the sources of information available to activists and non-activists, irrespective of the communities in which they lived or other factors, such as occupation. Activists ranked as their most important information sources union meetings (forty-four per cent), the picket line (twenty-four per cent) and the strike centre (eighteen per cent), whereas among non-activists the major source was television (forty-two per cent), although union meetings (twenty-four per cent) and the picket line or friends who had been picketing (sixteen per cent) were still important. Overall, using the ratio of activists to non-activists established in Chapter three, we can estimate that television (forty-one per cent) was the major source of information to Yorkshire strikers, followed by Union meetings (twenty-five per cent) and the picket line (sixteen per cent). The strike centre was ranked most important by eight per cent of non-activists but eighteen per cent of activists, suggesting nine per cent overall. Inevitably, owing to their cost, newspapers were less important but six per cent of non-activists ranked these their major source compared with one per cent of activists, suggesting six per cent overall. More activists (eight per cent) than non-activists (four per cent) regarded the *Yorkshire Miner* as the most important source of information and three per cent of each group ranked local strike bulletins their major source. Non-activists showed a greater reliance upon non-Union sources of information but even activists were more dependent on the television than the *Yorkshire Miner* or strike bulletins. Quite apart from the greater power of its images, the television was a 'passive' source of information, automatically available to most striker households (but not all – see Chapter four).

Adeney and Lloyd are undoubtedly correct in their assertion that 'the failure of the NUM leaders to swing public opinion, including that of their own members, decisively to their side' was a key factor in the miners' defeat.[167] However, these authors fail to acknowledge their own roles as key reporters on the strike, and offer little guidance as to how the NUM could have been more effective in the propaganda war. Three means were available to the NUM to exert some influence over media coverage of the strike: fraternal union action, academic intervention and the Union's own public relations.

Fraternal action by print-workers was important in preventing the worst excesses of newspaper coverage of the miners' dispute. Before the strike the *Sun* had published a 'ballot form' for miners on the overtime ban.[168] The NGA *Sun* chapel obtained an equal space equal prominence right of reply for Scargill, which the paper published a few days later.[169] When the *Sunday Times* gave MacGregor one half-page to outline NCB plans for the coal industry,[170] print unions won equal space for a reply by Scargill.[171] The *Daily Express* conceded a centre-page spread to Scargill following SOGAT intervention over the paper's 'Scargill confession'.[172] On the same day, however, under the caption 'Mine Führer' the *Sun* front page pictured Scargill with arm raised, implying a Nazi salute.[173] The print-workers refused to handle the picture or headline and the paper was printed without either.[174] The major weakness of the fraternal action by media unions was the lack of support among journalists and producers, particularly those involved in the television news.

Academics attempted to shift the media focus from Scargill, violence and democracy to the major issues of the strike. Apart from passing references to an assumed logic of closing uneconomic capacity, however, the media avoided discussion of the arguments surrounding pit closures. Most academic contributions took the form of corrections to detail through the letters pages of the quality press, which had a limited circulation. Serious criticism was invariably marginalised through being detached from the offending report, and many academics were unhappy with their treatment by media personnel.[175]

The NUM's press and public relations was one of the major weaknesses of strike maintenance. Whereas the Department of Energy employed specialist consultants to monitor media broadcasts,[176] the NUM used its normal channels and provided far fewer press releases than the NCB. There *were* journalists and press officers prepared to work closely with the NUM during the strike but, like the academics, these were not approached by the Union. Moreover, Scargill's periodic verbal attacks on the media, however justifiable, were counter-productive in that they vindicated hostile journalists and alienated those who were sympathetic. As for communication within the NUM, the inadequacy of the Union's maintenance in the form of strike bulletins (see Chapter four) is immediately apparent from the analysis of sources of information available to strikers.

5.5 NCB union-busting tactics

Before the miners' strike there were few warnings that American union-busting tactics were being introduced to Britain.[177] MacGregor's involvement in neutralising union influence at BL[178] and BS [179] attracted little attention. The popular view of MacGregor was as a 'geriatric American butcher' rather than an accomplished union-buster. Gillman reported that Amax, the US corporation which MacGregor controlled from 1967 to 1977, made its fortune by cornering fifty per cent of world molybdenum supplies and only moved into coal in 1969.[180] *World in Action* later revealed how Amax investment in coal increased after MacGregor defeated the United Mineworkers of America (UMW) in the newly-opened Belle Ayr mine at Gillette, Wyoming.[181] In 1974 MacGregor broke with other coal operators in refusing to renew the UMW contract; he directed propaganda against the Union and made pension credits transferable only after ten years' service. Letters to strikers announced the corporation's intention to reopen with or without union labour. Scabs were brought to the mine under armed escorts provided by the Wakenhut Corporation's 'Emergency Services Unit'. MacGregor told a striker who interrupted a shareholder's meeting, 'you'll never get a union contract at Belle Ayr mine'; it remained non-union and fourteen new mines, also non-union, have since opened in the same valley.

MacGregor's description of the use of the local sheriff and judge[182] is reminiscent of the labour relations warfare organised by the Harlan County Coal Operators in the 1930s.[183] However, MacGregor's union-busting pedigree goes back further than Amax. In 1953 he defeated the Teamsters' Union in a strike at Manning, Maxwell and Moore, using 'a publicity and letter-writing campaign to encourage our people back to work and to discredit the union leadership'.[184] In 1949 at Campbell, Wyant and Canning he was involved in defeating a United Auto Workers' strike by using white-collar workers to run the foundry.[185]

The Mohawk Valley Formula

The strike-breaking tactics MacGregor employed were developed in the US steel industry, which MacGregor joined in 1940. Four years earlier, in the Remington Rand strike, the steel employers devised a 'scientific' approach to strike-breaking which became known as the Mohawk Valley Formula. NCB tactics during the 1984 miners' strike conformed more closely to this formula than has employers' action in any previous strike in Britain. Steuben provided a detailed analysis of the formula and identified six key elements:

1. When a strike is threatened, label the union leaders as 'agitators' to discredit them with the public and their own followers . . .
2. Disseminate propaganda, by means of press releases, advertisements and the activities of the missionaries; such propaganda must stress the arbitrary demands of the strikers . . .
3. Align the influential members of the community into a cohesive group opposed to the strike. Include in this group, usually designated a Citizens' Committee, representatives of the bankers, real estate owners and businessmen . . .

4. Bring about the formation of a large armed police force to intimidate the strikers and to exert a psychological effect upon the citizens . . .
5. Most important heighten the demoralizing effect of the above measures – all designed to convince the strikers that their cause is hopeless – by a 'back-to-work' movement, operated by a puppet association of so-called 'loyal employees' secretly organised by the employer. The association wages a publicity campaign in its own name and co-ordinates such campaign with the work of the missionaries circulated among the strikers and visiting their homes . . .
6. Close the publicity barrage, which day by day during the entire period has increased demoralization worked by all of these measures, on the theme that the plant is in full operation and that the strikers are merely a minority attempting to interfere with the right to work, thus inducing the public to place a moral stamp of approval upon the above measures. With this, the campaign is over – the employer has broken the strike.[186]

In the early part of the strike, attempts to discredit the NUM leaders and the Union's demands were consistently made both by the NCB and the Government, occasionally with the support of right-wing academics. The importance of 'law and order' and the 'right to work' were emphasised by Government ministers rather than citizens' committees,[187] but were no less effective for that. At first the NCB did not attempt to undermine the Union by appealing directly to members rather than dealing with national, area or branch officials. In Yorkshire most pit managers were in regular contact with branch committees and reports concerning the state of the colliery or requests for safety work to be undertaken all passed through the normal channels, and were policed by the SCC. In Scotland, however, Area Director, Albert Wheeler, used safety work as a weapon against the strike, threatening to turn off pumps unless manpower was provided and thereby having enginemen available to wind down scabs.[188] In all areas, theft of or damage to NCB property by NCB employees was dealt with summarily by the colliery manager without the opportunity of raising the issue through the conciliation machinery.

The leaders of the back-to-work movement, who were also involved in legal action against the NUM, were probably recruited before the strike. Robert McGibbon and his wife Irene, who attempted to undermine the strike at Betteshanger colliery, Kent, in April, were viewed as under-cover management agents when it was revealed that they had organised an anti-union campaign in the BL Cowley plant in 1974 with the right-wing Freedom Association. Richardson has claimed that the Nottinghamshire Working Miners' Committee (Notts WMC) was established in October 1983 and briefed to ensure that Nottinghamshire would remain in production in the event of a national strike.[189] Certainly Chris Butcher was preparing a back-to-work campaign from the end of March, although the Notts WMC was only publicised during May. Butcher was financed and promoted as the 'Silver Birch' by the *Mail on Sunday* and toured the country with its Industrial Editor establishing contacts with other scab organisers like Tony Morris at Lea Hall and Jim Lord at Agecroft.

Since organised strike-breaking was not anticipated in Yorkshire, few theories were offered by miners before the strike about who would behave in this way. Even once strike-breaking had begun, there was no consistent view of who was involved, other than the NCB. In Yorkshire the back-to-work organisers were middle-aged at thirty-four per cent of collieries and old-timers at twenty-one per cent, while at thirty-six per cent they were thought to come form a wide range of age groups. They were newcomers to the industry at thirty-six per cent of pits and old hands at twenty-three per cent but both were involved at forty-one per cent of collieries. The organisers were drawn from all occupational groups at forty-three per cent of collieries, but craft-workers predominated at twenty-one per cent and surface-workers at fourteen per cent. Face-workers and other underground-workers were each cited as being the mainstay of the back-to-work movement at ten per cent of pits. Kinship was thought to be irrelevant at forty-eight per cent of collieries, while at twenty-nine per cent scab organisers were from alien families and at twenty-three per cent they were from mining families. At forty-eight per cent of pits they lived outside the mining community, and although at only fourteen per cent of pits they came predominantly from within the community, at thirty-eight per cent of pits they included both insiders and outsiders. The scab leaders were characterised as being overtly conservative at fifty-two per cent of collieries and as non-political at twenty-three per cent, while at twenty-five per cent of pits they included a diverse range of political opinion. It would have been difficult, therefore, for the Union to have anticipated who was most likely to organise dissent. Moreover, it was irrelevant to the NCB who was involved, so long as a few individual miners could front the movement for management.

Working miners and their supporters

Although the Notts WMC was supposedly independent of the NCB, Butcher admitted in July that a substantial amount of money came from 'wealthy business people who want the strike to finish'.[190] Rank-and-file miners were unlikely to have had the business contacts or have been in a position to reassure sponsors that their donations would be used appropriately. David Hart, an adviser to Thatcher and Brook Street property developer of extreme right-wing views,[191] was involved with the working miners at least from July and probably earlier. While the story of Hart's chance involvement after a visit to Shirebrook[192] is not entirely plausible, it is impossible to be certain when he was first involved and how he was brought in. Thatcher may have put MacGregor on to Hart (although Hart's brother had worked with MacGregor in America) as a person who could implement the strike-breaking formula MacGregor had outlined. One of the Union's national office staff believes that 'Hart came in MacGregor's kit-bag'. An effective back-to-work movement necessarily had to operate beyond Nottinghamshire. Under Hart's guidance, therefore, Clarke and Liptrott established the National Working Miners' Commmittee (Nat WMC). Extending the back-to-work movement from the major working area,

Nottinghamshire, into areas where the strike was solid conferred two tactical advantages: firstly, the early work in Nottinghamshire was unlikely to meet serious challenge and publicity could be built around the right-to-work (or to a ballot) theme rather than around strike-breaking and, secondly, the back-to-work movement would appear to be *gathering momentum*, a necessary condition for encouraging others to break the strike.

Hart operated from Hobart House, co-ordinating the legal action, finance and publicity of the back-to-work movement. Hart's legal strategy was designed both to prevent the Union from using its resources on the strike and to sever the connections between the officials and the membership.[193] The notion of independent action taken by rank-and-file miners is contradicted by MacGregor's admission that he sought Lawson's assistance in May to identify 'bright young lawyers' to take cases against the NUM.[194] Even the earliest actions of working miners used solicitors prominent in Conservative politics. Butcher put Taylor and Foulstone in touch with David Payne of Hodgkinson and Tallents, who were connected with Sarah Hogg, Economic Editor of *The Times*, for whom Hart was occasionally writing. Clarke and Liptrott, key members of Hart's Nat WMC, used David Negus of Ellis Fermor, who put their case in the hands of Michael Burton QC, barrister in the Taylor and Foulstone case. When Taylor and Roberts approached the NCB for advice on taking action against the NUM, they were sent to John Lewis of Shacklocks (Mansfield), who knew Negus. These legal actions were an integral part of the back-to-work movement, as each judgement reduced the credibility of the strike and made its organisation more difficult.

Hart raised finance for the legal action, estimated at £500,000, through business contacts. Sir Hector Laing, Chairman of United Biscuits, who appears to have recommended MacGregor to Thatcher for BS and whose home was used for secret meetings of that pair, was particularly enthusiastic.[195] Laing rallied capital against the NUM in a letter to over one hundred companies calling for donations to the working miners' legal costs: 'All that is needed for evil to triumph is that good men do nothing . . . They are in the front line fighting for what they believe in and they need to know they have friends'.[196] Lord Taylor, whose Taylor Woodrow opencast operations were in competition with the nationalised coal industry, was among those who donated to the fund. Bob Copping, the Barnsley winder who first chaired the Nat WMC, resigned on 5 September over Hart's involvement and leaked the story to the *Daily Mirror*.[197] According to Copping, Hart was 'the money man. He was handing over sums of up to £300 in cash as floats . . . Hart was not acting as an adviser, he was running the show'.[198]

Publicity was the key both to fund-raising and the back-to-work campaign. Hart brought in Tim Bell, from the Tory Party campaign agents Saatchi and Saatchi, and Gordon Reece, Thatcher's 'image moulder'. Together the three ran a publicity campaign involving media briefings, newspaper advertisements and letters to individual strikers. Press releases built upon the picture the media had already created of the strike (see above) and presented the Nat WMC as a membership reaction against a strike of this sort, rather than against strikes (or unions) in general. To be effective, the campaign had to appear independent of the NCB and not give the impression of being anti-union. Thus in an article in *The Times*, which Hart wrote to give credibility to the Nat WMC, he failed to mention his own involvement and claimed: 'It is a genuinely national committee with men from Wales, Derbyshire, Lancashire, Staffordshire, Warwickshire, Yorkshire and Nottingham. They are financed by collections at working pits and by contributions from ordinary members of the public sent in response to advertisements placed in the national press . . . They have shunned offers of help from 'big business' and even from Conservative miners.'[199]

The first advertisement Hart placed on behalf of the Nat WMC in the *Daily Mail*, *Daily Express* and *The Times* had the dual objectives of discrediting Scargill and generating finance. Under the banner 'Come off it Arthur!', the advertisement claimed 'children have been terrorised; property has been attacked' and called for donations to 'help us fight for freedom and democracy'. Once the back-to-work movement was underway, advertisements were placed directly by the NCB. The premise of the campaign was that 'the majority . . . were on strike out of loyalty to the Union or out of fear of it' and aimed to win 'the hearts, minds and pockets of those who neither supported the strike all out, nor had settled to return to work early on'.[200] The advertising strategy was one of 'saturation media coverage', establishing the broad principle that negotiations had ended so people might as well go back to work: 'the real selling of the idea was done at local level'.[201] To undertake this campaign, the NCB switched in October from its regular publicity agents to Lowe Howard-Smith, chaired by Hart's brother Tim, prompting strong objections from Geoffrey Kirk, Director of Public Relations.[202]

NCB propaganda

In June, following the breakdown in negotiations, MacGregor sent a letter to all miners in the industry, which was designed to generate rank-and-file pressure on the union leadership to agree to NCB proposals and end the strike: 'The NUM which called the strike, will end it only when you decide it should be ended . . . At the present time there are between 20 and 30 pits which are viable which will be in danger of never reopening if we have a lengthy strike'.[203] Uncertainty was created about the future of collieries which were in the short-term at least, safe, and the letter concluded with an appeal to strikers to return to work. Adeney and Lloyd attribute the June letter to Hart,[204] although MacGregor claims not to have met him until 2 July.[205] Some branches responded to MacGregor's letter. Brodsworth branch, for example, wrote to members warning that 'the Board are now trying to convey to you that our pit is in trouble and that we as your representatives don't care'. Ned Smith, head of industrial relations, was unenthusiastic about MacGregor's letter, fearing it would alienate moderates working within the NUM for a collective return-to-work and jeopardise future negotiations.

Responsibility for nurturing support among miners for an end to the strike was placed in the hands of NCB area and colliery management. Four Area Directors showed particular enthusiasm for the back-to-work campaign: Ken Moses (North Derbyshire), Albert Wheeler (Scotland), David Archibald (Northumberland) and Albert Tuke (Doncaster). The first priority had been to keep Nottinghamshire working; with this secured, the strategy entailed building upon Nottinghamshire. North Derbyshire was seen as the fulcrum for the Notts lever on Yorkshire. Shirebrook colliery, part of a major complex involving Warsop Main and Pleasley, was chosen to launch the back-to-work movement. Shirebrook is adjacent to the Notts Area and furthest from Yorkshire (and the M1 which the majority of pickets still used); a large proportion of its workforce lies in Mansfield where the majority of miners were working at Notts pits. There was already a nucleus of seven miners scabbing at Shirebrook, including Roland Taylor of the Nat WMC, whom Hart had visited when establishing 'cells of men who would act as the basis and the rallying point for local back-to-work movements'.[206] Moses had written to miners on 18 May and by the end of the month had persuaded fifty-two men back to work at Shirebrook. In June, while Hart was operating in Shirebrook, Moses gave a seminar for other Area Directors on the methods they were employing. The tactic involved building personal contact to break the strike by 'rolling it back from the edges'. A wall map showed the location of every miner's house, their attitudes towards returning to work indicated by coloured pins. When a small group of miners in one locality appeared ready to break the strike, arrangements would be made for their collection by armoured bus. Information was gathered by management teams to build up a profile of each miner, identifying those who had regularly worked overtime before the ban, owned their own houses, had particular financial difficulties or were opposed to the strike. Those living in outlying areas or next to scabs were targeted and a friend who had already broken the strike would be despatched to discuss their problems. By the end of October there were almost 200 back at work at Shirebrook.

Before talks were resumed in July, a number of letters were sent to individual strikers in the South Yorkshire Area. In July every Kiveton Park miner received from the manager a hand-written letter which was positive about the future of the pit and critical of the democratic processes of the NUM:

> The overtime ban and then the Yorkshire strike were imposed upon you without a ballot . . . unless union leaders have a change of heart the strike could go on for many months longer . . . make your concern known to your union officials . . . telephone the colliery and ask to speak to me or the senior BACM man on duty and let us know if you are interested in returning to work.[207]

At about the same time, all Treeton miners also received a hand-written note from their Manager, which expressed concern about the consequences for 'the pit and the community if we don't get back to work soon', and encouraged miners who might consider returning to work to telephone him regarding arrangements. Similar letters were sent by the Managers of Brookhouse and Manton. In July, Albert Tuke, Doncaster Area Director, wrote to all NUM members in the area informing them that they would all be receiving 'a letter from your own colliery manager setting out the problems at each of the collieries in some detail' and urging miners to make their concern over the strike 'known to colleagues and union representatives'.[208] Tuke also mentioned that a special office had been established to provide information for those wishing to return to work.

The SCC was so incensed at the new management tactics that they resolved to withdraw safety cover from some collieries.[209] The NCB responded by assuring Yorkshire Area NUM that, *while the present negotiations between the NUM and NCB were taking place*, no more letters would be sent to individual miners, and that the area officials would be informed of any letters to be sent in the future.[210] To combat the risk of members abandoning the strike, the Yorkshire Area officials sent a letter encouraging all members: 'victory is closer every day; we urge you to do all in your power to make it sooner rather than later'.

Before the national talks resumed on 18 July, MacGregor sent a letter to all miners outlining the points of agreement between the NCB and NUM, and the one sticking point: 'other reasons for pit closures . . . in view of the large amount of agreement we have achieved, it seems to me, it would now be in the best interests of all of us to go back to work and resolve this dispute. That is what the Board will be trying to achieve when the negotiations resume on Wednesday'.[211]

Patterns of strike-breaking in Yorkshire

The back-to-work offensive which the NCB had prepared in anticipation of the collapse of negotiations was launched in Yorkshire with meetings of 'moderate' miners to discuss ways of engineering a return-to-work. At Kellingley colliery all NUM members were contacted and urged to end this 'political strike' by a mass return-to-work. Members were encouraged to contact the *Manager*, not the 'moderates committee', if they were considering returning to work. Phil Bassett reported in the *Financial Times* that rebel miners in Yorkshire intended to try to secure 'a large scale return to work in that area' by concentrating their efforts on three pits where their support was strongest.[212] Their objective was to get a dozen men back at each of the three pits to trigger off a wave of similar action throughout the Yorkshire coalfield and nationally. The back-to-work organisers were prominent at Kellingley, Denby Grange, EmleyMoor, Kiveton Park and Manton. In August MacGregor again appealed to individual miners to pressure the Union to negotiate an end to the dispute, as if the NCB's proposals were rejected then it would be a 'considerable time' before an 'alternative basis' for negotiation could be found.

The first Yorkshire scabs went in during August and September to sixteen collieries evenly spread between the four areas. The pattern of *individuals* returning to work revealed little of the NCB's strategy, other than that the first scabs were generally at pits on the periphery of the coalfield like Denby Grange, Markham Main, Allerton Bywater and Kiveton Park. As each new colliery was breached, its whole commuity was often mobilised in an effort to deter a few scabs. In turn the police laid siege to pit villages, particularly in Doncaster and South Yorkshire.

By November new national talks and attempts at conciliation by ACAS had failed and the NCB launched a campaign with a leaflet to every miner's home which was aimed at producing an *organised* return-to-work.[213] The NCB expected that, by offering a large financial incentive ('£1,000 for facemen and about £800 for surface workers') just before Christmas when financial pressure was greatest, groups of miners might return together. In addition to letters and advertisements in newspapers, the back-to-work message was carried by *Coal News* which was posted to every miner's home each month. In October *Coal News* headlined 'miners now at 93 pits', while inside the paper described deteriorating conditions at collieries. As part of the 'November push', that month's *Coal News* proclaimed 'end of the talks' and gave further details of twenty faces lost because of the strike. November saw greater numbers returning to work in Yorkshire than in the previous eight months and by the end of the month fifty collieries were affected. Only three pits (two in Barnsley, one in Doncaster) remained scab-free into 1985.

The NCB and the working miners continued their efforts to break the Union and get men back to work in December. In Doncaster Area all miners received a message from the working miners exhorting them to return to work and attempting to divide the men from their Union. The letter, mailed by the NCB, implied that Union officials were not suffering the same hardships as the members by asking how many officials 'have had to rely on food parcels . . . have outstanding debts [and] . . . have sold their possessions to survive?'.[214] A similar attack on the Union was launched in South Yorkshire, where Treeton miners received a letter from the working miners implying that the Union was undemocratic, as there had been no ballot over a return-to-work, and alleging that the gap between officials and members was demonstrated by the Tenerife holiday planned for one Union leader at Christmas. In some collieries the anti-Union propaganda was accompanied by a personal letter from the Colliery Manager. Treeton's Manager wrote that 'a negotiated settlement to the dispute is as far away as ever and the only way back is for the men to decide for themselves to return to work'.[215] At Kiveton Park the Manager reminded miners of 'unpaid bills, mounting debts, no fuel and Christmas fast approaching'.[216] The managers' letters were accompanied by a leaflet ('Earn it or lose it') proclaiming how much miners returning by 10 December could expect to earn before Christmas; those who failed to return 'would lose this money forever'.[217] Again *Coal News* reinforced the message 'miners at most pits' and 'miners bring pits back to life;', while advertising 'back-to-work pay and extra cash'.

In December the preparations for Christmas acted as a brake on the return because people were preoccupied and had a milestone to reach, so fewer people gave up the strike. The psychological impact of the number of collieries affected was out of all proportion to the numbers who scabbed, since up to Christmas under five per cent of the Yorkshire miners had returned to work. South Yorkshire was worst affected with eleven per cent scabbing by the end of the year, compared with only one per cent in Doncaster. In Barnsley three per cent had returned to work and in North Yorkshire five per cent had gone back. The patterns of scabbing up to Christmas showed that the NCB had applied the strategy used in Derbyshire. In South Yorkshire the back-to-work movement was extended from neighbouring Nottinghamshire and Derbyshire via Manton and Shireoaks. In Barnsley the campaign centred on the collieries around Calderdale in West Yorkshire which are separated from the rest of the Area by the M1. In North Yorkshire the strike was undermined from the north-eastern edge through the Selby complex and Kellingley. The position by Christmas 1984 is illustrated in Fig.6. The strike was still solid in the heart of the coalfield, with fewer than one per cent at work in sixteen collieries and under five per cent at a further twenty-four. However, over twenty per cent had given up the strike at five collieries: Manton, Shireoaks, Denby Grange, Emley Moor and Ricall.

All the theories seeking to explain patterns of activism (see Chapter three) were repeated in reverse to explain patterns of strike-breaking. In Yorkshire, those who returned before Christmas were predominantly middle-aged at thirty-four per cent of collieries, family men at twenty-three per cent and old-timers at twenty-one per cent. Youngsters were prominent at only two collieries (three per cent) although there was a more even age distribution at eighteen per cent of pits. In terms of length of service, they were mostly newcomers at forty-six per cent of pits, old hands at twenty-three per cent and both at thirty per cent. They were drawn from all occupations at forty per cent of collieries, while surface-workers were over-represented at twenty-one per cent and elsewhere-below-ground workers at twenty per cent of pits. Face-workers predominated among the early starters at only nine per cent of pits and craftsmen at eleven per cent. At thirty-eight per cent of collieries they came from alien families, but at thirty per cent they were from traditional mining families and thirty-two per cent from both. They came primarily from outside a mining community at seventy-one per cent of Yorkshire pits, from within the community at eleven per cent and from both types of area at eighteen per cent of collieries. The scabs were characterised as non-political at seventy per cent of collieries and conservative at twenty-three per cent, although at eight per cent there was a mixture of political persuasions involved. In so far as it is possible to characterise the 'typical' scab at all, he was therefore unlikely to be a youngster or a face-worker, more likely than not to have joined the industry in the previous ten years and come from an alien family, and almost certain to live outside a mining community and be non-political or Conservative.

The pattern of strike-breaking and the background of individuals who gave up the strike before Christmas 1984 demonstrate the weakness of NUM strike

Fig 6 Yorkshire strike-breakers, Christmas 1984

organisation and maintenance. The early strike-breaking mostly involved the collieries on the periphery of the coalfield and individuals living outside mining communities. Had the NUM established community-based strike committees (see Chapter three), local information about likely weak spots would have been available to the Union and it might have been possible to take action to prevent people from drifting back to work. Certainly local strike committees would have made it more feasible to prevent the back-to-work movement penetrating the coalfield communities. The propaganda battle was conceded before the strike began because the NUM had made no provision for independent distribution of strike literature (see Chapter four). The NUM ought to have been in a position to circulate the *Yorkshire Miner* and strike bulletins to all members, to counteract NCB propaganda and media reports.

Just as the NCB recruited its agents to erode support for the strike within the coalfields, so the Union should have established a campaigning network of activists to monitor striker opinion and pinpoint areas where maintenance needs, both material and psychological, were most acute. The coalfield communities were ideally suited to the development of new methods to disseminate Union information and to gather intelligence. The political funds ballots held by several trade-unions, including, after the strike, the NUM, demonstrated the potential of new methods of campaigning. These campaigns, designed to encourage debate and discussion at workplace level, involved training activists as 'campaign contacts' and supplying them with speaking notes, leaflets and posters.[218] The dependence upon lay members to run the campaigns was without precedent and produced overwhelming support in membership ballots for maintaining the political funds, despite the decreasing proportion of trade-union members who had voted Labour in the previous four general elections. These methods would have vastly improved the *Campaign for Coal*, and campaign contacts could have formed the basis of a community network during the strike. Each activist could have been responsible for a part of a community, within which he could have briefed others to put across key arguments and monitor striker opinion. Such pyramid tactics could have reached every striker and would also have helped resolve the conflict between the different levels of strike organisation by integrating local initiatives with central strategy.

Chapter six

Strike settlement

This chapter examines how the strike came to an end. Most stoppages end either when one party capitulates rather than face the costs of continued action, or when a compromise is reached through negotiation. In the 1984 coal strike the Government was prepared to withstand enormous losses in order to defeat the miners, the parties failed to reach a negotiated compromise, and ultimately it was the costs to individual strikers which forced the NUM to end the action. The chapter is organised into four sections: in the first, factors influencing settlement by the Government, the costs of the strike and the level of coal stocks, are analysed; the second section examines the progress and failure of negotiations, and the involvement of third parties; thirdly, the erosion of the strike by the back-to-work movement is shown to have forced the Union's decision to return without an agreement and, finally, the attempts to obtain an amnesty for miners sacked during the strike is analysed.

6.1 Factors influencing settlement

In conventional economic models of strike activity, the duration of a strike is generally governed by the costs imposed on the parties. Although the inequality of the employment relationship ensures that a strike imposes greater personal hardship upon the workers involved (see Chapter four), employers are assumed to weigh the costs of strike action against the costs of concession to union demands with the object of maximising the present value of future profit.[1] Such models are formulated in terms of wage disputes where the parties may compromise as a strike imposes costs upon them. The conflict over pit closures, however, defied compromise: either surplus capacity was to be closed or a coal-based energy policy developed to maximise long-term coal production. This fundamentally irreconcilable *issue* determined that the conflict would be waged almost regardless of the costs incurred. While the strike was in progress, however, two factors were thought to influence the Government's willingness to promote a negotiated settlement: the unprecedented costs of defeating the miners, and the risk of interruption of electricity supply if coal stocks fell below the usual level of safety.

Cost of the strike

Adeney and Lloyd associate 'loss without limit' with Scargill's leadership, but it was Chancellor Lawson who told the Commons on 31 July 1984 that the £400m so far expended on the miners' strike was a 'worthwhile investment'.[2] Just how extensive an 'investment' was made to defeat the miners will probably never be public knowledge, but Lawson's £400m was certainly an underestimate of the costs even up to the end of July. Blazyca *et al.* put the total cost of the first twenty weeks at £1,400m, or £70m per week.[3] The figure comprised NCB net losses of £27–34m per week, additional CEGB costs of £20m per week from the increased oil burn, losses to BR of £4–5m per week and to BS of £5m per week, lost tax revenue from miners of £5–7m per week, social security payments of 1m per week, and policing costs of £3–5m per week. Later that month ITN *News* estimated total costs of £1,340m after twenty-four weeks,[4] and *Newsnight* reckoned on £1,500m after twenty-seven weeks,[5] each representing an average of about £56m per week. In the depth of the winter, the weekly cost of the strike was estimated to have risen to £85m.[6] As the strike was ending, *Panorama* anticipated that the NCB would incur an operating loss of £875m, of which £467m was interest charges.[7]

Further details of the major costs were only revealed when the nationalised industries presented their annual reports for the 1984–85 financial year. BS attributed £180m of its losses to the miners' strike, which included 700,000 tonnes of lost production.[8] For BR, the loss of revenue from coal traffic, some of which proved permanent, amounted to £400m.[9] At the end of July the NCB reported losses of £2,225m, compared with £875m in the previous year, of which £1,725m was the net loss due to the strike.[10] A few days later the CEGB reported that increasing the oil burn and other measures necessitated by the coal strike had a net cost of £2,020m.[11] To the combined costs to the nationalised industries of £4,330m must be added lost tax revenue of up to £400m, the costs of policing of about £200m and social security payments of over £50m. In total the strike cannot have cost central Government less than £3,500m; *The Times* estimated £6,000m,[12] and Scargill cited £7,000m at the conference which called off the strike.[13] Local government also bore costs to an extent which varied in proportion to sympathy for the strike. In areas like South Yorkshire, in addition to loss of rent and rate revenue there was the further cost of free school meals and public transport (see Chapter four).

The economic effects of the strike outweighed its immediate costs. The Government was expected to face costs of £600m in the financial year after the strike, as a result of the deterioration of collieries.[14] The oil balance of payments was adversely affected by about £2,800m,[15] interest rates were thought to be 0·75% higher because of the strike,[16] and GDP growth was retarded.[17] The Chancellor claimed that the strike had reduced the scope for tax cuts,[18] a view challenged by Tory 'wets'.[19]

Although it is difficult to assess the knock-on effects of the strike, these were enormous in areas where coal-mining employment was highly concentrated. Two different multiplier effects may be distinguished. Employment multiplier effects concern jobs created outside the colliery as a result of its activities, while expenditure multiplier effects refer to the impact on demand for goods and services through the purchasing power of miners (see Chapter seven). Employment multiplier effects will have been less severe during the strike because a large proportion of mining machinery is manufactured for export, where demand was unaffected by the dispute. Expenditure multiplier effects, however, could have cost the equivalent of 50,000 jobs during the year of the strike from the reduced spending power of strikers. Associated with this loss of employment was reduced tax revenue, and increased welfare expenditure from rent and rate rebates to unemployment benefit and social security. In September the Chairman of Wakefield Chamber of Commerce said twenty per cent of local businesses had gone bankrupt and others had been forced to diversify because of the strike.[20] The total costs of multiplier effects to central and local government could have been of the order of £140m.

The enormous costs of the strike led to the charge that the Government was behaving irrationally in economic terms by spending more on the strike than any savings which would accrue from closing high-cost capacity. Orme attacked Walker for his failure to meet the NUM: 'he prefers to spend £4 billion trying to defeat the strike'.[21] The Henley Centre for Forecasting, in its anlaysis for *Newsnight*, claimed that the longer-term economic gains would outweigh the immediate strike costs; break-even would occur in 1989.[22] The Henley calculation ignored social costs, costs to the NHS, and effects on housing and the local economy of pit closures. It assumed an average annual increase in miners' earnings of eight per cent, but ignored projected productivity improvements. It also assumed that the experience of steel-workers made redundant at Corby would be repeated in the case of miners and that the costs of unemployment would cease when miners became re-employed, a fundamental misconception exposed earlier (see Chapter one). Furthermore, the Henley Centre underestimated the cost of the strike in September at £920m, whereas previous estimates would have suggested twice that figure and after the strike a September total cost of £3,000m might have been considered more realistic. Taking the total strike costs to be £6,000m, even without challenging any of their other assumptions, on the basis of the Henley analysis the costs of the strike will not be recouped until 2005.

An investment which is only recovered after twenty years under the most favourable assumptions can hardly be considered attractive in economic terms. One must, therefore, assume that the 'worthwhile investment' related to the Government's wider political purpose. Ross argued that 'a sixty-day strike over 2 cents an hour may be irrational in the economic lexicon, but viewed as political behaviour it may have all the logic of survival'.[23] Equally it would be wrong to expect simple economic rationality of employer behaviour, particularly where a state-owned industry is concerned. From a political perspective, the Government would have paid any price to defeat the miners, not as revenge for 1974, but in order to further the interests of capital by inflicting damage upon the labour movement.

The economic cost of the strike would have been £1,000m higher but for the CEGB's policy of building coal stocks.[24] However, the strategic importance of the coal stocks was far greater than their contribution to containing strike costs, for they were the key to defeating the strike (see Chapter five). The Government had to ensure the continuity of energy supply, both to demoralise the strikers and to maintain public support for their stand against the miners.

Coal stocks

Before the strike began, the *Miner* reported that stocks of coal at power stations amounted to 22·5m tonnes and usable pit-head stocks represented a further 17·5m tonnes.[25] Power-station stocks would be exhausted after ten weeks and even if colliery stocks were moved these would only last a further eight weeks: power cuts were to be expected, therefore, by the first week of July 1984. When the strike began, miners had very little idea of how long it was likely to last. Our survey of 1,830 strikers in Yorkshire showed that fifty-five per cent thought the strike would be over within three months, and another thirty-four per cent thought it would last between four and six months; almost ninety per cent of strikers therefore expected to be on strike for six months or less. Only 1·5% thought the strike would last over a year. Strikers' estimates did not vary with their degree of activism, community or occupation. The continued supply of coal from Midlands collieries was expected to extend the duration of the strike but it could be won even without the support of Nottinghamshire, according to NUM leaders (see Chapter two).

There was little serious debate of the coal stocks question for several months, apart from the occasional observation that the end of winter was not a sensible time to begin a coal-mining strike. On May Day Thatcher assured the Commons that coal stocks would last 'for many, many months' and *The Sunday Times* accurately estimated power-station stocks to be 20m tonnes.[26] In June, *Labour Research* quoted Scargill's estimate of 16m tonnes held by the CEGB and calculated that power-station coal consumption was running at a million tonnes per week even in May.[27] To its rhetorical question 'when will the lights go out?' *Labour Research* gave the non-committal reply 'much sooner than the Government's attitudes suggest'. Imminent power cuts became an article of faith among the strike's supporters, but such belief was not founded upon serious analysis of the situation.

Weekend World estimated that from a July figure of 17m tonnes, power-station stocks would be down to 11m tonnes by November, necessitating power cuts, and would reach 5m tonnes by January 1985, bringing domestic cuts and short-time working in industry.[28] In the same programme, Geoffrey Goodman speculated that by February, with stocks at 3m tonnes, troops would have to be used to move colliery stocks and predicted widespread civil unrest. Walker denied that the Government was considering establishing a body like the 1926 Organisation for the Maintenance of Supplies and gave an assurance that coal stocks would last into 1985. The following day there was further media speculation over the use of civilian volunteers and reports that Philips and Drew, city analysts, had estimated coal stocks would fall to 4m tonnes by December.[29] Phil Wright estimated that, even with 300,000 tonnes per week from working collieries, coal stocks would be running down by 1·2m tonnes per week in the winter and predicted power cuts before Christmas.[30]

In September it emerged that deep-mined output had averaged 400,000 tonnes per week, reaching as much as 487,000 tonnes, while the 4m tonnes stocked at working pits was being moved at a rate of about 130,000 tonnes per week.[31] Power stations were therefore receiving 530,000 tonnes of coal per week. Nevertheless, coal stocks were thought to have fallen to 14m tonnes at power stations and 20m tonnes at collieries.[32]

Wright presented a detailed analysis of coal stocks at the Durham University weekend school on 22 September 1984.[33] Under the best possible scenario for the Government, coal stocks would only fall to 10m tonnes by January 1985, reaching a minimum of 7·5m tonnes in March before restocking. The best scenario for the NUM, however, would result in coal stocks reaching 7·4m tonnes by October and below the critical 4m tonnes level in November. An intermediate scenario, depending upon solidarity action to reduce coal deliveries, would bring coal stocks down to 7m tonnes by December and below 4m tonnes by February 1985. Such projections were made more difficult during the strike by the Government's policy of publishing details of coal stocks in *Energy Trends* one month later than usual. What was also unknown to all outside the CEGB and the Government, however, was the extent to which more sophisticated computer control of the grid system had increased the flexibility of power generation (see Chapter five). Over the year of the miners' strike, the CEGB supplied two per cent more electricity with only fifty-two per cent of the coal consumption of the previous year.[34]

Wright concluded that power-station stocks would have been below the critical level in September 1984 had all NUM members been on strike from the beginning of April.[35] Using the actual levels of stocks and consumption shown in Table 15 it is possible, retrospectively, to determine whether or not the strike could have been won. Throughout the strike (end of March 1984 to end of February 1985) coal deliveries to power stations averaged 530,000 tonnes per week; these were crucial in undermining the strike. Assuming that the coal actually consumed was the minimum to maintain the continuity of electricity supply, power-station stocks would have reached critical levels by August 1984 had the strike involved all NUM members from 6 March 1984 (confirming Wright's conclusion). If 17·5m tonnes of the 21·7m tonnes distributed stocks were suitable for power-station use, these too would have been effectively exhausted by January 1985. The exhaustion of power-station stocks would have necessitated a massive transport operation to move distributed stocks, but not one much greater than was actually used to move scab coal if deliveries and consumption are compared in Table 15. In December and January almost one million tonnes of coal per week would have to have been delivered, but Wilsher *et al.* claim that this figure was achieved in February 1985.[36]

Table 15 *Coal stocks and consumption ('000 tonnes)*

	Actual					*Total strike*	
	Total stocks	*Power station stocks*	*Undistributed stocks*	*Power station consumption*	*Deliveries*	*Power station stocks*	*Undistributed stocks*
February 1984	51,497	27,166	22,667	7,649	–	–	–
March 1984	46,178	23,014	21,723	9,113	4,961	18,098	17,500
April 1984	43,161	19,937	21,873	4,363	1,286	13,735	17,500
May 1984	41,913	18,413	22,116	3,191	1,667	10,544	17,500
June 1984	40,576	16,707	22,378	3,819	2,113	6,725	17,500
July 1984	39,485	15,930	22,316	2,659	1,882	4,066	17,500
August 1984	39,194	15,527	22,533	2,278	1,875	1,788	17,500
September 1984	38,954	15,564	22,100	2,989	3,026	–	16,299
October 1984	38,516	15,484	21,635	2,888	2,808	–	13,411
November 1984	37,845	15,323	21,085	2,991	2,830	–	10,420
December 1984	36,548	14,402	20,753	3,769	2,848	–	6,651
January 1985	34,708	12,799	20,570	3,918	2,315	–	2,733
February 1985	34,087	11,606	21,113	3,886	2,693	–	–
March 1985	33,437	11,916	19,942	6,019	6,329	–	–

Source Energy Trends.

Had the strike been complete, then some time between August 1984, when power-station stocks would have run out, and January 1985, when power cuts would have been introduced, a settlement favourable to the NUM would almost certainly have occurred. Such a settlement could have been concluded earlier in this period if the power supply unions had refused to accept coal moved from pit-head stocks. Wilsher *et al.* claim that Lyons insisted that 'his members would never co-operate in the use of troops to move coal from picketed pit yards',[37] but it is doubtful whether troops would have been required, given the volume of coal that was actually moved. However, had the NUM membership been united and coal stocks clearly dwindling at a faster rate, the power-supply unions and the rest of the labour movement would probably have been more supportive of the miners. Just as no-one likes to back a loser, no-one likes to miss backing a winner.

There is no escaping the conclusion that the NUM's analysis of coal stocks and the strategic importance of Nottinghamshire was seriously inadequate. At the Durham school in September, Kim Howells spoke of the 'ineptitude of not having discussed the situation of coal stocks before the strike' and claimed 'we were nineteen weeks into the strike before there was a directive from Sheffield to picket power stations'. Howells argued that the CEGB should have been the prime target rather than BS, which was of marginal importance: on the day of the largest Orgreave picket, Rugeley power station received 1,028 deliveries of coal. Howells called for national co-ordination of picketing directed towards bringing power stations to a halt. Although the NCC had experienced great difficulty in co-ordinating picketing of power stations, the next meeting responded:

> From the start of the dispute, national office had advised all areas to concentrate pickets on power stations and recently sent out further circulars containing up-to-date guidelines which took into account discussions between power station unions and the NUM.
>
> It would be useful if an experienced co-ordinator could be deployed into the Trent Valley area. South Wales were asked if Kim Howells could undertake this task but they said he was already heavily committed.[38]

For most of the strike the NEC appeared to have no serious discussion of coal stocks, although the NUM research department apparently prepared a report on pit-head and power-station stocks following Wright's revelations at Durham.[39] In contrast, in November TUC researchers had completed a confidential report on fuel stocks at power stations and concluded that there was little likelihood of power cuts or of an 'early settlement of the dispute'.[40] Shortly afterwards, Jack Eccles, TUC Chairman, publicly stated that 'with the strike dragging on for so long, a one hundred per cent victory is no longer possible'.[41] Whatever the source of the Union's inadequate assessment of the situation in the electricity supply industry, its consequences were disastrous. Effectively the strike could not be won without Nottinghamshire, so from the moment it was certain that that area could not be won over, the conflict was futile. Moreover, as official NUM predictions of electricity shortages failed to materialise, the Union's leaders lost some credibility with the members and outside supporters:

> the projection that Arthur often made that they'd so many weeks left and so on; I think the Government welcomed that in a way because they knew all the time that they'd got that tied up and when the strike went into the winter and there were obviously no cuts coming that was a boost for the Government which they rode on and it had an adverse effect upon the miners themselves.[42]

6.2 Progress of negotiations

When the strike began, neither the NUM nor the NCB appeared eager to negotiate. The NCB initially attempted to constrain the escalation of the strike through the courts but did not pursue this tactic (see Chapter five). The struggle within the NUM over the ballot was to the NCB's advantage; Heathfield warned that there could be no talks until the strike was one hundred per cent solid. Nevertheless, once the 19 April SDC voted in favour of strike action, the NUM wrote to the Board seeking an opportunity to discuss the future of the industry.[43] The divisions within the NCB were as great as those within the NUM at this time. While MacGregor ruled out the possibility of compromise with the NUM, Smith intimated there was room for negotiation over pit closures.

There were three phases of negotiations. The first meetings were held between May and July and were marked by the *absence* of serious negotiation. In September, a second round of talks brought NACODS, ACAS and the TUC into the discussion. In the third phase, which began in November as the back-to-work movement gathered momentum, the TUC became increasingly involved and the NCB sought total capitulation by the NUM.

No negotiation

The Government made no attempt to bring the two sides together; Thatcher claimed 'we are not going to intervene in the coal dispute. Our job is to see that the industry has proper management and adequate resources to secure its future'.[44] Walker emphasised the fact that coal stocks would last for months in an effort to demoralise the strikers. The Government's support for MacGregor's 'no negotiation' stance contrasted with the attempts made by Labour Shadow Minister on Energy, Stan Orme, to stimulate negotiations. In May, Orme asked Walker why he hadn't yet met the miners although the strike was in its third month. When Walker replied 'why don't you meet them?', Orme invited Scargill to address a meeting of Labour MPs. MacGregor contacted Orme and presented the NCB case with Cowan, and a meeting between the NCB and NUM was eventually arranged for 23 May. The Union stressed that the only way to resolve the present dispute was for the NCB to 'withdraw its pit closure/manpower reduction programme which was in violation of the *Plan for Coal*'.[45]

MacGregor walked out of the meeting and declined to comment on the Union's case, to the amazement of members of the NUM negotiating team. Jack Taylor told reporters 'we've wasted our time and money coming here today . . . he's insulted the NUM and the employees of the NCB'.[46] These early talks set the pattern for the negotiations throughout the strike as MacGregor played cat-and-mouse both with the NUM and with his own NCB staff.

During late May and early June three meetings were held. The first, in MacGregor's absence, was a cordial meeting at which the Union believed 'the Board had indicated their willingness to withdraw their closure programme'.[47] This progress was followed by a further meeting attended by MacGregor where 'a further measure of agreement was beginning to develop as to the means whereby the current dispute could be resolved'.[48] Orme, who had been instrumental in establishing these talks, told the Commons: 'they are taking place against a threat from the Government because they are trying to implement a victory over the miners, not a negotiated settlement'.[49] Before the third meeting could be arranged, MacGregor publicised the NCB's intention to eliminate uneconomic capacity, thereby destroying any hope of further negotiation and leading Scargill to claim 'MacGregor has clearly had his further instructions from Mrs Thatcher'.[50] Labour MPs similarly believed: 'Mrs Thatcher has broken her pledge that she would not intervene in the pits dispute', although Walker claimed that 'the Government had intervened only to ensure capital investment in coal and decent pay for miners'.[51] During the period of no negotiation MacGregor suggested that the NCB might conduct its own ballot of mineworkers and speculated that the outcome would be a majority in favour of a return-to-work. The NUM viewed the proposal as an attempt 'to usurp the constitution and function of the National Union of Mineworkers'.[52] When the *Guardian* reported MacGregor's ambition to see 'the leadership of the Union . . . thoroughly defeated rather than settling the dispute quickly',[53] the NEC called for his 'immediate resignation'.[54] The objective of breaking the Union precluded a negotiated settlement but Orme believed that one could be reached, given a change in Government policy. To maintain the isolation of the NUM, MacGregor had to present a *public* image of being prepared to negotiate. MacGregor contacted Orme and over lunch: 'he said "tell Arthur we'll arrange this meeting in Edinburgh; you fix it up . . . tell him to bring his toothbrush and his pyjamas and we'll stay till we've settled it." . . . I went out of Hobart House walking one foot off the ground: I thought we're going to do it!'[55] As a result, a series of talks was held in Edinburgh and London during the period 5–9 July 1984. Initially, progress was made in the negotiations and the national officials: 'had secured the verbal agreement of the Board to withdraw their pit closure programme announced on 6th March 1984, to retain five pits which were currently under threat and that future closures would be on the basis of exhaustion to be defined in accordance with the Plan for Coal'.[56]

The NCB refused to put the agreement in writing and MacGregor made his famous comment that he had 'an audience to satisfy', an allusion to the Government's involvement. On 18 July the NCB submitted a formal document to the NUM agreeing to withdraw the pit closure programme announced on 6 March since 'the loss in production due to industrial action made all their projections irrelevant'.[57] The NCB proposed to keep open the five collieries in dispute (Polmaise, Herrington, Cortonwood, Bullcliffe Wood and Snowdown) provided the NUM accepted clause 3c: 'The NCB and NUM agree that where a comprehensive and in-depth investigation by their respective mining engineers shows that a colliery has no further mineable reserves that are workable and which can be beneficially developed, such a colliery shall be deemed exhausted.'[58] The NUM negotiators refused to accept the term 'beneficially developed' since this was tantamount to accepting pit closures on economic grounds, and the NCB terminated negotiations.

With the breakdown of negotiations, the NCB launched its back-to-work campaign, but with little effect (see Chapter five). Although these 'anti-trade union practices' generated ill-feeling,[59] informal negotiations continued. Orme presented an amended version of the July document to MacGregor, who at first claimed not to have received it.[60] Following MacGregor's attempt to dismiss this conciliatory move by the NUM, the leaders of fourteen major unions wrote to him explaining that the NUM had reluctantly accepted the new wording suggested by Orme. Laurie Sapper called on MacGregor to 'stop playing politics and start behaving like an employer and get negotiating'.[61]

The NUM were reluctant to pass over any significant control to the TUC General Council, a decision which Will Paynter endorsed with reference to the 1926 General Strike.[62] Nevertheless, the TUC represented the only platform from which the wider support of the labour movement could be canvassed to pressurise the NCB into negotiating. Before Congress, the General Council met NUM officials and agreed a policy statement of overwhelming support for the miners, but made it clear that TUC support was not unconditional and that they wished negotiations to resume. Amid a blaze of *Daily Mirror* publicity, the newspaper's proprietor Robert Maxwell persuaded MacGregor and Scargill to renew talks during TUC week. MacGregor appeared to have made this arrangment without consulting his industrial relations staff.[63] Certainly Ned Smith believed negotiations to be ill-advised after the NUM leaders had received their morale-booster from the TUC, but the pretence of negotiation was an essential ploy for MacGregor to prevent TUC rhetoric from being translated into action: 'They were urged on by the anti-Scargill electricians and engineering managers, who were opposing the Scargill line at the TUC but none the less phoned the Electricity Council to warn that other unions in power stations might harden their position if talks did not ensue'.[64]

Third party intervention

Initial correspondence to establish a new round of negotiations entailed recrimination over the failure of the July talks and disagreement over what had already been agreed. When confirming the NCB's offer of talks, Kevan Hunt expressed the hope that 'the closure of uneconomic pits can be dealt with in accordance with the principles of the various Reports on the Plan for Coal'.[65] Heathfield, in reply, sought confirmation of the Board's verbal undertaking to withdraw the pit closure programme, keep open the five pits under threat and 'reach agreement with the NUM regarding the exhaustion of colliery reserves on the basis of Plan for Coal'.[66] The following day Smith contradicted Heathfield's statement, claiming that 'the Board did not, either in writing or verbally, agree to the course of action implied by the three points contained in your letter'.[67]

When negotiations began in Edinburgh on 9 September, Willis, Buckton and Basnett were on hand in case TUC intervention was required. Talks continued the following day and media reports suggested that the sticking point was the single word 'beneficially', although Heathfield had accused MacGregor of going back on a verbal undertaking to withdraw pit closures on economic grounds. On 11 September MacGregor and Smith flew to London, almost certainly to consult the Government. Negotiations were to be resumed near Selby the next day but media intrusion forced a change of venue to Doncaster, where the parties talked until 2 a.m. and were thought to be on the verge of a settlement. The negotiations were adjourned for the NEC meeting but the Union's Executive received little information:

> At the request of the Board, the national officials suggested that the details of the present negotiations be not discussed given that the negotiations stood adjourned and would be reconvened on 14th September and it may be prejudicial to any outcome that any current form of words under consideration were given a greater significance than necessary thus preventing the ability of either side to reconcile an interim situation with their own standpoint and the eventual outcome.[68]

The parties resumed negotiation on 14 September only briefly before any prospect of a settlement collapsed. MacGregor told reporters that the strike was over who managed the industry and predicted more miners would now return to work. Scargill blamed Government intervention and the 'political motivation of destroying the NUM'. The Board had appeared to respond to the TUC initiative without conceding the issue of uneconomic capacity: 'their purpose was to undertake a public relations exercise to the detriment of our membership'.[69]

When the September negotiations broke down, Scargill and McGahey immediately discussed the next move with TUC leaders and Willis criticised the NCB for breaking off the talks. Walker said it was now up to the TUC to resolve the dispute,[70] and they became increasingly involved. Orme made a further unsuccessful attempt to find a formula relating to collieries with reserves about to become exhausted, then concluded that it was necessary to involve an independent third party. Orme had kept ACAS 'informed of what was going on . . . secretly' but his suggestion that they should intervene was not welcomed by either side; MacGregor's was 'the most bitter objection to ACAS'.[71] Both MacGregor and Scargill met ACAS representatives in late August but the Service was first involved materially in negotiations in early October when there was a threat of a *national* strike by NACODS.

In the interest of long-term harmony underground, it had been accepted practice during NUM disputes that NACODS members would still be paid if they refused to cross NUM picket lines. In the 1984 strike, NACODS members in Yorkshire had made no effort to work since there was no-one to supervise and occasionally reminded NUM branch officials to mount a token picket to assuage a colliery manager who had become difficult about the understanding. In areas where the majority of miners continued to work, so did NACODS members. On 15 August Merrick Spanton, Director of Personnel, informed the NACODS Executive that their members would not be entitled to payment unless they crossed picket lines on the meshed buses to supervise miners who had broken the strike. By putting NACODS members in the front line, the NCB provoked a recommendation from their Executive not to supervise 'any workmen who try to return to work in defiance of a strike decision approved and endorsed by the NUM'.[72] The NUM had earlier attempted to persuade NACODS to make such a decision which could have halted coal production in Nottinghamshire.[73] When the recommendation of strike action was put to membership ballot, NACODS General Secretary Peter McNestry predicted that they would secure the necessary sixty-six per cent majority.[74] The NUM Executive learnt that NACODS officials 'were optimistic that strike action would ensue' and that other unions representing power-workers were considering ways of increasing their support for the NUM in line with TUC policy.[75] Although the power workers' support did not materialise, NACODS members voted eighty-two per cent in favour of strike action.

The NCB immediately entered into negotiations with NACODS to avert a national stoppage. With the active involvement of the TUC, LP, ACAS and NACODS, the control of negotiations inevitably drifted away from the NUM. Three documents were presented to the NUM and NCB, but ACAS failed to produce a formula which satisfied both parties, even though at one point the difference was between a review of collieries 'in line with Plan for Coal' (NUM) and 'in line with the principles of Plan for Coal' (NCB). The difference was more fundamental than the terminology suggested, however, and the talks again collapsed. Sir Pat Lowry, ACAS Chairman, thought it 'difficult to contemplate bringing the two sides together again in the near future'.[76]

Negotiations with the NCB failed to resolve the NACODS dispute and on 16 October the Executive announced its intention of calling a strike the following week. Support for the strike was even strong in Nottinghamshire and the Midlands where the deputies could alienate the miners with whom they would afterwards have to work. MacGregor used ACAS to reach agreement with NACODS and separate them from the NUM. After ACAS mediation,

NACODS withdrew its call for strike action on 24 October, satisfied with the Board's assurances that deputies would continue to be paid and that the colliery review procedure would be modified to include an independent review body (see Chapter seven). Once the NACODS crisis was resolved, the Board renewed its effort to break the NUM strike with the 'November push' of the back-to-work campaign (see Chapter five).

Surrender terms

At the end of October, Eccles said that the TUC should increase its involvement in the strike. Scargill replied that there was 'no chance' of the TUC taking over their strike and Basnett explained that Eccles was expressing a personal view.[77] Nevertheless, the miners' leaders approached Willis in November for TUC support in persuading the NCB to negotiate. Willis condemned NCB tactics of promoting a return-to-work in place of negotiation, but also indicated that he hoped to encourage the NUM negotiating team to be more 'flexible'.[78] The NEC were concerned at the number of miners returning in some areas and agreed that the Union would: 'welcome any initiatives by other bodies, including the TUC, Labour Party and Church leaders, which were designed to bring the Board to the negotiating table prepared to resolve the dispute in the industry in line with those proposals submitted by the Union at the last meeting with ACAS'.[79]

The General Council established a Liaison Committee comprising Willis, Buckton and Basnett, who had monitored the NUM dispute from the start, along with Eccles, Moss Evans (TGWU), Bill Keys (SOGAT82) and Gerry Russell (AUEW). On 5 December the committee met Walker and King, without the knowledge of the NUM Executive or the TUC General Council, and found that the Government would not countenance any settlement which could be construed as a victory for 'Scargill'.[80] Undeterred, the committee produced a formula to resolve the strike which included an immediate return-to-work, the withdrawal of the capacity reduction, keeping open the five named collieries and a review of *Plan for Coal*, which they put to Walker on 14 December. Walker rejected the committee's formula and complained that the TUC had failed to bring about 'a change in the NUM attitude'. While Willis insisted 'we shan't be backing away from . . . finding a negotiated settlement', he also feared that, without one, the strike 'could go on for years'.[81] With victory in sight, the Government would accept nothing short of total capitulation and saw no role for the TUC unless they could deliver the NUM.

In the new year Michael Eaton put the NCB line that because 'NUM leaders are not moving ground . . . we have to rely on individuals returning to work'.[82] MacGregor appeared to contradict Eaton by suggesting that only a negotiated settlement could resolve the situation. At the next NEC meeting the President suggested that the Board should confirm its willingness to negotiate and it was agreed that the full NEC should attend future negotiations.[83] The NCB proposed an informal meeting to reconvene negotiations during the CIBTO/CISWO Annual Meeting on 16 January. Heathfield and Windsor developed an agenda with Smith and Hunt on 21 January; at a meeting with ACAS the next day it was agreed that a formula could be found which would be acceptable to both sides. The formula contained the necessary face-saver for an irreconcilable conflict: the NCB would retain its right to close uneconomic collieries, while the NUM could raise its objections pit by pit. The NEC endorsed the report of these meetings and confirmed the Union's 'willingness to enter into negotiations without preconditions in which the full NEC will participate'.[84]

When Scargill publicised the NEC decision, he learnt from reporters that an NCB press release that day had insisted on a *written* undertaking from the NUM to discuss the closure of uneconomic capacity as a precondition for the resumption of talks. Simultaneously Thatcher told the Commons that the NUM must make its position clear: 'there must be no fudging'. MacGregor and Eaton suggested that Smith had 'gone too far' in accommodating the NUM but the day's events were seen as 'Thatcher's wish for a total *political* victory over the NUM'.[85] A letter from the NCB now required a 'written undertaking that the Union is prepared to . . . co-operate with the essential task of closing loss-making pits', which Spanton confirmed with Heathfield on 29 January.

The NUM Executive reiterated its willingness to negotiate without preconditions and asked the national officials 'to maintain liaison with the TUC and their efforts to facilitate the resumption of negotiations'.[86] The Board's preconditions appeared to contradict their agreement of the previous October with NACODS, but there was little prospect of NACODS mobilising at this late stage. The way was now open for the NCB and Government to use the TUC to bring the NUM to negotiate on their terms. When Willis addressed the NUM Executive, he stressed that the role of the TUC was to seek a 'procedural route to resumed negotiations' and 'not to be a party themselves to such negotiations'.[87] Ned Smith retired in February as a result of MacGregor's sabotage of the January talks and expressed the hope that the miners would obtain a negotiated settlement: 'it was not a British thing to do, to grind men back to work'.[88] However, doubts were growing within the NEC that a settlement would ever be negotiated and in South Wales the possibility of a return-to-work without an agreement gained support: 'better no agreement than a bad one', ran the argument.[89] The only alternative to an organised return-to-work was the remote possibility that Willis would be able to arrange new negotiations.

As the only avenue available, the TUC became inexorably drawn into active negotiation with the NCB: 'they got sucked in . . . they didn't set out to negotiate . . . but in the end they came up with a proposition that the Coal Board agreed to, which the miners had not agreed to'.[90] After discussion with the NCB, Willis requested a meeting with the NUM Executive and presented a document, the terms of which had been 'prepared solely by the NCB . . . endorsed by the

Government and were not open for discussion or negotiation'.[91] After a heated debate, the NEC agreed to accept the NCB document provided certain amendments were made. The NUM wished to delete references to the closure of pits on economic grounds without discussion through the review procedure modified by the NACODS agreement. Willis conveyed the NEC view to the NCB but the Union's amendments were rejected. During their discussions the NCB issued a press release which implied that the document had been drafted by the TUC; Willis assured the NEC that the document was entirely of the Board's making and 'deplored the slur against his offices and the TUC'.[92]

The following day the NEC met at the ACAS offices and asked the Service to establish a committee of inquiry if they failed to persuade the NCB to resume negotiations. At the same time NACODS officials, fearing that the NCB document jeopardised their agreement, met Cowan and Spanton. The NCB stressed that the document was not intended for NACODS and that the reference to the use of 'existing procedures' was included because the NUM had not agreed to the modified review procedure. Later at the TUC offices the NACODS officials told the NUM that they believed the NCB would accept the amendments NACODS proposed if submitted by the NUM.[93] When Willis presented the amendments to the NCB on behalf of the NUM, however, these were again rejected and NACODS leaders believed that their agreement had been destroyed 'as a result of Government interference'.[94]

On 19 February, in an effort to overcome the Government's objections, the TUC Liaison Committee met Thatcher, who instructed Walker to raise the TUC's points with the NCB. The following day Willis met Walker and MacGregor, who produced an amended NCB document. Willis presented the document to the NUM and explained that there could be no further changes. The NEC agreed that these latest proposals were 'infinitely worse than those presented last weekend' and proposed the SDC adopt their earlier amendments to the document as a basis for restarting negotiations, and accept the NACODS agreement on the modified review procedure.[95] The TUC Committee made no further attempt to intervene in the dispute; with the Government they had lost prestige, while with the NUM their credibility, if not their integrity, was in question.

At the SDC on 21 February Scargill described in detail the progress of negotiations since 16 January and delegates were provided with copies of the various versions of the NCB document. Jack Collins (Kent) launched an attack upon the TUC for negotiating with the NCB and Government instead of 'organising other people in the struggle' and called upon the national officials for 'correct working-class leadership . . . if you run to the bosses you will never stop running'.[96] Successive speakers argued that the Union should seek an amnesty for members sacked during the strike, particularly in areas where the strikers were in a minority. The SDC reaffirmed the amended NCB document of 17 February as a basis for talks, declared its willingness to accept the

NACODS agreement, and called upon 'the wider labour and trade union movement to implement the TUC Congress decision of last September and not to leave the NUM isolated'.[97]

At a meeting with NUM national officials on 25 February, the TUC Finance and General Purposes Committee strenuously denied that the TUC had any input to the NCB document. The Union sought 'further means of implementing Congress policy' and asked the TUC to establish a trust fund to enable the NUM to continue to operate but 'the outcome of the TUC's lengthy deliberations were not considered by national officials to be entirely helpful'.[98] In a final attempt to reach a negotiated settlement, the NEC agreed to 'accept the NACODS agreement in its entirety . . . [to] enable discussions to take place on all other outstanding matters'.[99] On 28 February the NEC discussed a Northumberland motion for an immediate return-to-work and another from Kent calling for the Union to stand firm but decided to convene an SDC on 3 March and invite areas to submit motions for its consideration. Orme, meanwhile, urged MacGregor to use the NACODS agreement as a basis for settlement of the NUM dispute: 'You can get it now, you can get the NACODS agreement; take it. "Mr Orme", he said, "the TUC's proposal is on the table; take it or leave it."'[100] The TUC's involvement was therefore more damaging than was immediately apparent since the Liaison Committee provided the Government with an excuse to deny the NUM even the safety net of the NACODS agreement.

6.3 The return-to-work

Although the back-to-work campaign had intensified before Christmas, its effect in areas like Yorkshire had been minimal (see Chapter five). Ned Smith, who had opposed the tactic, believed miners would rejoin the strike once they had earned enough to buy their families presents.[101] Christmas posed a challenge for strike maintenance which the mining communities met with great ingenuity (see Chapter four), and had represented a target for which to strive: 'we'll stay out till Christmas if necessary' was often heard during the summer months. With the holiday past, however, maintenance activities were relaxed and strikers experienced an anti-climax as neither power cuts nor negotiations were in sight. The absence of a tangible target (except Christmas 1985 as some suggested, half-seriously) deflated the strikers' morale to an all-time low and individuals pre-empted the collective decision to end the strike. Such a crisis of morale was predictable: 'Critical periods of the strike occur most frequently in connection with Sundays and other holidays. At these times the strikers are especially inclined to desert because of the absence of picketing or other mass activities and because of the sight of privations at home.'[102] The NCB predicted a 'surge' back to work after Christmas but Union officials were relieved when few people returned during the three working days that constituted the first week of 1985.

No further to go

The back-to-work campaign had continued with letters to individuals from colliery managers. At Brodsworth the Manager wrote: 'we have already lost one of the five faces, with a second one at risk. Repairs are required now before it is too late.'[103] To the imminent risk, the manager added the encouragement that over forty people had already returned to the pit in response to an earlier circular.[104] *Coal News* in January announced 'all pits ready for new year return' and detailed production and recovery work already underway. The publicity barrage intensified at the weekend and on Monday 7 January the NCB claimed a record 241 'new faces' in Yorkshire. While Eaton linked the return to the NUM leaders' unwillingness to negotiate on NCB terms, Jack Taylor reminded members that the 'best way to get negotiations going is to stick together'. The following Monday was less spectacular, but a fortnight later on 21 January the return of 571 Yorkshire miners promoted fears that the return-to-work was accelerating. Even so, towards the end of January the scabs still only represented nine per cent of the Yorkshire miners. The scabs at this time were heavily concentrated in South Yorkshire, where twenty per cent had returned, while in Doncaster and Barnsley only three per cent had given up the strike, and North Yorkshire matched the overall average.

The increased rate of return on 21 January possibly persuaded the Government that it was unnecessary to negotiate. Successive breakdowns in negotiations probably caused more strikers to return than would have done in the absence of any negotiation. Therefore, after the NCB refused to resume negotiations on 1 February without the NUM's prior agreement to discuss 'uneconomic pits', a record 586 Yorkshire miners returned on the Monday (4 February), despite the efforts of about 5,000 pickets. It was during this week that the NEC discussed a return-to-work organised by the Union and without any agreement with the NCB. Clearly the NUM would suffer a rupture from which it might not recover if the rate of return continued. On the other hand, without NCB agreement not to victimise activists the NUM could be just as emasculated by the removal of key members. The passions aroused in discussions of the tactic obscured the conflict between the twin objectives of keeping the Union together and sustaining effective opposition to pit closures. Moreover, media treatment of the debate served to encourage *individuals* to return, the opposite of the effect intended by those who had first raised the issue in Yorkshire.

At the weekend the NCB maintained their back-to-work campaign with advertisements in major newspapers announcing '83,000 miners have decided the strike is over'. February's *Coal News* claimed '5,500 Yorkshire pit men go back to work' and ran features on the seven collieries where sufficient numbers had returned for coal to be produced. Although the TUC had arranged a regional day of action for Monday 11 February, a substantial number of miners again returned. The next day the High Court ruled that at eleven Yorkshire collieries (seven of them in South Yorkshire) mass-picketing was illegal. The

Area Executive and Council accepted the ruling and disbanded the SCC.

The abortive attempts by the TUC and ACAS to restart negotiations resulted in Eaton's declaration on 17 February that the NCB's 'final position' would not be modified.[105] The next day the week began with 326 new scabs in Yorkshire, limited damage considering the small number who continued picketing. When the SDC of 21 February rejected the NCB ultimatum, Walker asserted that this was the last offer: 'now all that's left is for NUM members to return to work'. Walker's message was repeated over the weekend and reinforced with NCB advertisements in the Sunday papers declaring that there was 'No further to go'. In South Yorkshire the NCB placed advertisements in the *Dinnington and Maltby Trader*, a free paper delivered to every household in the area and probably the only newspaper seen by most strikers. The advertisement claimed that twenty per cent of the workforce had returned at Kiveton Park and seventeen per cent at Dinnington. Another advertisement on 22 February again encouraged miners to 'Beat the taxman' by returning before 25 February and claimed that 7,500 miners were already back in Yorkshire, 3,500 of them in the South Yorkshire Area.

There was a massive demonstration of support for the miners in London on Sunday 24 February, but Jack Taylor confirmed that the strike was rapidly losing momentum: 'you can only grind men down so far . . . there is terrible hardship in coal mining areas'.[106] Over 1,000 new scabs returned to work in Yorkshire the next day and with, for the first time, every colliery affected, Taylor spoke of miners being starved back to work. Officials were clearly concerned to preserve the unity of the Area Union in the light of increasing numbers returning to work, many of whom had been active in picketing earlier, and avoided denouncing the latest blacklegs. In this way Union officials gave the appearance of legitimising the return-to-work, thereby causing further disillusionment among those less committed to the strike and attracting recrimination from the 'bitter enders'. Another 500 miners returned on the Tuesday and by Wednesday 27 February the NCB was claiming twenty per cent of Yorkshire miners were back at work.

The strike was not expected to last much longer but the Union had the invidious task of determining its end while there was still sufficient cohesion to make a corporate decision: 'Defeat may be registered by withdrawal of demands, return to work without negotiations, discharge . . . or by a shut-down of the establishment. The strike may end through gradual defection and collapse of group control or through formal decision and collective resolution. As the stresses of conflict increase, morale may weaken to such an extent that the strike disintegrates.'[107]

The hunger scabs

By the end of February, the strike was weakening even in the heartland of the Yorkshire coalfield and the branch meetings which had mandated their delegates to vote for a return had in many cases insisted that the Union lead them back the following week as they had 'had enough'. Despite these decisions, the strike had not deteriorated so much as the media and the NCB claimed. The NCB had only succeeded in getting five per cent of Yorkshire miners to give up the strike before Christmas 1984. Before the formal return-to-work on Tuesday 5 March 1985, eighteen per cent had given up the strike, the vast majority of them during the last week of February. This figure includes the considerable number who returned on Friday 1 March in the mistaken belief that this was necessary to secure holiday entitlements, many of whom rejoined the strike on the Monday. Moreover, the return had been heavily concentrated in South Yorkshire (thirty per cent) and North Yorkshire (twenty-two per cent) and even in these areas several collieries were still relatively solid. Barnsley Area had only twelve per cent back at work and in the Doncaster Area the comparable figure was seven per cent. The position during the final week of the strike is illustrated in Fig. 7. The strike was still relatively solid in the heart of the coalfield, with under five per cent at work in eighteen collieries and under ten per cent at a further seven. However, over fifty per cent were back at nine pits: Manton, Shireoaks, Kiveton Park, Dinnington, Denby Grange, Emley Moor, Ricall, Stillingfleet and Gascoigne Wood. The strike was broken but only in a limited sense even to the end in Yorkshire, although for how much longer that position could have been maintained is open to question.

Most strikers who returned during this period put themselves in a different category from those who had returned before Christmas. As Union officials had already acknowledged, many had formerly been active in the strike but were starved back to work. In several collieries fights broke out between recently returned strikers and the early scabs. While the strikers who stayed out to the end treated all scabs with equal contempt at the time, most eventually accepted a distinction between the 'hunger scabs' who had had enough and the scabs who gave up before it was obvious that the strike was lost.

The theories rehearsed to explain patterns of activism (see Chapter three) and strike-breaking (see Chapter five) reappeared in rank-and-file explanations of who became a hunger scab. In Yorkshire those who returned after Christmas were predominantly middle-aged at twenty-nine per cent of collieries, old-timers at twenty-five per cent and family men at twenty per cent. While twenty-seven per cent of collieries reported all ages involved to the same extent, none reported youngsters as being overrepresented, despite the hardship experienced by single miners (see Chapter four). In terms of length of service, the hunger scabs were predominantly newcomers at thirty-four per cent of collieries and old hands at twenty-five per cent, but forty-one per cent of pits regarded them as drawn evenly from both groups. All occupations were involved at fifty-five per cent of collieries, while elsewhere-below-ground workers predominated at eighteen per cent of pits. Face, craft and surface workers were each held to be overrepresented at about nine per cent of pits. At

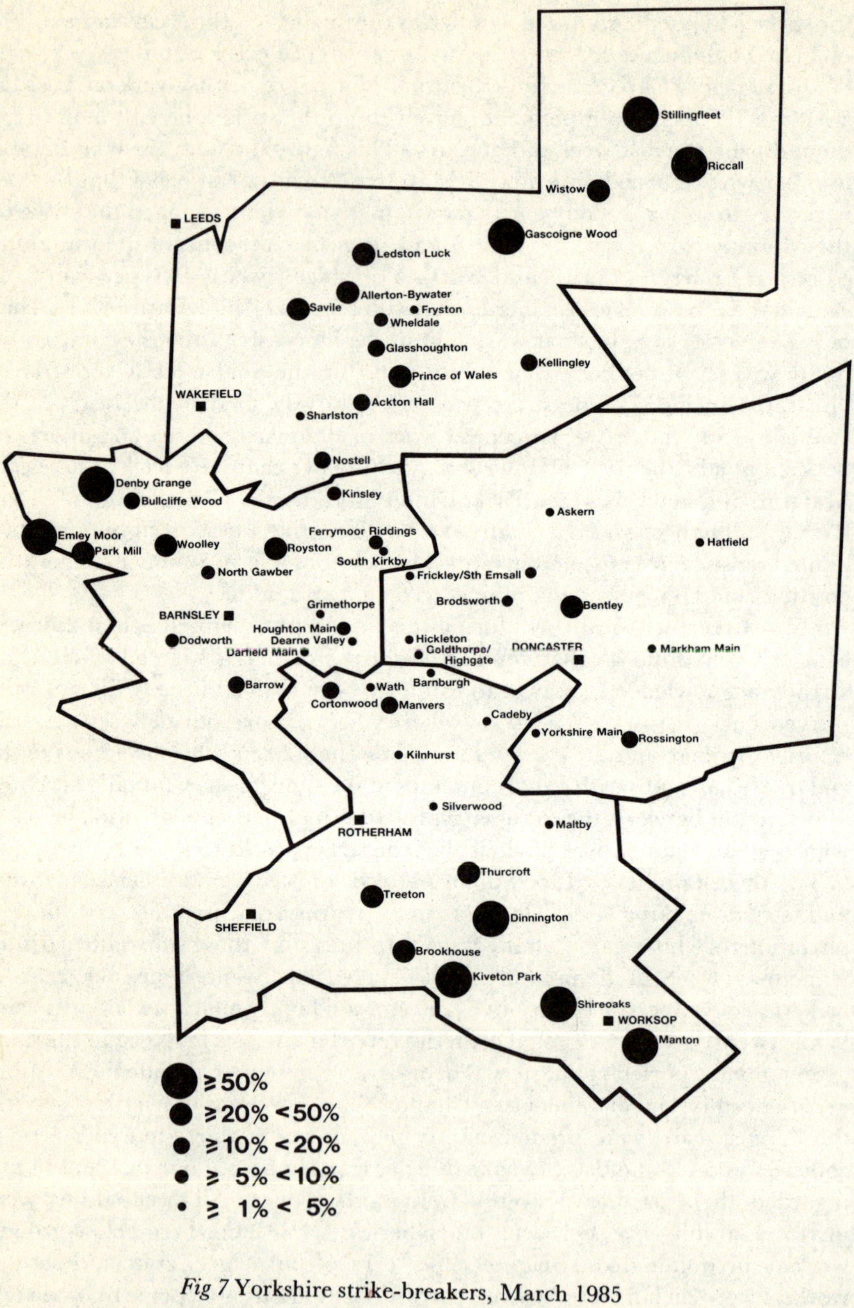

Fig 7 Yorkshire strike-breakers, March 1985

forty-five per cent of collieries they came from traditional mining families in the main, while at twenty-seven per cent they were from alien families and at twenty-nine per cent from both. They mostly lived outside a mining community at forty-eight per cent of collieries, but were from within the community at twenty-five per cent and from both at twenty-seven per cent of pits. Their politics were characterised as non-political at eighty-two per cent of collieries and Conservative at eleven per cent, while at seven per cent the whole spectrum of pit politics was involved. It is difficult to characterise the typical hunger scab, therefore, except that he was unlikely to be a youngster or a radical and more likely to live outside a mining community.

Had the strike continued any longer, more miners would have given up on an individual basis. Hiller argued that 'the striking group normally has a margin of indifferent members . . . from whose ranks the strikebreakers or scabs are recruited'.[108] Our respondents attempted to identify those whom they thought objected to the strike and would have returned had it continued. At thirty-eight per cent of collieries these objectors were drawn from all age groups while they were predominantly middle-aged at thirty per cent; family men and old-timers each predominated at fifteen per cent of pits. In terms of length of service, newcomers and old hands each predominated at twenty-five per cent of collieries and they were evenly involved at the remaining fifty per cent. Objectors included all occupations at fifty-seven per cent of collieries and each of the four job categories was cited as overrepresented at roughly eleven per cent of collieries. They were from traditional mining families at thirty per cent of collieries and from alien families at twenty-seven per cent, while both were involved at forty-three per cent of pits. Objectors most often lived outside a mining community at thirty-six per cent of pits and within the community at fourteen per cent, while at fifty per cent of pits they were from both categories. They were regarded as non-political at thirty-nine per cent of pits, and Conservatives at twenty-three per cent but included a range of political attitudes at thirty-six per cent of pits. As with the hunger scabs, therefore, the objectors can only be characterised in broad terms: they were unlikely to be youngsters or radicals and more likely to live outside a mining community.

The defining characteristics of the early scabs, the hunger scabs and the objectors confirm Hiller's view that: 'the strikers' distance from the group centre, their traditional attitudes and degree of devotion to the group are factors which influence the rate of defection from the corporate pursuit'.[109] Moreover, the differences between the three groups show that, as the strike dragged on, the defecting group became less easily distinguished from the activist core.

The formal decision

The NEC meeting on 28 February called an SDC for Sunday 3 March to allow areas to reach a decision on whether to return without an agreement. The following day Yorkshire Area Council delegates voted 42–22 to continue the strike but to reconvene on Saturday after branch meetings had been held. The reconvened Council meeting voted to return 38–31 on a show of hands but this was reversed on a card vote with the slender margin of 557 to 561. Yorkshire Area was to oppose a return without an amnesty for sacked miners but would abide by the national conference decision.

The NEC met before the SDC to consider eight motions submitted by the areas. The President ruled that the NEC were bound by the previous SDC decision, so the Committee agreed: 'that the motions could be put to Conference without a specific recommendation from the NEC, although the President would explain the fact that the Executive were already committed to previous policy decisions of Conference'.[110] Scargill gave the NEC report to the SDC and asked the delegates to allow the NEC to make no recommendation until the end of the debate. Des Dutfield (South Wales) accused the NEC of 'abrogation of their responsibility', and warned 'it is no good trying to get off the hook: we have been on the hook for twelve months'.[111] Sammy Thompson (Yorkshire) agreed: 'if you lead us, you lead, and if you don't lead somebody else will lead for you'.[112] Thompson's motion that the NEC should reconvene to agree a recommendation was carried 72–11.

The NEC reconvened and the representatives of four areas agreed to withdraw their motions in favour of other motions on the agenda, which were then discussed. The South Wales motion proposed an organised return-to-work without agreement, with an amnesty for sacked miners to be a negotiating priority once work was resumed. Durham Area supported this motion conditional upon the establishment of a trust fund for victimised members. Kent proposed that there should be no discussion of any other motion until an amnesty was agreed, while the Scottish motion was for a return-to-work conditional upon an amnesty. Yorkshire proposed that Conference should reaffirm the previous position until the interests of members at the five named pits and those dismissed during the strike could be secured. The Kent and Scottish motions were defeated while the NEC split 11–11 on both the South Wales and Yorkshire motions.[113] The NEC rejoined Conference and the President explained that in each case they had either rejected the motion or failed to agree, so the SDC would have to determine policy without a recommendation from the NEC.

Terry Thomas moved the South Wales motion:

> What stands before us is whether the men . . . return to work with leadership or whether they return to work in chaos . . . I do not believe that the members . . . should have the loyalty they have shown this Union abused by allowing this Union to be destroyed . . . You either give them leadership and repay the loyalty they have given us, or you sit back with your blindfold on and you let the strike collapse around you.[114]

Jack Collins moved the Kent motion with an impassioned plea not to desert the sacked strikers: 'if you are encouraging a return to work without these blokes going in you will be guilty of encouraging scabbing'.[115] John Burrows

(Derbyshire) disagreed; if the SDC failed to organise a return 'we shall simply be turning first-class trade unionists into scabs'.[116] Moving the Scottish motion, George Bolton argued that the Union must not only stand firm, but also 'conduct an enormous national campaign' if they were to obtain a general amnesty.[117] Sammy Thompson accepted the arguments about an amnesty but reminded delegates when he moved the Yorkshire motion of the pits they came out on strike to protect, two of which were in the Yorkshire coalfield where 'we have more men sacked now than what we went out on strike for'.[118] John Lockett (Midlands) pleaded with the areas where the strike was still strong to continue the struggle for a *national* amnesty for the sake of areas where strikers were in a minority.[119] Nigel Ashfield (Power Group), however, told delegates that the minority on strike in his area 'cannot keep it up any longer . . . they want to return to work with some sort of co-ordinated leadership'.[120] Peter Fenton (Group No. 2) believed that the Union's credibility depended upon negotiating an amnesty as a condition for the return-to-work, otherwise every NUM official would be 'branded a traitor'; he reminded the platform of the number of times they had stressed 'we will not go back to work until we see those people that are sacked get their jobs back'.[121] Thomas feared that if they atempted to hold out for an amnesty:

> and the drift back continues until you have 10,000 or 12,000 men still out on strike in the British coalfield. Then you have got the Coal Board into a position where you won't be talking about 700 jobs. You will be talking about 10,000 or 12,000 jobs . . . when they reach the figure they are satisfied with . . . they will have no qualms in sacking the hard core . . . Of course, we can still have a platform chanting 'Here we go', but comrades, while we are in the Conference discussing this our members are back in the pit in ever-increasing numbers.[122]

Mike Beddows (Cokemen) reiterated: 'if we don't lead the men back they will go back on their own . . . and we will be in no position to bargain for our colleagues who have lost their jobs'.[123]

The President wound up the discussion and suggested that the SDC was 'probably one of the most difficult if not the most difficult in the history of the NUM' and praised the 'courage, strength and solidarity' of NUM members and the Women's Support Groups. Scargill concluded:

> A lot of people will look at this struggle and try to pick out what advances have been made and to do that would be to miss the point. The greatest achievement is the struggle itself, because we have already shown that provided we are prepared to fight against their policies we can prevent their implementation . . . their struggle will inspire workers not only in this country but abroad, who have a completely new outlook as far as the fight for jobs is concerned.[124]

When the votes were taken, the South Wales motion was carried 98–91, while the Yorkshire motion was lost 91–98. Both the Kent and Scottish motions were defeated 19–170. The President confirmed: 'That means the National Conference will be calling for an organized return to work on Tuesday, 5th March, without any signed agreement, that the NEC will be calling for negotiations with the Board on a national basis for an amnesty for those men dismissed in the dispute.'[125] Conference then approved a motion from the NEC to establish a trust fund for the benefit of members experiencing acute hardship or victimisation (see below). Scargill told miners lobbying outside Congress House of the SDC decision: 'we go back together and the Union goes on fighting for pits, jobs and communities'. There were shouts of 'traitor: you've sold us out' and a group chanted 'we're not going back'. Many of those whom the NCB had dismissed were never able to go back, but the rest returned over the next few days.

The organised return

The return-to-work was as confused as the mobilisation (see Chapter two). Nationally about 1,100 more miners went back to work on Monday, 4 March, but the remainder waited for the organised return the following day and some who had gone back on the Friday came out again for the day. At Kiveton Park 100 men who had broken the strike walked out when three men were sacked for 'gross industrial misconduct'.[126] Against the advice of area officials, Scottish delegates voted 7–6 to continue the strike until an amnesty was agreed. In a separate meeting SCEBTA delegates voted 10–4 against continuing the strike but would respect NUM picket lines. In Kent 2,000 miners voted almost unanimously to stay out for the reinstatement of forty Kent miners, at least until a further meeting to be held by 17 March. Branch meetings at Markham Main and Hatfield Main voted not to return until the victimised men were taken back. These two Doncaster pits and Betteshanger in Kent deployed token pickets through the Yorkshire coalfield on the day of the official return.

The return was highly emotional; branches went back behind Union banners and colliery bands. In a final display of loyalty, Women's Support Groups and children accompanied the miners to the pit gates while local traders, relieved that business would soon resume, looked on. Many branches decided to return in the daylight at 9 a.m. instead of the regular shift time. At Ledston Luck the manager locked the gates and refused to allow the miners to return at an unorthodox time. Management responded similarly at Cadeby and Kilnhurst collieries and at Carcroft and Birdwell workshops, so all the workers at these units returned home. Hickleton branch voted to stay out for the rest of the week to protest at a new shift system introduced unilaterally. At Manton, where over fifty per cent of the membership had returned before the end, ninety per cent came out for another week over sackings and suspensions. Kellingley deputies walked out when they discovered management had allowed NUM scabs to undertake their duties. Cortonwood branch voted not to cross a token Kent picket, although a few individuals ignored the decision. Barrow strikers, accompanied by Scargill and women workers from a local clothing factory, refused to cross a picket of Armthorpe miners and returned home again.

Armthorpe pickets were as far afield as Bold, where embarassed branch officials voted not to lead the men back while pickets and police were present. The Yorkshire pickets agreed to leave once the police had dispersed. In South Wales miners at Celynen South and Merthyr Vale walked out when management refused to move scabs. In Northumberland, the Manager of Ellington refused to allow the men to start until later in the day. There was heavy picketing in Scotland where almost half the workforce had already given up the strike. Pickets were also out in Kent, but relatively few miners went back in the area. The NCB estimated that nationally eighty per cent of miners were now back at work, but this included those who had temporarily joined the strike in its early months. The 36,000 still on strike (25,000 of them in Yorkshire) constituted almost thirty per cent of the numbers on strike before the November push. After six months of systematic back-to-work campaigning and an official decision to return, still only seventy per cent of strikers had defected.

On 6 March Markham Main and Hickleton were still out and Kent pickets brought Barnburgh out again. However, Yorkshire Area NUM instructed branches that the Kent pickets were unofficial and they failed to prevent a return at Wath, Manvers and Cortonwood. As Cortonwood miners walked silently past a lone picket from Betteshanger, he called after them: 'I never thought I would see the day when Yorkshiremen crossed picket lines; we came out for you lads and now forty of us have been victimised'.[127] At the end of the second day back 10,000 miners were still on strike: 6,000 in Scotland, 2,000 in Kent and 2,000 in Yorkshire. The Scottish Area delegate meeting was brought forward and voted 10–5 to call off the strike without an amnesty in view of the scale of the return and on 7 March Markham Main voted to go back. On the Saturday, miners in West Fife and Kent voted to return the following Monday. On 11 March as Kent and the rest of Scotland went back, Polmaise miners, who were already out on strike when the dispute over closures began, voted to return the next day. At Betteshanger, miners walked out again after management refused to allow sacked miners to be elected workers' safety inspectors.

The strike was over but the magnificent endurance of the miners had ensured theirs was not a simple defeat. The ability to resist for so long offered an inspiration to other workers and was widely viewed as an opportunity to damage 'Thatcherism'. Nevertheless, beneath the pageantry of the colliery bands and the rhetoric of marching back with 'heads held high' was the grim realisation that the strike had been called off without securing the original demands or even the minimal assurances of no victimisation: 'the return to work is marked by dejection; for readjustment must now be made to a lowered status before the victorious employer and curious onlookers. The workers, therefore, seek solace in ideal values, which, though varying in character, always pertain to the wish for the integrity and vindication of the group and for greater devotion to the ultimate "goal of the labour movement".'[128]

Management's reception to returning miners was reported to be less friendly where strikers had been in a minority.[129] In Yorkshire, however, the miners' initial reception owed more to the character of the area directors and individual colliery managers. For example at Wheldale, where fewer than ten per cent had given up the strike: 'the manager had us all lined up across the road on the first day back and said "you've got the worst record in North Yorkshire, you're a bunch of animals; if anyone steps out of line they're sacked" '.[130] At Gascoigne Wood, however, where only the Union officials and two others stayed on strike: 'Before we went on strike the attitude of management, deputies and under-officials was hardened but when we went back it was completely different; it was like working at Butlins. I could come up when I wanted, go home when I pleased, no-one said anything to me.'[131] Gascoigne Wood, the hub of the Selby complex, was behind schedule, so management were eager to avoid disputes in a location of such strategic importance. However, at neighbouring Ricall where over fifty per cent had also returned to work, the management 'treated everyone like dogs'.[132]

On the eve of the official return, Scargill had predicted guerilla warfare in the pits. MacGregor wrote to every miner: 'This long strike is now over. It has been severely damaging to the relationships between all of us who work in the industry . . . This strike was not of management's choosing nor of our making. Now that it is over, every effort must be made to restore harmony within industry.'[133] Doncaster Area Director, Tuke warned: 'we will manage the industry properly and even-handedly but we're not going to have a guerilla war; we have inspectors underground'.[134] South Wales Area Director Philip Weekes similarly said that he would not tolerate action against those who had given up the strike. Most managers warned against intimidation. At Kiveton Park, where over fifty per cent had given up the strike, the Manager wrote: 'Firm but fair discipline will be imposed as it always has been. Violence of whatever form, including fighting, physical provocation and verbal abuse or intimidation, will be considered an act of gross industrial misconduct carrying the usual penalty. All men will be treated equally.'[135] The Manager of Brodsworth, where under ten per cent had returned, similarly warned: 'violence or intimidation of any man or his property, no matter which side he took during the strike, *will not be tolerated* and could result in instant dismissal.[136] In fact, the execution of this policy similarly varied with the strategic importance of collieries and the attitudes of individual managers (see Chapter seven).

6.4 The sacked miners

The political significance of the miners sacked during the strike far outweighed their number: 'the demands on the leadership are nowhere near as great when a strike is won, or partly won, as when it is lost . . . In the event a strike ends in defeat, the union's first duty is to take care of those courageous workers who suffered most – the jailed, the blacklisted, the hungry.'[137]

As early as June the Notts representatives feared that the minority of strikers in their area might be victimised. The NEC agreed that it should be 'an essential prerequisite of any settlement of the current dispute that all members who have been victimised would be reinstated'.[138] Concern grew as the strike continued and the number sacked increased. In the absence of progress towards a negotiated settlement, the Union was unlikely to gain reinstatement for sacked miners. In November the NEC agreed that miners who had been unfairly dismissed should apply to industrial tribunals within the qualifying period, although the Union intended to deal with the issue in future negotiations.[139] However, the Chairman of the tribunals in Yorkshire ignored ACAS and the Area Union, and insisted on hearing a number of cases; the Union suspected that the NCB 'was pressing to have cases heard in order to breach the collective agreements'.[140] The Union was unable to manipulate the tribunal system or to persuade the NCB to negotiate over an amnesty. At the informal meeting of 21 January 1985, NCB representatives insisted that a resolution of the dispute on closures would be followed by a discussion on 'other related matters', including dismissals.[141] This final initiative, however, was sabotaged (see above).

At the February SDC Peter Fenton (Group No. 2) argued that the proposal to resume negotiations without preconditions neglected 'the boys who have been sacked'.[142] Nevertheless, the SDC agreed to 'resume negotiations immediately without preconditions'.[143] The Chairman emphasised that the amnesty was 'all part of the negotiating process' but many delegates remained unconvinced when a return-to-work without agreement was considered.

Financial support

Jack Collins failed in his bid to have sacked miners admitted to the March SDC, which approved the South Wales motion for a return-to-work without an agreement, and unanimously adopted the following motion from the NEC:

> The experiences of the dispute over the past twelve months has demonstrated the need for an established Trust Fund enabling the Union to fulfil its obligations in accordance with Rule 42 both in the current circumstances and on any future occasion. Conference therefore instructs the NEC to urgently examine this question and to take such steps as are necesary to establish a Trust Fund for the purpose of hardship and victimisation benefit.[144]

The NEC therefore recommended: 'that a special levy be called from members . . . As this levy would be for other than industrial membership it would require to be approved by the membership.'[145]

It is difficult to determine the precise number of miners sacked because of the strike, as the total fluctuated even after the strike with people being reinstated and others sacked for related misdemeanours, particularly actions against scabs. At the end of the strike the NCB put the total number sacked at 718 and indicated that they were prepared to re-employ sixty-seven who had been convicted only of minor offences.[146] Later the NCB put the total at 900 sacked, of which 280 had been reinstated; the number still sacked had therefore fallen from 651 to 620.[147] The NCB put the Yorkshire total at 318 but the total of those cited by our respondents at every Yorkshire colliery amounted to 464 miners sacked. Two months after the strike ended the national figure settled at 700 but none had been taken back in Scotland and Kent.[148]

In the national ballot held on 20–22 March, the majority of members failed to vote in favour of the levy; only forty-six per cent were prepared to pay 50p per week to the Miners' Solidarity Fund. The NEC therefore agreed that those who had voted in favour should be invited to make voluntary contributions.[149] National officials were dismayed at the ballot result but activists criticised the absence of a campaign on the issue and argued that there was insufficient time to 'prepare the ground'.[150] Moreover, some branches had already introduced a local levy for the sacked miners and were not prepared to commit themselves to an additonal national levy. Branch officials found it difficult to win support for a levy because the media had stigmatised the sacked men and most miners were critical of theft or violence:

> what people had to realise was that these things only happened because of the strike. They wouldn't have done it otherwise. We had one sacked and the Union disowned him . . . he wasn't a full shilling . . . because he went to the pit looking for scrap and there's only one live cable to the shaft and he cut through it . . . we nearly lost the pit. He got no support but rumours went round that he was getting £45 per week.[151]

Similarly, at Woolley, one of two miners sacked for stealing coal on a commercial scale had his hardship money reduced because he had stolen for financial gain rather than to meet personal heating needs.

Just as they had during the strike, the National and Area Unions struggled for control over the finances generated in support of sacked miners: 'areas should be requested to divert all future contributions received for hardship and solidarity to the Miners' Solidarity Fund in order that there could be a uniformity in the treatment of all cases'.[152] The sacked miners had most support in the heart of the coalfield. At Frickley, for example, an additional branch levy brought the money received from the hardship fund (£35 for single miners, £45 for married men) to £100 per week. While such local initiatives guaranteed a more acceptable income for sacked miners, they were far from universal, so generating inequities between collieries and emphasising the parochialism of post-strike relations.

Towards reinstatement

There were no national negotiations on reinstatement, although the NCB did discuss the issue at area and pit level.[153] On 14 March 1985 the all-party Employment Committee of the House of Commons announced that it would convene a committee of inquiry 'to consider the criteria governing the non-reinstatement of those who had been dismissed' and 'to consider whether those criteria had been applied consistently and fairly on a national basis'.[154] Evidence was heard from the NCB and the NUM. MacGregor told the committee that the reinstatement of sacked miners was an area issue and each case was to be considered individually, but the Board's philosophy was that: 'areas should not reinstate anyone who had been dismissed following conviction for acts of serious violence, intimidation or vandalism, or dismissed on good evidence of involvement in them'.[155] Hunt, head of NCB industrial relations, 'agreed that the process of assessing each case at area level was a form of 'summary treatment', a view which MacGregor disputed, even though the sacked miners had been denied trade-union representation at the review hearings. MacGregor accepted that inconsistencies in the area reinstatement policies 'are bound to occur when you have a fully decentralised policy' and argued that an individual could complain of unfair dismissal to an industrial tribunal. However, MacGregor: 'declined to give the committee an assurance that the NCB could necessarily accept such a finding if an industrial tribunal upheld a complaint of unfair dismissal and ordered reinstatement'.[156]

The NUM appealed to the committee for an amnesty for all sacked miners 'similar to the amnesties offered after the coal miners' strikes of 1972 and 1974', although Heathfield accepted that an 'appeal procedure' could at least resolve anomalies. The committee supported the idea of a special review to re-examine the cases of miners who had not been reinstated; Scargill viewed this as 'a very helpful step', and MacGregor would 'not rule out completely such a possibility' but preferred to keep his options open. In summary, the committee recommended:

1. It should be standard practice after an industrial dispute involving strike action, for the provisions of the ACAS Code of Disciplinary Practice and Procedure to be reactivated by any employer once normal working has been resumed.
2. The NCB should (a) institute a national level review into the cases of the dismissed miners whose reinstatement has been rejected at Area level and (b) ensure that if there are any Areas in which no review of dismissals has yet taken place, such a review precedes any re-examination at national level.
3. The terms of reference of any review should be consistent with the spirit and guidelines of the ACAS Code of Disciplinary Practice and Procedure.[157]

The NCB rejected the Employment Committee's recommendation that they reconsider their refusal to take back 599 sacked miners.[158] However, the NCB was persuaded to review the cases of 203 miners sacked in Scotland after a meeting between McGahey and John Lowden, the new Area Director,[159] and by December seventy-five Scottish miners had been re-employed. Many of the miners who had been sacked for criminal offences were acquitted in the summer of 1985 (see Chapter five), but acquittal did not guarantee reinstatement.

Although cleared of criminal charges, the miners were guilty of being active members of the Union. Dave Nellist drew attention to the discrimination against Union activists when he asked in the Commons: 'Is the Secretary of State aware that in Scotland seventy per cent of those who were dismissed were lay officials or elected members of strike committees? That is a clear example of victimisation of strike organisers.'[160] Thatcher described the committee's findings, published on 19 June, as 'shabby' and criticised their failure to condemn violence and intimidation against working miners.[161]

The Employment Committee's recommendations would ensure the reinstatement of most miners charged with less serious offences, so enjoyed broader support than the idea of an amnesty, which the leadership obviously regarded as a liability. The Shadow Home Secretary was opposed to an amnesty because:

> I don't believe that you can ask for an amnesty for people who've committed certain criminal offences and not others . . . How do you distinguish? It's one thing for a miner who has been arrested for obstruction and fined. It's quite another thing – look at the two men – who although in my view were mistakenly prosecuted for murder, undoubtedly did do something very wrong indeed that can't conceivably be justified by its context of a normal industrial dispute.[162]

Amnesty campaign

The NUM ambition of a general amnesty was taken up by Benn, who argued: 'one of the first tasks of the next Labour government must be to proclaim complete amnesty for . . . all those miners who are now in Jail . . . and all the trade union funds that have been sequestrated . . . must be returned with interest'.[163] Benn presented the Miners' Amnesty (General Pardon) Bill to Parliament on 28 June 1985, seeking: 'to provide for a general amnesty for all those who were fined, imprisoned or otherwise punished by the courts, the NCB or others, for acts done or offences committed during the 1984–5 strike by the NUM; to secure that full pardons, reinstatement at work, and compensation shall be offered to all the persons involved; and for purposes connected therewith'.[164] There was little support for the Bill but it served to publicise the amnesty campaign, which had two purposes: 'First of all it is directed at the wider audience who say after a struggle you want conciliation . . . But, of course, its more immediate objective is to commit the TUC and the LP conferences to an amnesty.'[165] The NUM motions to the TUC and LP conferences called upon the next Labour government to review all cases of miners jailed, granting free pardons wherever appropriate, to reinstate and compensate those dismissed, and to reimburse the Union's funds lost in fines. Despite the General Council's opposition, the TUC passed the NUM motion by 4,649,000 votes to 4,585,000.[166]

Kinnock similarly made clear his opposition to a blanket amnesty, winning Walker's approval a second time: 'the Opposition are divided on the matter. Some . . . are in favour of reinstating all miners, no matter how guilty of violence they are, and, indeed, of taking miners out of prison. I am glad that at least the Leader of the Opposition condemned that.'[167] In a heated debate at the LP Conference Kinnock made a vitriolic attack upon the NUM, criticising the Union for striking at the wrong time, failing to hold a ballot and adopting mass-picketing tactics which he claimed had caused a particular style of policing.[168] Miners in the gallery shouted 'scab' and 'traitor' but Kinnock was supported as vociferously by about half the delegates for a speech that was seen to be avenging Scargill's success of the previous year. Benn had been optimistic that Conference would support the NUM motion, claiming that he could 'not think of a constituency party that won't do it, except perhaps one or two in the Notts area'.[169] Conference did support the NUM resolution by 3,542,000 votes to 2,912,000, but since this was less than a two-thirds majority, Kinnock's prediction that the resolution did not stand 'a chance in hell' of finding its way into the LP manifesto proved correct.[170]

In order to generate support for the practical application of the NUM motions, the National Amnesty Campaign (NAC) was formally launched at a joint meeting of Trade Union Briefing and the Campaign Group at the LP Conference. The NAC sought to raise support and finance for the victimised and imprisoned miners, and to promote a second amnesty bill sponsored by the Campaign Group. The Justice for Mineworkers Bill introduced by the Campaign Group in 1986 redrafted the wording of Benn's Amnesty Bill to bring it in line with the resolutions of the 1985 TUC and LP conferences. Although the new Bill provided for a review of all cases, rather than a general amnesty, the Campaign Group were aware that it had 'no prospect of being passed by the present Parliament' but saw its main value as 'part of the nationwide political campaign in support of the miners'.[171] The Justice for Mineworkers Bill provided a national focus for the NAC, but co-ordinating the various local amnesty initiatives was at least of equal importance. The NAC steering committee met on 27 October 1985 and established a panel of speakers from the NUM, WAPC, support groups and Campaign Groups to support and build local campaigns.

Groups like the South Yorkshire Defence Campaign (SYDC), established in October 1984, became part of the NAC network. The SYDC had been active from the end of the strike, campaigning for sacked miners, monitoring the progress of cases against miners and publicising their outcome. Ultimately it was at local level, pit by pit, that progress was made towards the reinstatement of some of the sacked miners.

Where the local union organisation had survived the strike with the minimum of damage, action was taken in an effort to obtain reinstatement, but many branches were too weak for this to succeed. Silverwood struck for a week over the issue of sacked men and brought a resolution to Council for a policy of guerilla disruption at the most profitable pits, and days of action over the issue of the sacked men: 'Hatfield supported it, five or six others from Doncaster, but it was massively beaten, not just beaten; actually delegates expressed great dismay at the very idea of taking any sort of action'.[172] At Thurcroft, 'branch officials raised the dismissals every day with the manager' and by May 1985 had obtained reinstatement for nine of the twenty miners sacked at the pit.[173] In the Doncaster Area, management agreed to reinstate two miners at Bentley and Markham Main and restored original shift times, in return for the Union's agreement to the introduction of contractors at Frickley. At Fryston, twenty men had been sacked in connection with the attack on strike-breaker Michael Fletcher; the branch succeeded in gaining reinstatement for all but those who were sent to prison. In an effort to break the strike in Yorkshire, management offered to reinstate those they had sacked, provided they became scabs. At Woolley one of two men dismissed for identical offences was taken back on this basis, while the other refused to break the strike and remained sacked. At Ricall the branch actually instructed two men to return to work in order to secure their jobs. At Bilston Glen 200 miners struck in support of two miners who had won their cases at industrial tribunals but whom the NCB refused to take back.[174]

The majority of those who were reinstated had their jobs back in the first six months after the strike. The NCB was determined to exclude the hard-line militants and activists wherever it had an excuse. Inevitably the loss of these men has added to the Union's weakness following the strike, and their continued victimisation serves as a warning to others who may wish to take their place.

Chapter seven

Strike aftermath

The dispute was not resolved by the ending of the strike, and many of the issues at stake only became apparent afterwards. This chapter examines the aftermath of the strike which, like the period before the conflict formally began, represents an essential part of the coal crisis and its attendant struggles.

Six interrelated themes are examined. Firstly, the pattern of post-strike industrial relations is explained in relation to management objectives at the heart of the conflict: moving to a form of production entailing a higher organic composition of capital and an increased rate of surplus value. The new management style established nationally is shown to have been translated at pit level into an offensive against union organisation. Secondly, we trace the emergence of the Union of Democratic Mineworkers. The separation of the Notts Area, actively encouraged by the NCB, represented the most serious damage inflicted on the Union. Thirdly, we assess changes in the politics of the NUM since the strike. Within Yorkshire, the women's action groups and activists' movements, which had been crucial to the strike, were unable to exert any lasting influence over Union officials and their rank-and-file foundations rapidly disintegrated. In the fourth section we examine the development of a modified colliery review procedure, designed to resolve the issue at the centre of the strike. Having failed to prevent the restructuring through strike action, the NUM attempted to resist pit closures using the agreement with which the NCB divorced NACODS from the NUM dispute, but found the procedure deficient. The fifth section analyses how local authorities, concerned at the impact of pit closures on local economies, established the Coalfield Communities Campaign. Although launched during the strike, the initiative was deliberately kept separate from the miners' struggle, and ultimately degenerated into lobbying for regional assistance. We conclude with our assessment of the prospects for British coal-mining.

7.1 The new industrial relations

Well before the strike, the new-style NCB industrial relations was evident in management provocation during the overtime ban and MacGregor's 'off your butts' productivity drive. The pit conciliation machinery, which had often been short-circuited in the past, was enforced by managers refusing to negotiate until men were working normally and the Union was back in procedure. However, it was during the strike, and partly through the opportunities afforded by the strike, that the major changes were instituted. The back-to-work campaign was particularly significant: 'A study of strikes in which employers developed or attempted to develop back-to-work movements leads to the conclusion that this tactic is generally used when the employer's objective is to break not only the strike, but also the union'.[1] Kirk, Public Relations Director, and Smith, Industrial Relations Director, strongly objected both to the back-to-work campaign and the involvement of Hart and Bell (see Chapter five). MacGregor recruited Area Directors as well as Tory agents to usurp the two professionals' roles. Moses and Wheeler were used as industrial relations battering-rams, and for a short while Eaton became a public relations figure who was prepared to toe the MacGregor line. After questioning MacGregor's public relations strategy, Kirk was forced to take early retirement and was killed shortly afterwards in a boating tragedy off Skye.[2] Smith and MacGregor had a heated argument over NCB industrial relations strategy in front of leading capitalists, including Sir Terence Beckett and Sir Michael Edwardes, at a Hobart House dinner.[3] Smith walked out over MacGregor's handling of the NACODS dispute and, after making further attempts to bring about a negotiated settlement, also took early retirement. These clashes were not about personality, although MacGregor was obviously a difficult person to work with, but about policy and *ideology*.

MacGregor was appointed to bring the perspectives of international capital to the NCB. Kirk edited a book of Schumacher's writings on energy,[4] which were contradictory to MacGregor's perspectives and the policy these entailed for the coal industry. MacGregor's management reorganisation after the strike included the replacement of people like John Mills (Deputy Chairman) and Philip Weekes (South Wales Area Director) who were committed to the British coal industry, with part-timers representative of international capital: David Donne (Dalgety), David Newbigging (Jardine Matheson), Colin Barker (ITT) and Sir Melvyn Rosser (Deloitte, Haskins and Sells).[5] If international capitalist objectives were to be pursued, the power of the NUM had to be weakened through the imposition of a unitary philosophy hostile to trade-unionism. Smith was firmly attached to the sophisticated pluralist style of NCB industrial relations, causing MacGregor to comment: 'He thought many of the things I suggested would cause trouble . . . our differences in approach were fundamental. I saw the need to hold the line to re-establish the right to manage as paramount; he thought the need to compromise was paramount.'[6]

At the end of the strike MacGregor had appealed for a restoration of harmony, but a management crackdown was widely expected. Kirk feared the consequences of American-style management, while Smith argued that the abrasive style and apparent need to defeat the miners was bad management.[7] Heathfield and Wilson, General Secretary of BACM, believed authoritarian management could not work because a greater degree of co-operation was required in mining than in other industries. Within a week, MacGregor's message to activists was clear: 'people are now discovering the price of insubordination and insurrection. And, boy are we going to make it stick'.[8] It was obvious that the Government wished to make an example of the miners when Thatcher commented that 'the miners' strike taught the facts of life to union leaders'.[9]

After the strike had been called off, the NEC paid formal tribute to the strikers and their supporters, disbanded the NCC and agreed to convene a meeting of the Union side of the JNNC to consider negotiating priorities.[10] The NCB made the lifting of the overtime ban a precondition for negotiation, and the ban was cracking in the same parts of the coalfield where the strike had broken. Manton and some of the Selby pits had worked weekends since the end of the strike and Shireoaks had voted to lift the ban. Although Yorkshire, Durham and Kent were opposed, the NEC recommended to an SDC that the ban be lifted.[11] On 2 April the SDC voted 122–74 in favour of lifting the ban. Some Yorkshire activists argued the ban was their only bargaining lever and feared this decision was a further signal to management of the Union's weakness. On 19 April 1985 agreement was reached on the outstanding pay claims for 1983 and 1984: two increases of 5·2% on grade rates, payable from 1 November 1983 and 1 November 1984, and the consolidation of a bathing and changing allowance. Significant backpay was owing to those who had broken the strike, attracting the label 'scab bonus' from Yorkshire activists. However, the settlement did little to alleviate the financial hardship of those who had remained on strike beyond Christmas 1984.

Scabs and activists

The two groups most affected by the end of the strike were those who had returned to work earlier, the scabs, and those victimised by management, who were mostly activists. In the immediate aftermath, management was concerned to prevent the harassment of scabs, while the Union sought the reinstatement of sacked activists.

In areas where the strike had remained solid, the few who had broken the strike were treated with contempt by the returning strikers. The media highlighted Paul Watson, the only scab at the Aberdare phurnacite plant, whose car was attacked and who found excrement in his socks. When five men were suspended for allegedly pouring oil over him, the workforce walked out but Union officials advised them to return and acknowledged that the men were sacked as an example to the community.[12] In September Watson left Aberdare for a 'secret address in the Midlands'.[13] At Betteshanger thirty scabs were reported to be afraid to return to work after their names and photographs had been posted around the pit.[14]

In Yorkshire the treatment of scabs varied according to the proportion who gave up the strike. At Selby and the Calderdale pits, where the strike was effectively broken, there was no harassment of scabs nor of the minority who remained on strike, although some of the latter were victimised by management. In the South Yorkshire collieries where over fifty per cent had returned to work the superscabs were still the object of abuse, often from the hunger scabs who felt humiliated at having given up the strike. In the heart of the coalfield the few who had broken the strike were abused and feared violence or 'accidents' at work. At North Gawber, for example, where one team setting rings (girders) in roadways included a scab, the others all let go of the ring on a pre-arranged signal, leaving the strike-breaker to collapse under the weight. The activists never knowingly spoke to scabs. Often, however, a scab was *safer* working in a team of activists; if anything happened to a scab, the militants would get blamed, so they would look after a strike-breaker while refusing to engage in conversation. At several collieries the scabs were forced to ride in the lower deck of the cage while those on the top deck emptied their water bottles or urinated on them. At South Kirkby, miners lined up at the pit bottom and in the baths to threaten, abuse and physically assault the scabs. On 30 April, in an effort to control this behaviour, management dismissed four men for alleged intimidation. The pit and neighbouring Ferrymoor Riddings went on unofficial strike; by the end of the week 3,000 miners were involved, but the Barnsley Panel instructed them to return to work, the Yorkshire Area Executive having refused to make the strike official. A month after the strike ended, forty people one night besieged the house of Thurcroft scab Keith Mettam, who fled with his family to the refuge of Maltby police station.

Many isolated scabs asked for transfer to Nottinghamshire, or at least to collieries where more had returned to work. As pits closed and miners were transferred, it became increasingly difficult to identify scabs and hence maintain their ostracism. Ackton Hall miners, notoriously implacable towards scabs, were split up when the pit closed and sent to collieries like Allerton Bywater and Kellingley where they were forced to work with strike-breakers. The Ackton Hall men resorted to scratching their cheeks or the backs of their hands to indicate to colleagues that another miner was a scab. Within nine months, attitudes had softened towards the hunger scabs, but not towards the early starters. Over a year after the strike ended, 400 miners walked out at Hatfield Main when they discovered that they had been working with a scab since the strike; the NUM had assumed that all twenty scabs had been transferred.

MacGregor promised that the NCB would tackle the 'distressing problem of intimidation in the industry',[15] but it was clear that the strike-breakers were now of only marginal importance to management. Although many scabs were moved, the Board did not agree to all requests for transfer and Eaton claimed that 'a decision to return to work was a personal decision on the part of the individual'.[16] Where isolated scabs were not transferred, they often took sick leave and eventually left the industry. Just as the Union was accused of deserting the sacked miners, the NCB was accused of deserting the scabs. The balance of power between management and Union hinged upon control over the majority of miners, who were neither victimised activists nor intimidated scabs. In an effort to rehabilitate the scabs, some pit managers negotiated over sacked miners. At Savile, for example, men were reinstated in return for Union agreement to accept scabs from Fryston and Wheldale.

Management often used sacked miners as a lever to gain Union agreement over introducing contractors:

> The management are saying, if we take a sacked lad on with these contractors then the militants are less likely to kick up against it and twelve months later we can put that lad back on the dole . . . The Union are off-loading these men because there's that many of them on £35 and £45 a week, but they're selling these lads down the line . . . they didn't come to us and say 'would you rather pay a levy or set them on with contractors?'.[17]

At the end of January 1986 there were still 143 men sacked in Yorkshire,[18] and by August 1986 the national total of sacked miners had only fallen to 500.[19] MacGregor's replacement as BC chairman in September 1986 by Sir Robert Haslam (who had also succeeded him at BS) did not affect the pattern of labour relations, despite Haslam's declaration of wishing to heal the rift left after the strike. Nevertheless, Haslam promised to review 400 cases of sacked miners;[20] two years after the strike ended, 135 were re-employed (though not reinstated to their former jobs),[21] but there were still 350 sacked miners nationally, 134 in Yorkshire, and little prospect of their being re-engaged.

Activists were naturally a target for management reprisals. Six months after the strike, Rossington colliery came out for two weeks over management's refusal to uphold the Doncaster Deployment Agreement relating to manning levels and the allocation of work by rota. According to branch officials: 'the colliery management stated that they would not give face work to "activists of a disruptive nature" . . . The manager wants a market system where only yes men get the job. But we forced him to rethink his decision. He didn't think the men at Rossington would take action'.[22] At Barnsley Main, management tactics similarly provoked a stoppage in the first year after the strike:

> we have always had a custom and practice that if a team had to come off a face, providing it had been on the face for six weeks, we would put the checks in a hat and draw out for that team, which is fair to all concerned . . . The undermanager refused to do this and when we came to work on the Monday . . . the team was picked.[23]

The undermanager wished to remove unilaterally a team who had taken a hard line against scabs and, despite the local strike being solid, it was settled by the parties' agreement that the decision on which team to remove should alternately fall to management and the branch. At Frickley, 'they isolated all the militants . . . there were four of us in our team; they kept sending the others on the pit top and keeping me underground. The manager told us they had files on us and picket-line photos . . . they split us up'.[24] At the same pit another activist was warned that a young miner's deputy training would be stopped if the activist was seen again in his company. This was no idle threat since activists elsewhere (for example, at Dinnington) had found their deputy training terminated after the strike. The clamp-down on activists even extended to traditional political activities: at Ricall the manager instructed a miner not to bring Labour Party literature onto the premises.

Management objectives

The new industrial relations reflected longer-term managerial objectives, the attainment of which was facilitated by the altered balance of power between the forces of production. The limited recovery of former relations at pit level occurred against a back-drop of a new order imposed from national level. Much of the strategy can be traced to the Miron Report (see Chapter one), which appeared to have influenced the Ridley Report (see Chapter five). In September 1985 the NCB released details of stringent cost objectives, which were prepared by Moses without trade-union involvement.[25] The Moses Strategy defined a cost ceiling of £1.50/GJ, roughly translated into £38/tonne by the NUM, although the equivalent cost per tonne is crucially dependent upon the calorific value of the coal being produced. Coal of high calorific value produced above £38/tonne can fall below the £1.50/GJ ceiling, whereas low-calorific-value coal has to be produced at below £38/tonne to meet the £1.50/GJ criterion. The implications of the Moses Strategy were interpreted almost exclusively in terms of pit closures, yet, given the commitment to 90m tonnes per annum output in the Strategy, *other* management plans for reducing costs and increasing productivity were of more immediate importance. Ultimately, productivity increases would of course lead to pit closures unless cost reductions increased the market for coal.

A plan formulated by Wheeler, the former Scottish Area Director who took responsibility for the key Notts Area, was endorsed by Haslam as BC strategy for the future.[26] The Wheeler plan outlined a number of means to attain the £1.50/GJ target which amounted to lengthening the working day, intensifying supervision and reducing safety standards, an approach reminiscent of the coal-owners' strategy after the 1926 lockout. The Miron Report recommended the reintroduction of the 1965 Bevercotes Agreement relating to ROLF (see Chapter one), which allowed seven-day working under continental shift patterns. Under the Wheeler Plan, working time is to be increased by producing coal six days per week, with miners working four continental shifts of up to nine hours per day underground and shafts winding at full capacity for twenty-three hours per day.

The problem of decreasing returns is endemic to coal-mining, and a major cause is the reduction of productive time as coal workings become more remote from the shaft. In the Durham coastal pits, travel time has such an effect that BC calculates the optimum shift length (to allow continuous production) to be in excess of nine hours. Four nine-hour shifts would increase machine-available time by thirty per cent. For some time BC has considered asking the Government to amend the Coal Mines Regulation Act 1908 (as amended by the Coal Mines Act 1932), which limits the underground shift length to 7.5 hours plus one winding time. The Durham Area resisted pressure form BC to allow longer shift times, having recognised that Union acceptance of a shift length of eight hours in that Area would have been used by BC to advocate removing statutory limitations on the length of the working day for all miners. If

BC were able to introduce optimum shift lengths, NUM researchers estimated, 50,000 miners' and deputies' jobs would be lost nationally, 4,000 of them in South Wales.

The reduction of statutory limitations on hours of work underground would undoubtedly have an adverse effect upon miners' health; evidence on health and safety was used to support the Coal Mines Regulation Act 1908. Other aspects of the Wheeler Plan are also likely to worsen conditions. The Board wishes to introduce new working practices from America, including the use of diesel-powered free-steered vehicles and roof bolts in place of conventional supports. There are plans to reduce the number of deputies and officials, partly as a result of automation (see Chapter one), and to alter their function from safety to supervision. In order to introduce these changes in working practices, BC has been pressing the Health and Safety Executive (HSE) to relax statutory standards. Under the Health and Safety at Work etc Act 1974, the HSE have drafted general regulations to replace the detailed provisions of the Mines and Quarries Act 1954. The 1974 Act was enabling legislation designed to promote the more effective application of existing provisions and to replace these progressively with detailed regulations which would 'maintain or improve the standards'. The NUM and NACODS argued that the new general regulations constitute a dilution of standards and in 1986 withdrew from consultative discussions in protest:

> The changes to existing Regulations do not just flow from the Health and Safety at Work Act, but are inspired by the NCB's new strategy for production. It is clear from Wheeler's proposals that the Board see existing Regulations as a constraint on the technological revolution. It is not merely a question of replacing out-of-date statutes. From the Board's viewpoint it is as much a question of establishing highly flexible working practices, *in all parts of a mine*.[27]

Having failed to alter working hours in Durham, BC made the development of a £900m super-pit at Margam conditional upon Union acceptance of six-day working, in breach of the national Five Day Week Agreement.[28] National Union opposition was stifled and after a four-hour meeting on 7 March 1987, two years after the strike ended, South Wales delegates agreed to local negotiation on the issue. Area President, Des Dutfield, claimed that the decision was taken because of the 'responsibility to school leavers and the unemployed' in South Wales, while Scargill accused the area of violating NUM rules.[29] Kinnock wrote to the South Wales Area, congratulating them on their independent stand at Margam.[30] Flexible working was bound to dominate the 1987 NUM Conference, and a few days earlier BC reiterated that Margam would not produce coal until six-day working was agreed.[31] The NEC split 10–10 on the issue, which was therefore debated by Conference delegates without an NEC recommendation. South Wales Area delegates supported flexible working, while opposition was led by delegates from Durham and the Midlands. After extensive debate, delegates rejected flexibility, including six-day working, by more than two to one, and agreed to hold an individual membership ballot with a recommendation to support the Conference resolution.[32]

Management tactics

Aside from the immediate issues of scabs and sacked strikers, many pit managers made unilateral changes to working arrangements, even before the strike ended. At Gascoigne Wood, miners who had gone back to work rejoined the strike when management introduced continental shifts.[33] Although only seven per cent had gone back before the end in Doncaster, Area Director Tuke showed most enthusiasm for initiating the post-strike régime. Several concessions were withdrawn throughout the Doncaster Area, including the practice of paying craftsmen one half-hour's overtime for completing reports required under safety legislation.[34] Craftsmen's report money was unaffected in Barnsley, where twelve per cent had gone back to work. After several months the Union negotiated back a portion of report money in Doncaster. At Woolley, where thirty per cent gave up the strike, 'from day one management stopped all concessions, water money and so on, which we never got back'.[35] Six months after the strike ended, Houghton Main miners walked out over management's refusal to implement the Yorkshire water agreement. At Frickley, where only one per cent had broken the strike, management altered shift times unilaterally and dropped rates for outbye workers to fifty per cent bonus in contravention of the agreement; shift times were restored within six months but it took over a year for the branch to get outbye workers back on sixty-five per cent bonus. At Wheldale, where about seven per cent had gone back to work colliery management gradually eroded demarcation, using workers' increased flexibility to reduce manpower. The need to earn bonus money to clear strike debts removed any possibility of concerted restriction of output, but there were still disputes over productivity as management attempted to increase the rate of production. At Dawdon, for example, there was a two-day strike following a warning letter from the pit manager that face-workers had not met performance targets and would have to improve.

Although the *extent* of the crack-down depended upon local management attitudes and the strategic importance of the operating unit, rather than upon the incidence of strike-breaking, it was nevertheless the breaking of the strike and the damage inflicted upon union organisation which made it possible. The damage was apparent before the strike ended when eleven miners at Maltby rejoined the strike after management proclaimed 'there's no union here now'.[36] The first four months after the strike were regarded by many branch officials as a 'punishment period'. Management put branch officials back on three shifts and allowed only one official to remain on the surface. This practice was designed to reduce the Union's effectiveness and expose Union officials to the inevitable complaints about the deterioration of working conditions caused by the stoppage. Branch officials fulfilled the role of personnel officers smoothing out local difficulties and acting as a brake on the militants. Management were so successful in reducing time off for union business that the NUM Head Office circulated details of trade-union rights to local officials.[37] Roy Mason (Lab., Barnsley C.) argued in the Commons that 'union officials should be allowed to work on regular day shifts and that NUM branch secretaries should have two days on the surface to enable them to deal with trade union affairs'.[38]

The pattern of industrial relations at pit level was established from the first day at some collieries:

> we were waiting about for ages while some of the lads sorted some scabs out . . . by the time we were going down we realised we wouldn't be underground while eleven o'clock, and the manager says, 'I'm not paying you till you get down pit', so some of us wanted to kick back. Union says, 'oh no, we've just had twelve months, we don't want any more bother lads'. Since then all we get from the Union, at least in the first nine months, whenever there's a dispute, is 'well we lost the strike'. And now when we go to the Union they say, 'well they'll not even talk to us'.[39]

In many pits management refused to talk to branch officials; this policy was adopted at Allerton workshops, where a majority returned to work, at North Gawber where under twenty per cent had returned, and at Frickley, which had remained relatively solid: 'they'd only meet one branch official at a time and then they'd not turn up nine times out of ten; they were trying to belittle us'.[40] Selby management would similarly only see Union officials individually after the strike but branches ruled that individuals cannot make agreements and officials now create situations where two of them see management together. It took twelve months for workers at Allerton workshops to organise strike action in order to re-establish normal negotiations.

As well as restricting Union officials and discriminating against activists, NCB management continued the union-busting tactics begun during the strike, using the law and appealing directly to the men without involving the Union. These tactics were particularly evident in the Selby complex. In January 1986 miners at Stillingfleet walked out because management stopped pay after men arrived one minute late owing to adverse weather conditions. In response, the Manager wrote to the Branch Secretary, Ted Scott, claiming that he had 'deliberately caused men to take strike action' and warning that any future action would be regarded as 'a serious breach of the existing machinery'.[41] The branch instituted an overtime ban to pressurise management into repaying the men and agreed to organise a strike ballot if Union officials were disciplined for carrying out trade-union duties. Within an hour of the Manager seeing a copy of the ballot paper, an area official had telephoned the Branch Secretary advising against the ballot as Area Director Tuke was seeking a court injunction. At Whitemoor miners were encouraged to take their problems to management without involving the Union. At Ricall chargehands' meetings were held every month as part of a 'quality circle' approach, but these were abandoned after chargehands began using them as an opportunity to air grievances to management.

Eighteen months after the strike there was something of a return to former industrial relations at Ricall, probably because management wished to avoid disruption to the development of the spine tunnel to Gascoigne Wood, which was behind schedule. Nevertheless, management used the incentive scheme to enforce discipline at Selby. In February 1987, almost two years after the strike ended, management stopped a week's bonus pay for 'restricting output' after workers in the Selby complex had walked out over visits from local Conservative MP Spencer Batiste and Bob Siddall (son of Sir Norman), who had reputedly told miners to 'eat grass' during the 1984–85 strike. On 16 February all 3,000 Selby miners staged a week-long protest strike which was unanimously supported, even by those who were labelled 'super-scabs' for having given up the strike before Christmas. However, the Union did not expect management to reverse their decision on bonus pay and activists feared that BC would automatically stop bonus payments for a week at any colliery where there is a stoppage of work: 'if they can get away with it at Selby, they can do it anywhere'.[42]

This management offensive on union organisation was only a partial success for three reasons. Firstly, the centrally-directed NCB hard-line management introduced in conjunction with a process of decentralisation created a paradox: pit managers continued to enjoy the autonomy to decide whether to adopt the new approach. Secondly, sufficient cohesiveness remained in the core of the coalfield to resist the most flagrant attempts to assert managerial prerogative. Thirdly, even where the strike had been broken, the strategic importance of locations like Selby, Denby Grange and Manton still demanded the co-operation of the workforce to achieve development rates and productivity targets.

A further challenge to union organisation came with management's unilateral introduction of a new disciplinary code.[43] The new code sanctioned summary dismissal for 'gross industrial misconduct', including acts committed away from the workplace, so represented a codification of some of the deficient practices criticised by the House of Commons Employment Committee. The code also gave management control over the representation of offenders; pit managers could refuse to allow union representatives to attend members' disciplinary interviews on the vague grounds that they were 'not acceptable'. Shortly after the 1987 NUM Conference, the use of the new code provoked a stoppage at Frickley colliery, deflecting attention from the flexible working issue which had dominated the Conference. The stoppage began after five Frickley miners, who had allegedly left work early on the eve of their holiday, refused to attend a disciplinary interview with the colliery manager and were suspended.[44] The colliery's 900 miners walked out and the next day picketed out six other Doncaster pits.[45] A day later pickets spread the strike to the remainder of South Yorkshire, so sixteen pits and 14,000 miners were involved.[46] The colliery Manager wrote to all Frickley strikers warning that they would be sacked unless they returned to work; copies of the letter were sent to other strikers in South Yorkshire.[47] After the NEC agreed to hold a national ballot on the disciplinary code, the Yorkshire Area Executive instructed the strikers to return to work pending the outcome of the ballot. Frickley Branch Secretary, Steve Tulley, was dismayed by the decision: 'for the first time since the strike there is resistance to the management's dictatorship and the executive are telling fifteen thousand miners to go back to work'.[48] Taylor attended the Frickley branch meeting and urged them to return to work, but the strikers voted overwhelmingly to continue.[49] Although other collieries had agreed to abide by the Executive's instruction, Frickley pickets stopped nine South Yorkshire pits and spread the strike to the two Barnsley pits in the neighbouring village of South Kirkby.[50] The following day, in the interest of area unity, Frickley strikers voted to return to work. The Frickley dispute presented the Union with a classic dilemma; they chose the legitimacy and unity of an official dispute in preference to the gathering momentum of a spontaneous strike movement.

While the Frickley dispute brought the entire South Yorkshire coalfield to a standstill, a simultaneous disciplinary case brought the threat of a strike throughout North Yorkshire. Stillingfleet's Branch Secretary Scott, who had been the subject of earlier management discipline, was dismissed for instructing miners at Wistow colliery not to shear at the weekend. This was potentially a far more dangerous situation for BC than the Frickley dispute, for three reasons: firstly, the dispute linked the issues of the new disciplinary code with flexible working; secondly, management was disciplining a branch official for attempting to uphold a national agreement which, for all the rhetoric of area leaders, was being steadily undermined at a number of Yorkshire collieries, and thirdly, management clearly wished to avoid a stoppage at Selby. The Frickley dispute pre-empted action over Scott's case and fragmented the Yorkshire Area. Frickley pickets were asked not to picket North Yorkshire collieries because this was 'bull week' when miners worked exceptionally hard to increase their holiday pay packets. South Yorkshire collieries had already taken their annual holidays so were more willing to take action. Its timing suggested that the Frickley dispute was deliberately provoked by management to divide North and South Yorkshire. At the height of the Frickley dispute, Area Director Tuke wrote to all North Yorkshire miners warning that if they struck over Scott's dismissal, their involvement in the action would be recorded in their personal files.[51] A later letter informed miners that Scott had been offered re-employment at Wheldale (which was expected to close within six months), and that any industrial action would guarantee the closure of seven named collieries.[52] Tuke also made extensive use of press releases to reinforce the message.[53] The North Yorkshire Panel discussed the possibility of a 'rolling strike', involving groups of collieries taking action alternately, but activists expected Scott's case to become subordinate to any national programme of action against the code. The ballot on the new code was organised during August and sought authority for the NEC to call 'various forms of industrial action' if BC failed to withdraw the code.[54] Whilst the ballot was underway, the code provoked another strike when six men were sent home at Deep Navigation in South Wales. Following a three-hour meeting with the NUM and the involvement of ACAS, BC agreed to re-examine the code, but not to withdraw it.[55] Although Yorkshire activists were critical of the inadequate campaign mounted by the Union, the membership ballot authorised industrial action by a majority of 77·5% nationally, eighty per cent in Yorkshire.[56] In response, BC made the minor concessions of agreeing not to veto Union representation and to accept the decisions of industrial tribunals.[57]

The NEC viewed the concessions as inadequate and instituted an overtime ban with effect from midnight on Sunday 20 September.[58] In response, Haslam wrote to all miners individually, warning that the Five Day Week Agreement would be suspended to lay off miners in the event of a backlog of maintenance work.[59] The action was designed not to restrict maintenance, so disruption was minimal and Yorkshire activists criticised the adoption of ineffective sanctions after the membership ballot had given such unequivocal authority for action. Miners lobbbied the October NEC meeting but failed to convince the leadership to extend the action, while Yorkshire Area Council postponed its proposed ballot on the issue.[60] The indecision of the NEC was exemplified at the November meeting, where three motions in succession were rejected: one calling for an end to the overtime ban; one calling for its escalation, and the third for maintaining the *status quo*. Faced with such inconsistency, the President ruled that the ban should continue and announced his resignation.

In January 1988 another strike spread through the Yorkshire coalfield when three development workers were redeployed by the manager at Bentley colliery for allegedly making insufficient progress. Management had suspended the conciliation machinery so disputes were no longer referred to a pit umpire. In anticipation of an assertion of managerial prerogative reinforced by the new disciplinary code, Bentley miners struck on 4 January and picketed out other Doncaster pits. The colliery manager wrote to all strikers warning that the dispute could only be resolved once the pit was working normally.[61] On 7 January, when twelve pits and 10,000 miners were involved, the South Yorkshire Panel voted to continue the unofficial action. The following day, with seventeen collieries in South Yorkshire and two in North Yorkshire at a standstill, the South Yorkshire BC Area Industrial Relations Officer sent a copy of the Bentley letter to every miner's home.[62] The spontaneous action and management's immediate direct appeal to the workforce surprised Area NUM officials, who feared BC would seek an injunction to stop secondary picketing. The stoppage demonstrated that the sanctity of the picket line was re-established in Yorkshire, but Bentley miners were persuaded to call off the action. A week later the Bentley day shift struck again when ten miners were threatened with dismissal after management had identified them from video tapes as flying pickets.[63]

7.2 Union of Democratic Mineworkers

The Notts miners were the centrepiece of NCB strategy during the strike since the continued production of coal in the area ensured that the NUM could not win the strike (see Chapter six). Similarly, divorcing the Notts Area from the

National Union was a major ambition of the NCB, enabling them to implement the new strategy after the strike. The formation of a union in competition with the NUM was, for the NCB, the most attractive means of attaining the objectives outlined in the 1973 Miron Report (see Chapter one). The back-to-work movement (see Chapters five and six) should be interpreted in this wider context, rather than being viewed simply as a mechanism for breaking the strike.

Origins of the UDM

In seeking to mobilise the Notts Area, the NEC wished to prevent members of the Union from defying decisions of the NEC or SDC through suspension or total exclusion from the Union.[64] In August the NEC called for the resignation of Jack Jones (Leicestershire) because of his 'participation in a well-publicised function with the chairman of the NCB and remarks attrributed to him concerning the NUM National President'.[65] The vote to exclude Jones had little effect upon the Notts Area NUM, whose leaders were determined *not* to support the strike. In the summer the Notts membership elected thirty-one delegates, twenty-seven of whom were opposed to the strike. Anticipating NUM action, MacGregor and Moses had assured working miners that exclusion from the Union would not mean loss of employment, thereby announcing the end of the closed shop agreement. The obvious support and encouragement of management increased the confidence of working miners to pursue independent policies which would maintain production in Nottinghamshire and weaken the control of the National Union. Adeney and Lloyd viewed the passage of Rule 51 (see Chapter two) as 'the most potent recruiting agent for the working miners they were to have'.[66] The rule was intended to provide a means for disciplining scab leaders: 'the working men of Nottinghamshire were going to be targets of this new "star chamber court" – especially Roy Lynk'.[67]

The Notts Area responded by withdrawing from its members who were on strike all financial support 'in respect of legal fees and fines incurred whilst engaged on ... the pursuance of the current industrial action'.[68] The NUM was then forced to support Notts strikers from national funds, with the help of Yorkshire Area in particular. Liptrott, Branch Secretary of Sherwood, told strikers that 'they had ceased to be financial members on the grounds that their contributions were in arrears for more than eight weeks'.[69] The NEC instructed Liptrott to write to each member informing them that they 'remained fully financial members of the Union', and that failure to execute this instruction would result in disciplinary action.[70] To avoid open warfare with the NEC at this stage, Liptrott agreed to implement the NEC directive, but sought legal advice.[71]

Notts working miners waited until the middle of December before confronting the NUM with their intention to delete, from the Area rule book, Rule 30 subordinating the Area rules to those of the National Union.[72] Faced with this threat to their autonomy, the NEC agreed: 'the NUM opposes the proposed changes of Rules currently being considered by the Nottinghamshire Area of the NUM because they constitute a threat both to the NUM and the proper conduct of business within other areas'.[73]

The National Union was denied an injunction preventing the rule change,[74] and Rule 30 was deleted at the Notts Area Council by a vote of twenty-nine delegates to two. Before this decision was taken, the Notts Area representatives refused to meet the NEC. Lynk was obviously prepared for more than just a bid to re-establish area autonomy: 'he said that he and his colleagues were ready to run a wholly autonomous union if they had to: "once you have started down a road, you have to be prepared to follow it through".[75] The NEC called an SDC to decide whether to exclude the Notts Area from the Union, to prepare a strategy to ensure that Notts miners who supported National Union policy remained part of the NUM, and to empower the NEC to take legal action to secure the funds and assets of the Union which were under the control of the former Notts Area. The NEC sought to prevent the Notts Area from splitting from the NUM, agreeing: 'that the Secretary communicate with the Notts Area seeking an urgent meeting between the national officials or the NEC in order to discuss the present situation and to seek a solution in the interests of the Union that will obviate the need for the matter to be referred to the Special Conference'.[76]

Notts Area Council endorsed the rule change and suspended Richardson for advising Notts miners to maintain NUM membership.[77] Richardson's suspension was important, as George Bolton argued: 'we've never had . . . any significant, powerful, organised left progressive leadership in the Notts coalfield . . . the left had just begun to emerge in Nottingham over the course of the last two years, just before the strike. But it tended to be at a certain level, at a branch leadership level, or in the shape of Henry Richardson'.[78] Lynk's appointment as General Secretary in place of Richardson exemplified the reassertion of right-wing dominance in Nottinghamshire.

The Notts Area Council voted to lift the overtime ban before the strike ended and, on 11 March 1985, Richardson was dismissed by the Area Executive. Although two days later Richardson obtained a High Court injunction against the Area Executive ordering them not to act on their decision, Justice Mann was not prepared to impose on a trade-union an official in whom the members had no confidence. The National Union refused to accept the Notts decision: 'Mr Richardson was and would continue to be an employee of the NUM and the Secretary had directed the NUM Nottingham Area to forthwith withdraw the purported notice of dismissal of Mr Richardson'.[79]

The High Court granted an injunction preventing the introduction of Rule 51, but the NEC continued with the revision of the rules. One objective was to introduce proportional representation to NEC voting, in line with the procedures of the national conference; this change would give the left a greater majority on the NEC.[80] South Derbyshire Area was guaranteed to lose individual representation on the NEC through the rule changes; Toon expressed his support for the Notts breakaway and announced that the Area was also proposing to abandon the National Union.[81] Scargill met the South Derbyshire Area Executive in an attempt to avert a further split and was thought to have promised South Derbyshire leaders that their NEC representation would be secure.[82] Nevertheless, the fifty-four-man Area Council voted unanimously to introduce the rule change, which was supported by a seventy per cent majority of the membership in a ballot.[83] Toon claimed that the implementation of the rule would depend upon whether the National Union attempted to discipline South Derbyshire.

Disciplinary action could only be implemented by Area Councils; in Durham at the time of the March SDC, 200 scabs had already been expelled and another forty were in the pipeline.[84] A breakaway Colliery Tradesmen and Allied Workers' Association (CTAWA) had been formed in Durham by George Hunter, who claimed 1,500 members.[85] There were also threats that COSA would leave the NUM if the proposed rule changes went ahead.[86] In an area ballot, seventy-five per cent supported mandating Notts officials to oppose the rule changes.[87] Before the ballot, area officials conducted a vigorous campaign against the proposed rule changes and prepared a document which included a solicitor's comments for circulation, both within Nottinghamshire and to other areas through contacts who were formerly involved in the Nat WMC. The document purported to explain the new rules but its major message was against the NUM and Scargill in particular.[88]

In May the NEC recommended that Lynk, 'who opposed Union policy', should be dismissed, and Prendergast reprimanded.[89] The July 1985 Annual Conference voted eighty-two per cent in support of dismissing both Lynk and Prendergast from the NUM payroll. The case against the two officials was that they pursued policy against the NUM *before* the area policy had been officially determined.[90] Lynk and Prendergast were still NUM members and area delegates to the NEC, but they were no longer to receive their £18,000 salaries and expenses from the National Union. Anticipating NUM policy, both had signed contracts with the Area NUM when first appointed in order to safeguard their positions.

On 4 July 1985 delegates voted by a seventy-five per cent majority to adopt the new rules and the Notts delegation was led out of the Conference by Lynk and Prendergast. Two days later Notts Area delegates voted by 228–20 to leave the NUM, which was unanimously approved at an emergency meeting of the Area Council.[91] Scargill urged Notts miners to stay within the Union; the NUM announced that all Notts members would remain members of the NUM unless they resigned and instituted legal proceedings to regain the ownership of the Berry Hill offices. Out of a total Notts NUM membership of 28,000, about 6,000 were estimated to have retained NUM membership.[92] Lynk held Scargill personally responsible for the break-up of the Union, while Walker told reporters that it was a great sadness for the labour movement when a union started disintegrating because of the behaviour of its President.[93]

Notts Area President, Chadburn, refused to lead a breakaway union and Lynk claimed that he would have to find another place to work.[94] Chadburn was locked out of his Mansfield office and protested: 'I haven't resigned; how can you quit something you don't belong to?'.[95] In August Notts Area Council voted to dismiss Chadburn on the grounds of 'gross misconduct', making reconciliation between the breakaway and loyalist groups in Notts more difficult. When Richardson and Notts loyalists sought an injunction to prevent the Area breaking away, Justice Stuart agreed that the Area was acting as a breakaway but ruled, in an inversion of the *status quo* principle, that Nottinghamshire should be beyond the jurisdiction of the NEC until a ballot could be held on the rule change. The Area therefore had legal independence and officials could campaign among the membership without competition from the National Union. Activists from Yorkshire and even loyalists within the Notts Area were prevented by the NCB from distributing pro-NUM literature at Notts collieries. In public Lynk requested the NCB to end the ban on NUM propaganda, thereby appearing to favour democratic debate and to be independent of the NCB. However, Lynk also sent a letter to branch officials instructing them not to 'co-operate in any way to arrange . . . meetings or support any action not in line with Nottingham Area policy'.[96] This directive effectively ruled out any possibility of NUM speakers addressing Notts Area branch meetings, because Rule 14(c) expressly forbids persons other than members attending a branch meeting without the permission of the Area Agent.[97]

Other areas considered joining the breakaway 'Nottinghamshire Union of Mineworkers'. In July Notts officials discussed with officials of sympathetic areas the establishment of a national federation to rival the NUM.[98] Toon discussed South Derbyshire's withdrawal: 'it is a sad day, but we have had enough', he told reporters.[99] Compared with Nottinghamshire, there was less support for a breakaway in South Derbyshire, where Area President Kevin Richards resigned because the Executive had moved to the right and no longer represented the members.[100] Breakaway groups from Northumberland and Durham met MacGregor with a view to seeking recognition,[101] and soon after the CTAWA discussed forming a federation with Nottinghamshire.[102] The white-collar area, COSA, also consulted its members on the same question. Scargill predicted that the breakaway Union would not be recognised by the LP or TUC.[103] After Kinnock called for miners to remain in the NUM at the Durham Gala, Lynk accused the Labour leader of being afraid to stand up to extremists: 'I would advise him that Notts miners represent the truer body of the labour movement and not the extremist'.[104] In October 1985 the breakaway areas formed the Union of Democratic Mineworkers (UDM).

The breakaway had serious implications for the LP because there were five marginal seats in Nottinghamshire; the constituencies of Sherwood, Notts N. and Notts E. being particularly problematic. The LP was in no position to expel UDM members if called upon to do so.[105] The UDM established a political fund at its inaugural conference, but there was no indication as to which party would be involved. Since the LP would not recognise the UDM and Kinnock refused to attend the conference, delegates doubted whether the funds would be used to support the LP.[106] In November Lynk announced that the UDM was running its own candidate in opposition to the candidate backed by the NUM, Alan Meale, who was labelled 'extremist' for his membership of the Campaign Group. The UDM had an impact upon the 1987 general election in Nottinghamshire, but it is doubtful whether its influence was decisive. The Conservatives held Sherwood, Notts E. and Notts S., while Labour gained Notts N. Meale held Mansfield for Labour by a majority of fifty-six, there having been a 2·37% swing from Labour to the Conservatives.

The TUC response to the breakaway was similar, its leaders urging Notts miners not to desert the NUM. The TUC sought an urgent meeting to discuss mineworkers' unity immediately after the formation of the UDM, recognising that the implications for the TUC were no less serious than for the LP.[107] Hunter, UDM Vice President, confirmed that discussions were underway with other TUC affiliates regarding the possibility of forming a breakaway TUC.[108] The TUC was clearly aware that the only future for an organised workforce would be through reuniting the UDM and NUM, so TUC leaders urged miners not to desert the NUM. However, discussions between the TUC and NUM in November reached no conclusions on how to deal with the UDM. An internal report to the TUC Finance and General Purposes Committee emphasised the majority support which the UDM had in Nottinghamshire and raised doubts over the NUM's ability to mount an effective campaign against the UDM.[109]

The NCB and the UDM

From the outset, relations between the NCB and the breakaway Union betrayed management objectives. Having eased the separation during the strike, management sought to maintain the division in the workforce, and use it to their advantage. Leicestershire Area leader Jones, who had frequently criticised Scargill during the dispute, confirmed that the breakaway was the result of long-term planning by the NCB. Cowan apparently attempted to persuade Jones to get the Leicestershire Area to join the breakaway:

> [Cowan] said, we need you for the final breakthrough in Warwickshire, including Coventry – and from there through Staffordshire up to Lancashire. If we get them he said, we have captured the motorway block . . .
>
> I could . . . be put in touch with Tyrrell, the Board's solicitors and all my troubles would be taken care of financially . . .
>
> The plans for the break away were all done, signed, sealed and delivered even before the NUM Conference took place – because there is no way they could have produced all that documentation, virtually signed and sealed by solicitors in that short time between the end of our Conference and them announcing they were setting up a breakaway. And the Coal Board is involved all the way in it.[110]

The Notts breakaway leaders sought recognition and sole bargaining in the area from the NCB on 8 July 1985. The NCB announced it would recognise the UDM twelve days before the Notts, South Derbyshire and CTAWA members in the North-East were due to vote on the formation of the UDM. Scargill warned that the NCB were clearly trying to influence the outcome of the ballot in order to divide the Union.[111] The day before the ballot the NCB offered miners a financial incentive to vote for the UDM with the promise of a pay increase for UDM members.[112] Before the result of the ballot was known, Lynk predicted 'we will be the new national union of mineworkers'.[113] In Nottinghamshire seventy-two per cent voted in favour of the new Union, while in South Derbyshire the split was carried by only twenty-six votes.[114] The following day the 1,600 members of the CTAWA voted to join the UDM and Notts Area Director Wheeler immediately indicated that the NCB would recognise only the UDM in the breakaway areas.[115] Further support for the UDM came from individual branches voting to secede from the NUM. Miners at Daw Mill and Agecroft, along with the COSA membership at Wearmouth, voted to join the UDM.[116] While the UDM was granted sole bargaining rights where it was the majority Union, in Durham, where it was the minority Union, the NCB nevertheless afforded it the representational rights denied the NUM in Nottinghamshire. Clearly NCB action was motivated by a desire to promote a *particular form* of trade-unionism and not merely by a wish to avoid the problems of multi-unionism at pit level.

The NCB used pay increases, both as a way of promoting the UDM and as a lever on the NUM. In 1985, the NCB offered an increase of 6·7% to the UDM, which the leadership recommended members to accept.[117] Individual UDM members who broke the strike before Christmas were to receive backdated payments of £100, which Bell claimed was a further attempt by the NCB to promote the UDM. The NCB argued that it was the NUM negotiators who had refused the agreement containing the 'good attendance' bonus.[118] The NUM leadership was under enormous pressure from the members to negotiate a pay increase because of the continued burden of strike debts. Scargill claimed that the NUM had a comparable offer from the NCB, but MacGregor denied this.[119] The NCB refused to settle unless the NUM would give a written guarantee that it would co-operate with productivity arrangements, and negotiations broke down after five hours.[120] Activists in Yorkshire feared that if the NCB held out against granting an increase to the NUM and offered the UDM terms to individuals joining the breakaway, a substantial number of miners would leave the NUM. In November the NEC refused to sanction preconditions for pay negotiations,[121] but the following month the NEC gave the NCB a written guarantee on incentives.[122] Negotiations proceeded but reached an impasse over the question of the deficit of £120m in contributions to the mineworkers' pension scheme as a result of the strike. The NCB proposed to reduce benefits for those whose contributions were in arrears owing to the strike, unless the Union agreed to members so affected paying an additional one per cent over five years. If agreement was secured by the end of April 1986, the offer was to have been backdated to November, but in the absence of agreement the NCB withdrew the offer on 1 May 1986. At the UDM Conference MacGregor warned that the NUM might have to accept lower pay rises than UDM members.[123] One of the first acts of the new Chairman was to offer the NUM the same basic increase as the UDM but, in the case of those who had been on strike, to withold the back-pay owing for 1985–86 to offset some of the deficits in pension contributions for 1984–85. The amount of back-pay received was to depend upon the number of weeks' strike-breaking, rewarding scabs once more. The NUM rejected the offer but BC nonetheless imposed it unilaterally from 1 September 1986.[124] After further negotiation and the threat of industrial action, the back-pay issue was resolved just before Christmas 1986.

In June 1987, BC imposed another pay settlement previously agreed with the UDM involving £4 per week to all grades backdated to November 1986, £1 on standard performance incentive pay and a further settlement payable from November 1987. The settlement included a quarterly 'attendance bonus', aimed at reducing absenteeism, and a 'conciliation bonus' to discourage unconstitutional strikes. Heathfield described the settlement as 'an interim payment on the 1986–87 wage claim'.[125] Negotiation had been delayed by NUM resistance to the new conciliation scheme, whereby representation was available only to the majority union in each colliery.

BC also used UDM acceptance of the principle of six-day working to apply more pressure to the NUM over Margam.[126] The problem was compounded by the admission of the TGWU in South Wales that they would organise Margam, with agreement to flexible working, in order to keep the UDM out.[127] Sole bargaining rights at the proposed new Hawkhurst Moor colliery near Coventry were granted to the UDM, both as the majority Union in the area and because of agreement to six-day working.[128] So long as the division between the UDM and NUM remains, it will continue to be fully exploited by BC management.

While the relative size of the UDM is less significant than the fact of the division, from the outset BC management sought to promote membership of the UDM. For example, management granted check-off facilities not just in Nottinghamshire and South Derbyshire, but in any colliery, enabling miners to subscribe on an individual basis.[129] Once the UDM won official certification, the NCB revised its negotiating procedures,[130] and in June 1986 the High Court ruled that the NUM no longer had an exclusive right to represent mine-workers: the 1946 agreement was no longer valid.[131] Government support for the UDM was incorporated in the Coal Industry Bill 1986, which was published on 14 November and gave the UDM statutory negotiating rights.

For miners who had remained loyal to the NUM in Nottinghamshire, the administration of area assets and funds by the UDM represented a serious financial penalty: 'People who have been in the industry for years . . . are losing every penny we ever paid in contributions because we won't go with the breakaway. Yet our lawyers say that there is very little we can do.'[132] This fear helped to maintain the level of UDM membership. Inertia also explains the continued membership of the UDM since it requires an effort to deregister and register, which 'has led to moderates staying with the UDM because it was easier'.[133] The NCB refuses to recognise NUM officials and 'keeps a "holding list" of miners who belong, as yet, to no union. Their names are held in job applications; only those who indicate a desire to join the UDM are taken on'.[134]

MacGregor was keen to promote the UDM; on his retirement he expressed the hope that his successor would 'retain that commitment' and not be 'seduced by arguments about bureaucratic "untidiness" caused by having two miners' unions'.[135] The evidence so far suggests that BC is no less enthusiastic under Haslam. For the Government, the UDM will inevitably make the transition from a nationalised to a private coal-mining industry smoother. It is therefore in the interests of the Government and BC to foster the divisions within the coalfields and maintain their hold on the high production areas controlled by the UDM.

Despite the support of BC management, there are indications that the UDM is having difficulty in maintaining membership, let alone increasing it. The total number of miners working in Nottinghamshire has declined since the strike; a national officer of the NUM estimated that there were 23,000 miners in Nottinghamshire in March 1987, compared with 30,000 at the beginning of the strike. Pit closures and the introduction of new technology have considerably reduced manpower in Nottinghamshire as in other areas. The NUM estimated in March 1987 that UDM membership in Nottinghamshire was as low as 17,000, because 6,000 retained their NUM membership. In pits like Ollerton, Welbeck and Clipstone, NUM membership was approaching fifty per cent, while isolated UDM members in other areas had, like the strike-breakers, mostly left the industry. From parliamentary questions asked of Coal Minister David Hunt by Terry Patchett (Lab., Barnsley E.), the NUM estimated in June 1987 that UDM membership in Nottinghamshire amounted to less than 15,000, with a maximum 3,000 further members in other areas.[136]

BC management are clearly concerned about declining UDM membership and appear to have doubts about the quality of UDM leadership: 'the feeling among senior officials developed that Lynk was not the man for the job, but he could scarcely be replaced. Hart, who continued to take a close interest in the UDM, thought that a man was wanted who could project a moral intensity as white-hot as Scargill's – but of course in defence of a different cause'.[137] BC may yet be embarrassed by the internal government of the UDM, which is at variance with its proclamation to be democratic. Six of the eleven seats on the UDM National Executive are occupied by three men. Lynk has one seat as National Secretary and another as Notts representative, Toon has one seat as President and another as Derbyshire representative, and Hunter has one seat as Vice-President and another as Durham representative. The three figureheads of the UDM continue to dominate it as all Executive members are elected for life, unlike the NUM Executive members who face re-election every five years. Democratic practices at branch level were also questioned when the UDM was formed: 'Former NUM branch president Tommy Gay is now UDM branch president. Former NUM Area delegate Gus Johnston is now UDM delegate. NUM secretary Dick Emery is now UDM secretary – and so on. And since it is the Union of *Democratic* Mineworkers, not one of them has yet been elected by anyone.'[138] The relationship which developed between the UDM and the EETPU caused rumours of a merger, although the leaders of both unions denied this.[139] It was irrelevant to BC management whether Notts miners belonged to the UDM or the EETPU, so long as they remained, in Miron's words 'outwith the NUM' (see Chapter one).

7.3 The politics of the NUM

The dramatic emergence of the UDM overshadowed other important developments within the NUM following the defeat of the miners' strike: 'Defeat . . . is . . . frequently accompanied by a loss of the organisation, including the polarity which was temporarily developed for corporate action . . . The defeated group . . . is subjected to peculiar strains; recriminations and dissensions threaten its integrity, if not its very existence.'[140] The 'strains and dissensions' were displayed in the disunity of the left leadership at all levels, and in the struggles of women's groups and rank-and-file organisations. Each is considered in turn.

Left leadership

The leaders of Yorkshire Area and of the National Union owed their positions to the campaigning of the Miners' Forum (see Chapter one). The Forum had operated in the late 1960s as a rank-and-file movement riding on a wave of wage militancy. Once the left was in power in Yorkshire, however, the role of the Yorkshire Miners' Forum altered from that of a rank-and-file ginger group to one of a co-ordinating caucus to maintain left control. When the AIS fragmented wage bargaining the Forum became moribund as a rank-and-file movement. In 1983 a miner who had formerly attended its meetings explained why he had grown sceptical of the Forum:

> It's become simply an election machine to be wheeled out when there's a vacancy . . . Because we've captured the key Area positions, because we've supposedly got a left-wing leadership, they argue there's no need for the sort of organisation they had in the 1960s. What they mean is they don't want a challenge from the rank-and-file. Now if you challenge the leadership you're going against the Left.[141]

In the mobilisation of the 1984 strike, the Yorkshire Left provided the communication network for left-wing Executive members and branch officials (see Chapter two). Throughout the strike, however, the Yorkshire Left failed to provide the sort of campaigning lead which had typified the Barnsley Forum because over the years when Forum members gained control at area level they had not recruited sufficient lay activists to maintain a rank-and-file foundation for the Left. Nationally, the Broad Left met occasionally in Birmingham during the strike to consider the latest position in the areas and maintain left unity within the NEC. The Communist Party miners' group also met during the strike. Such meetings were essential because: 'you could never discuss problems on the NEC; many had members who were working. There was no real debate of the issues in the NEC because of leaks . . . People went back to being a parochial union. The problem for the NUM is that it is a federated Union.'[142]

Throughout the strike, the tension between the area and national leaders, exacerbated by differences over strike strategy, opened divisions within the left. The divisions were most visible when South Wales Area leaders argued in favour of returning to work without agreement in opposition to other left leaders (see Chapter six). The break-up of the national Broad Left after the strike was apparent in the two left candidates nominated to succeed McGahey as Vice-President in March 1987. The Yorkshire Left promoted Sammy Thompson, while Eric Clarke was nominated by Eurocommunist South Wales and Scotland. In Yorkshire, Thompson was portrayed as *the* left candidate but Kellingley, where a substantial proportion of the workforce has Scottish ties, nominated Clarke after Jimmy Miller was brought from retirement in Scotland to address the membership. The distrust between areas even affected the judgement of the activists at pit level. Although Scargill appeared to favour Thompson, differences between the national and Yorkshire Area leadership were almost as great as those between Yorkshire and the two other major left areas. Yorkshire activists generally supported Thompson on the grounds that he was reliable (and would attract parochial votes as a Yorkshireman), even though some believed Clarke to be the more able. Thompson was elected Vice-President with fifty-seven per cent of the vote against Clarke's forty-two per cent which was seen as an endorsement of the policies promoted by Scargill.[143]

Towards the end of 1987, Scargill announced his intention to honour a manifesto commitment to stand for re-election. Scargill compaigned on resisting pit closures, opposing the new disciplinary code and flexible working, and bringing the Notts miners, but not UDM officials back into the NUM. The election was contested by North Yorkshire Area Agent John Walsh, who advocated 'new realism' policies of halting the overtime ban, negotiating over flexible working and reuniting the UDM with the NUM. Walsh enjoyed considerable media support and was conveniently able to negotiate substantial compensation for Ted Scott, transferred to Kellingley from Stillingfleet, only days before the election.[144] The UDM leadership openly campaigned for Walsh as the candidate to reunite the Union and it was rumoured that Thatcher had instructed Conservative MPs not to voice their support for fear of costing Walsh votes.[145] Scargill argued that the opposition of BC management and the Government to his re-election was the best reason miners had to vote for him. Inevitably Scargill's majority was less than his exceptional 1981 victory, and more like previous presidential elections; he received 53·8% of the votes cast against Walsh's 46·2% in a poll of about eighty per cent. A more substantial victory would have given greater legitimacy to the policies which the National Union was struggling to maintain in the face of area weaknesses and management hostility. Moreover, although left control of the National Union was confirmed, the narrowness of the margin conferred considerable credibility on the right-wing candidate to run for the secretaryship upon Heathfield's retirement.

Since the strike, the major issues confronting the Union have had different implications for the central areas and the peripheral ones; the voting division is now more often between central and peripheral, the basis of the separation of Yorkshire from Scotland and South Wales, than the traditional left-right divide. The future restructuring of the industry will inevitably alter the balance of power between the factions within the Union. Miron (see Chapter one) believed reductions in the number of miners were as likely to occur in moderate areas as in militant ones and argued: 'we do not want . . . an industry of fewer men in total if the control of these men is still wholly or mainly in activist hands'. However, our analysis of the restructuring suggests that the left areas will lose their majority in terms of membership. Including the UDM membership and counting Lancashire and North Derbyshire alongside the Midlands as moderate areas, the proportion of membership in the left-wing areas would fall from sixty per cent in 1984 to forty-nine per cent in 2000 under Scenario II (Labour) and forty-six per cent under Scenario III (Conservative). Yorkshire, however, will increase in strategic importance; from twenty-nine per cent in 1984 to thirty-six per cent in 2000 under Scenario II and forty-six per cent under Scenariio III. Under Conservative policies, Yorkshire would constitute the entire left membership by 2000 and would increase its proportion to fifty-eight per cent of the total by 2010. Aware of the direction, if not the scale, of these developments, Taylor has spoken of a relatively secure future for the Yorkshire Area, with the Union altering its emphasis to that of a 'quality Union', defending the interests of a small, highly-paid workforce.[146]

At pit level in Yorkshire, the left also lost prestige following the strike, particularly when key individuals like Davy Miller from Kellingley took redundancy. Similarly, when Steve Widdrington from Ricall, who had played a key role in the SCC, became a training officer for the Board, the left at Selby lost crediblity for a time. Nevertheless, throughout Yorkshire the strike was widely viewed as having politicised miners and their wives. Activists repeated the cliché that 'things will never be the same again' and predicted an increasing role for the women's groups and rank-and-file movements in campaigning and mobilising communities against pit closures.

Women's groups

The view that the strike had 'changed irrevocably women's consciousness of themselves, challenged male working-class culture and gender relations'[147] proved optimistic, and others had expected that, once the strike ended, the women's 'bargaining power in relation to men [would] no doubt diminish'.[148] Nevertheless, there was evidence of changing attitudes within the NUM,[149] and of empathy among NUM leaders towards women's rights.[150] Some groups therefore expected the relationships developed with union branches during the strike to continue.[151]

Most women's groups declared their intention to continue to meet, if only for social purposes, but the return-to-work and the need to clear debts imposed domestic burdens which militated against women's involvement: 'the women can't get out because the men are working over'.[152] Many women wanted life to return to 'normal', but a minority resisted the re-establishment of pre-strike gender relations.[153] The number of individual groups dwindled as they merged to remain viable. By January 1986, Sheffield WAPC had amalgamated with the remaining members of several smaller groups to form South Yorkshire WAPC, although Kiveton Park Women's Group continued to function independently.[154]

A major factor in the disintegration of the women's groups was the absence of a focus for activity: 'there doesn't seem to be any real purpose for the local groups'.[155] Groups often continued when there was an issue like the amnesty or threatened pit closure affecting the local community. In some groups, the women who remained active were primarily 'sacked miners' relations . . . or the women who were involved on the central committee'.[156] When the Thurcroft WAG stopped distributing food parcels to sacked miners 'the group . . . folded up without a real decision to this effect having been taken'.[157] In Nottinghamshire women's groups campaigned to keep the area within the NUM, organising meetings and distributing the *Notts Collier* and the *Miner* outside local pits.[158] Darfield Women's Group constituted the prime mover in the campaign against the colliery's closure, with Marsha Marshall from Barnsley Miners' Wives Action Group becoming Secretary of the Darfield Campaign.[159] Barnborough Miners' Wives Support Group continued to meet for educational weekends at Northern College and directed its energies to the support of *Silent Night* strikers.

The National Conference of Women's Organisations held in November 1984 in Chesterfield had established the following objectives:

1. to ensure victory to the NUM in their present struggle to prevent pit closures and protect mining communities for the future;
2. to further strengthen the organisation of women's groups which have been built up during the 1984 miners' strike;
3. to develop a relationship between the NUM and the women's organisations at all levels;
4. to campaign on issues which affect mining communities – particularly peace, jobs, health and education;
5. to promote and develop education for working class women;
6. to publicise all the activities of the National Women's Organisation at all levels.[160]

The objective of strengthening the links between the women's groups and the

NUM centred upon associate membership of the Union which would have allowed women to attend NUM conferences and participate in educational events, but would not have given speaking or voting rights.[161] Most Area Unions were less than enthusiastic:

> we were told that certain areas of the NUM threw up their hands in horror when they first heard about it and weren't prepared to look at the details of how or whether it could work . . . the thought of women actually being involved or actually part of the NUM was too much. Yet we *are* part of the NUM, we are *bound* to be part of the NUM.[162]

The Yorkshire Area amendment to the proposed rule changes, deleting associate membership, was carried by a two-thirds majority at the NUM Conference on 4 July 1985.

The 1985 Conference decision destroyed any hope that the women's movement within the coalfields would remain effective. Women felt betrayed by the Union, as Sheila Capstick put it: 'they want it both ways. They want us there when they need us and when they don't need us they want us to melt away'.[163] The rejection was also felt at local level: 'I felt used. I felt I'd served my purpose. I looked after them while they'd been on strike and now they'd gone back to work they won't need us any more'.[164] At the National WAPC Conference held in Sheffield on 17 August 1985, 468 women delegates hotly debated the NUM Conference decision. McGahey was optimistic about affiliation and told delegates: 'the Scottish Area had voted for associate membership, the Scottish Area would have it, national policy or not'.[165] The women therefore resolved to pursue affiliation on an area basis with those areas that had been in favour. The national WAPC office was maintained in NUM headquarters but the surviving women's groups remain constitutionally outside the Union. It was later agreed that the women could apply *individually* for honorary (non-voting) membership of the NUM at an annual cost of £10. Adeney and Lloyd explain the area unions' resistance to affiliation in terms of the fear that women's groups would 'dilute the Union and bring in outside political and sectional interest groups not subject to the discipline of the Union'.[166] For many area leaders a more serious risk was that the *national* allegiance of the women's groups would undermine area autonomy and provide the national leaders with an independent network for information and influence. The same consideration affected the response of area officials to the development of rank-and-file movements.

Rank-and-file organisations

The meetings of the left leadership had no direct impact upon the young activists for whom the 1984 strike was their first major involvement in the NUM. In the absence of an existing rank-and-file organisation, the activists were attracted to various left-wing groups seeking influence inside the NUM. Trotskyist splinter groups made most progress because the Communist Party, preoccupied with divisions between the pro-Moscow *Morning Star* supporters and the Eurocommunist *Marxism Today* intellectuals, was associated with the

Forum and 'union bureaucracy'. Militant, the Socialist Workers' Party, Workers' Revolutionary Party, Revolutionary Communist Party, Socialist Organiser, Socialist Challenge and Socialist Viewpoint gained some support in the Yorkshire coalfield during the strike. All these organisations provided resources to help the strike, but they failed to win a substantial following because their sectarian perspectives appeared more important than the priorities of the strike. The activists needed a network through which to organise local campaigns to maintain the strike. Unfortunately, such a network did not develop until the strike was already beginning to crack up.

A rank-and-file movement grew out of the immediate needs of activists seeking to hold the strike together at local level (see Chapter four). The membership of this group was fluid and included supporters of all political groupings found within the coalfield. Unfortunately, *Socialist Worker* inaccurately described the group as the 'reconstituted Barnsley Miners' Forum', prompting denunciations of Trotskyist infiltration from the Area leadership. The first leaflets which the group produced were published on the authority of a few branches, particularly Ferrymoor Riddings, Armthorpe, Kellingley and Ledston Luck, whose representatives attended most regularly. By February, however, when the group was more directly concerned with strike tactics than with promoting the NUM case, the leaflets gained the support of the Barnsley and South Yorkshire Panels. Yorkshire Area officials who had themselves used the panels as a springboard for the Forum offensive against the right wing, became increasingly alarmed at the role of the new rank-and-file group, despite reassurances from Area Executive members who had attended the group meetings.

The leafleting group met after the strike in the Junction at Barnsley and, to prevent the activist network from being dissipated, constituted the Miners' Campaign Group (MCG): 'a broad-based, non-sectarian body within the Union. We seek to strengthen the left of the Union and help, through educational meetings and other initiatives, to develop and carry out socialist policies in the best interests of the Union as a whole.'[167] The monthly MCG meetings only succeeded in holding together part of the core of Yorkshire activists. Only thirty people attended the average MCG meeting, one-third of the number who had regularly attended Northern College out of the 200 involved in the group over the last three months of the strike.

The first MCG meetings after the strike were concerned with developing a strategy 'to roll back the NCB offensive and make the Union strong once again'. Campaigning for the victimised miners and against the UDM were obvious priorities and the group became involved in organising resistance to the closures of Darfield Main and Kinsley Drift. The regular attendance of Clapham and Feickert from national office, however, raised fears among the area leadership that the MCG, like the women's groups, provided Scargill with an autonomous information network and a lever on Yorkshire Area. When the Broad Left met in December, Sammy Thompson accused Scargill of having:

backed a new, ultra-left, 'rank-and-file' miners' group in Barnsley hostile to the Yorkshire Area leadership, now publishing its own newspaper and responsible, according to the Yorkshiremen, for fights breaking out over the conduct of the recent ballot on political funds. Scargill, they said had spoken on the group's platform and was suspected of assisting them to organise: the group itself had few members in the pits and is backed by such Trotskyist groups as Socialist Challenge and Socialist Organiser. The accusation, in short, was that the NUM president was organising an alternative left in Yorkshire.[168]

After the strike most of the ultra-left groups lost the bulk of their following in the coalfield; their support had been largely conditional and instrumental. The WRP retained some following through its campaigning for an amnesty. The RCP attempted to maintain influence by organising against the UDM but lost credibility through advocating nuclear power. It was also RCP members who opposed the political fund and caused uproar at an MCG meeting. Militant attempted to gain influence in the local committees fighting closures and anonymously launched a National Miners' Broad Left in January 1986; a second conference was held in Chesterfield a year later.

A genuinely autonomous rank-and-file movement seldom arises because effective organisation requires an existing power base within the union. The three left groupings currently active within the Yorkshire coalfield are subject to different external influences seeking to promote their own interests. The Yorkshire Left now represents a left-wing LP and Communist caucus of officials, lacking the rank-and-file involvement that typified the Forum. The Miners' Campaign Group is a broadly Trotskyist gathering of younger activists heavily influenced by national officials. The Miners' Broad Left amounts to a front organisation for Militant. None of these groupings has succeeded in keeping together the activists, the majority of whom have returned to the Labourist politics so firmly rooted in the coalfield.

7.4 Independent review procedure

The strike was triggered by the proposed closure of twenty pits employing 20,000 miners, but the scale of the coal crisis is far more serious (see Chapter one). The Union's central strike demand and the objective prioritorised by the majority of strikers and activists, was for an end to colliery closures other than through exhaustion of reserves (see Chapter two). Negotiations eventually focused upon the terminology of the criteria and the procedure to be followed when closure was proposed (see Chapter six). Changes in the criteria, procedure and rates of closure are therefore important indicators of changing power relations in the aftermath of the strike.

Origins of the review procedure

Pit closures have such a dramatic effect upon coal communuties that the procedures for closing collieries have always been problematic. Under the Coal Industry Nationalisation Act 1946 (s.46), proposed closures were dealt with through consultative machinery at district level agreed by the CINCC. In 1949 a national level appeal was added to the procedure. Baldwin cites an example of resistance to the closure of uneconomic pits under this procedure as early as 1951.[169] Nevertheless, the rationalisation undertaken in the first decade of the nationalised industry, and much of the contraction from 1958, was achieved with this procedure. In 1967, before the second wave of closures affecting Yorkshire, a new Administative Procedure for Colliery Closures was negotiated as part of a review of existing procedural agreements. The new procedure entailed discussions at pit and area level, during which the Union was entitled to undertake a technical inspection and had access to information on past performance and future production plans. The national level of the NCB was responsible for all decisions on closure, partial closure and merger of collieries where more than fifty workers were involved, and was required to notify the National Union three months in advance of the date of proposed closure. The NUM could request a national meeting within one month of receiving notice of the proposed closure but the NCB was entitled to begin redeployment of the colliery workforce two months before the closure date. Effective union opposition could be pre-empted if the NCB delayed meeting until redeployment was underway, and there was no early warning system that would allow union involvement in alternative strategies to maintain colliery production. However, the NCB could not close a colliery until the national appeal had been heard and had a duty to notify proposed closures to the Regional Economic Planning Council, whose Chairman could advise deferral of closure on socio-economic grounds. In such cases, the DEA would take up the issue with the Ministry of Power and, if the closure was deferred, NCB losses were reimbursed under the Coal Industry Act 1967.

In September 1972 the three mining unions agreed to a new Colliery Review Procedure (CRP) for a quarterly review of performance of all collieries at area level in order to identify pits with particular problems and seek ways of resolving these. In a circular to area secretaries, Daly explained that the purpose of the CRP:

> is to improve results [but] . . . recognised that some pits would have to close, either through exhaustion or because of heavy losses or changing markets, and the area director . . . would have to explain that he could not justify keeping the colliery open.
>
> The Board hope that in such cases agreement will be reached locally that the colliery could not continue in production and the unions nationally would be informed of the decision.[170]

The new procedure required the Area Director to express concern over a particular colliery at a routine review meeting and to hold a reconvened meeting with special reference to that colliery three weeks later, at which area officials would be joined by branch officials. Where local agreement was not reached, the Area Director would report to the NCB one month after the initial review meeting that continued operation of the colliery could not be justified. The NCB would then inform the NUM nationally and the union(s) would be entitled to make a technical inspection and present arguments to a national appeal held two months after the routine review meeting. Two weeks later the NCB would 'consider the points made at the appeal meeting and decide whether to close and if so announce it will take place in three months' time'.

The new procedure was accepted by the NEC in December 1972 but Daly's circular prompted Denby Grange branch to send objections to Area Secretary Sydney Schofield that the acceptance of closure through heavy losses or market deficiencies was 'totally against Union policy'.[171] After discussion at Area Council and Executive, Scargill (acting General Secretary) wrote to Eaton, Area Director, explaining Yorkshire Area's 'total opposition to the new colliery review procedure being put in operation . . . I am afraid I shall not be able to confirm our attendance at the proposed meeting on 27 February 1973'.[172] On 27 February 1973 the first Barnsley Area colliery review meeting went ahead without the NUM attending. Scargill confirmed on 9 May that 'the Union's attitude on the colliery review procedure is still the same' and the NUM failed to attend both the second routine review on 17 May and a reconvened meeting over the closure of Barley Hall colliery. By refusing to participate in the procedure, the Area NUM was unable to argue against closures and Jack Leigh (acting General Secretary) informed Eaton that the Area Union 'is now in a position to accept meetings regarding colliery review procedures'.[173] Area officials attended the second reconvened colliery review meeting over Barley Hall pit on 22 August 1973 and participated fully in the procedure thereafter.

The major weakness of all the procedures, and the source of union objections, was the fact that the national appeal was a consultative meeting after which the NCB would take the final decision. The NUM proposed independent arbitration in 1967 and 1981 but the NCB rejected this as incompatible with their statutory responsibility for the efficient working of the coal-mining industry. The closures brought about in restructuring under *Plan for Coal* were facilitated by the CRP, even when the rate of closures accelerated in 1981. Clearly the CRP did not offer a forum for the NUM to mount effective opposition to the closures announced on 6 March 1984.

Modified Colliery Review Procedure

As part of the settlement of the dispute with NACODS which began in August 1984, the NCB agreed to give consideration to altering the existing colliery review procedure in the following respects:

(i) An early warning procedure where local members believe their colliery is deliberately, by various means, being run down as a prelude to closure investigations.

(ii) Inside the colliery review programme, greater availability to branch officials, when preparing a case to oppose closure, of time and facilities to information.

(iii) Any final notice of closure to be issued would allow national officers time to consult the membership.
(iv) Any final closure notice not agreed between the NCB and the Association to be subject to an appeal to an independent body.[174]

The NUM resisted signing an agreement to closure on economic grounds since this would have undermined the NACODS agreement (see Chapter six).

The strike itself had reduced coal production to such an extent that the cutbacks announced on 6 March were unnecessary a year later, at least in the short term. A number of colliery closures were attributed to problems which arose during the strike: Ackton Hall, Fryston and Frances collieries had suffered underground fires, while Polkemmet had flooded. The first spate of closures announced after the strike affected South Wales. Celynen South and Bedwas were each closed with the agreement of the branches, although NACODS initially voted to fight Bedwas through the review procedure.[175] The closures of St John's and Maerdy, the last pit in the Rhondda, were announced within a month.[176] The next month the NCB announced the closure of two collieries in the North-East, Bates and Horden, and of Whitburn workshops.[177]

Within a month of the end of the strike the NCB had announced twelve pit closures, which McNestry claimed was 'in breach of review procedures'.[178] A NACODS delegate conference discussed strike action over pit closures and a membership ballot voted sixty per cent in favour of an immediate overtime ban in May. The ban delayed the start of work at most collieries as safety work had to be carried out before production could begin. Kellingley NACODS walked out on 24 May because management had stopped all bonus payments to those involved in the overtime ban. In Northumberland, Union officials sought an injunction to prevent the NCB closing Bates colliery before it had been through the review procedure; management had offered redundancy which two-thirds of the workforce had accepted.[179] After meetings between NACODS officials and the NCB on 28 May and 4 June, the Association agreed to call off its ban immediately, McNestry claiming 'we are satisfied that the NCB have come back into line operating the existing colliery review procedure'.[180]

Two months after the strike ended, the first closures were announced in Yorkshire; over a few months fourteen individual units were to close and manpower was to be reduced at other sites. In Barnsley, Emley Moor and Darfield Main were to close, while Bullcliffe Woood and Calder Drift were to merge with Denby Grange, and North Gawber with Woolley.[181] In Doncaster, Area Director Tuke announced 2,500 jobs would be shed by the end of the financial year.[182] In North Yorkshire the closures of Glasshoughton and Savile had been agreed before the strike, and Ledston Luck was added on grounds of exhaustion.[183] In South Yorkshire the closures of Cortonwood and Brookhouse were announced, along with further cuts in manpower at three other collieries and the closure of Carcroft and Birdwell workshops.[184] A month later Fryston manpower was cut back owing to an underground fire and ultimately closure was agreed locally.[185] Cutbacks among craft workers at Yorkshire Main were followed by closure with the agreement of the branch membership.[186] Local agreement was also secured to the closure of Penrikyber and Abertillery.[187] The closure of Bold and Betteshanger, both of which had incurred continuing heavy losses according to the NCB, were the last to be announced before agreement was reached on the new review procedure.[188]

Once the NUM had called off the overtime ban, a CINCC meeting was arranged for 11 April 1985 which established a sub-committee to negotiate details of a Modified Colliery Review Procedure (MCRP). The difficulties experienced by NACODS led Scargill to claim after the June NEC meeting that the review procedure had been shown to be 'no more than a sham', and later the NUM accused the NCB of altering the procedure without the unions' agreement.[189] After eleven meetings, agreement on the MCRP was finally reached between the NCB and the three unions on 30 October 1985. The existing procedure relating to proposed closures was extended to include 'major decisions which would significantly shorten the prospective life of the colliery'. A final stage was added to the procedure: an Independent Review Body (IRB) presided over by a Chairman drawn from an agreed panel of QCs and retired judges. The IRB would meet at the Union's request, two months after the NCB's decision from the national meeting, consider evidence from the parties and present its report. Within two weeks of the IRB recommendation, the NCB would make its final decision, having given 'full weight to the view expressed by the IRB' and, if the decision was to proceed with closure, the date of final closure would be announced.

When the MCRP was agreed, it was reported that since the end of the strike sixteen pits had been closed.[190] However, the NCB had actually announced its intention to close or merge twenty-five collieries and four workshops in addition to those agreed before the strike, to undertake a review of five more pits in South Wales and to reduce manpower in other areas. The NCB had already obtained local agreement to the closure of nine of the collieries, so these would not be considered through the MCRP. Branch meetings voted not to oppose the closure of Brookhouse and Cortonwood after management announced that those applying for redundancy would lose money unless they were on notice by a specified date. Peter Rose, Brookhouse NUM Secretary, accused the NCB of 'industrial blackmail',[191] and an Area Executive member later publicly condemned the role of an Edlington official in precipitating the same result at Yorkshire Main.

The new procedure in operation

In the absence of a Union victory on the question of energy policy and not merely colliery economics, the MCRP could not provide a safety net for closures. If the proposed closure of one colliery was withdrawn without an increase in coal demand, another would be at risk as a result. Nevertheless, the

MCRP was viewed as a last-ditch defence against closures and the NUM proceeded to fight those cases where the membership had not accepted closure. The preparation of cases for the MCRP was given new impetus by speculation that the NCB intended to close fifty more pits with a loss of 50,000 jobs. The NCB denied that there was any substance to the claim,[192] but it almost certainly arose from leaks of the 'Moses Strategy' which entailed the maintenance of 90m tonnes per annum output up to 1990 (see Chapter one) and the elimination of 'hopelessly uneconomic pits'.

In the summer of 1985 Scargill had regularly referred to the need for further industrial action to prevent pit closures.[193] Towards the end of July, a meeting was convened at NUM headquarters involving economists sympathetic to the union who would prepare the economic and social cases against closure to be taken through the MCRP. Those involved worked closely with the Union's mining engineers and NUM officials at area and branch level. At the first meeting of the economists group, Scargill argued that if the IRB recommended keeping a colliery open but the NCB went ahead with closure, this would provide the basis for mobilising the membership in strike action again. The same argument was repeated by Taylor to those of us preparing cases for Yorkshire collieries. At pit level the majority of miners did not seriously contemplate widespread action against pit closures; the industrial battle had been lost and those who were prepared to fight closure saw the IRB as an alternative 'political' arena.

The first three pits expected to be considered under the MCRP were St John's, Horden and Bates. Despite a publicity campaign which included an unofficial public inquiry, a branch vote agreed to allow St John's to close. Horden was considered at the first IRB hearing on 14–16 January 1986. Stuart Shields QC upheld the NCB's decision to close Horden on economic grounds and could not consider the social consequences since it only had potential for four to five years' production. At the next IRB hearing on 21–23 January 1986, Peter Bowsher QC accepted that Bates colliery would not be profitable but could, on the basis of the NUM's production plan, make a contribution to overheads from the third year. As the closure of Bates would sterilise substantial reserves of coal and have grave socio-economic consequences, Bowsher recommended that the NUM plan should be evaluated over a two-year period and, if proved viable, the pit should remain open. The NCB rejected the IRB recommendation and chose to announce their decision on 20 February, just fifteen minutes before Orme was to address the first conference of the Coalfield Communities Campaign, causing the Shadow Energy Secretary to alter his speech hastily. A week later Justice MacPherson ruled that the injunction preventing the NCB from preparing the run-down of Bates should be lifted because the NUM was unable to give an undertaking to meet any financial loss the NCB might incur as a result: 'full weight must be given by the NCB to the view or views of the IRB and to the whole context of the report. Nothing more than that can be or is required. The IRB report is not binding in any way.' The NCB response to the IRB over Bates and the MacPherson judgement reaffirmed that the MCRP was inadequate. MacGregor believed NACODS officials 'thought they had won a form of independent arbitration of our actions on pit closures, only clothed in other words . . . this was not the case'.[194]

The merger of Eppleton into the Hawthorne complex was considered by the IRB on 4–5 March 1986. Anthony Diamond QC supported the merger on the grounds that otherwise the whole complex, rather than simply Eppleton surface, could be closed. On 6–7 May 1986 the IRB recommended that Polkemmet, which the NCB had allowed to flood during the strike and effectively to close, should not be reopened. The cost of dewatering Polkemmet, £18–20m, was not justified for the potential seven years of reserves left in the colliery, particularly given that the pit was unlikely to regain its Ravenscraig market. On 10–12 June 1986 the IRB upheld the NCB's decision to close Kinsley Drift. Lord Fraser rejected the Union's argument that major investment of recent years had been directed at reserves amounting to 12m tonnes. Quite apart from the low quality of coal from Kinsley in recent months, geological factors made future mining impracticable. The closure of Cadeby colliery was considered on 9–11 September 1986 by the IRB. Anthony Diamond QC concluded that closure on strict financial grounds was reasonable as there would be substantial losses in mining the 8m tonnes of reserves. However, BC had a moral obligation to reconsider its decision in view of the wider socio-economic consequences of closing a colliery in an area of high unemployment. In November BC announced its intention to close Cadeby against the advice of the IRB.[195]

Bates and Cadeby were the only cases of the six considered by the IRB in which the Union's position was supported. In both cases BC chose to disregard the advice of the IRB. Three other proposed closures were withdrawn before the IRB stage. The closure of Betteshanger was announced on 18 October 1985 but it was reprieved on 11 February 1986 on the understanding that productivity would be raised at the pit and neighbouring Snowdown. In June 1986 the future of Tilmanstone was similarly described as being dependent upon productivity increases.[196] In October 1986, BC announced its intention to close the pit and a threat of industrial action proved irrelevant when deteriorating face conditions made closure inevitable.[197] BC intended the location of closure in Kent to be determined by the extent to which workers would subordinate their activities to managerial goals of increased productivity, although geology played a more important role in the final decision.

A similar trade-off occurred between collieries under threat in Yorkshire. From the time the closure of Darfield was proposed on 16 May 1985 to its discussion at national appeal on 7 November 1985, the three other Yorkshire contenders for IRB hearings, Cortonwood, Brookhouse and Yorkshire Main, had agreed to closure. During the week before Darfield was considered at national appeal, Tuke implied to the Area Union that the pit would be reconsidered provided two others which were causing concern, Fryston and Kinsley Drift, were allowed to close. Fryston was not expected to resist closure because the average age of the workforce was high and face conditions were poor; although development rates were exceptional, Tuke argued that they were 'racing towards failure'. Kinsley was also not expected to resist because many miners were not locals from Fitzwilliam but transferees from Rockingham colliery who had been promised jobs near to Wombwell at Houghton Main and Darfield. Fryston agreed to closure on 26 November 1985, but the Kinsley membership voted to oppose closure and a campaign was run by the local branch and Miners' Campaign Group up to the IRB decision. Darfield was in a better situation than Kinsley in terms of the quality and quantity of reserves and mining conditions and a month after the Darfield national appeal the NCB reconsidered its position and agreed to keep the pit open.[198] Although celebrated as a victory for the NUM, the reprieve was on the basis of BACM's proposals for the pit's continued operation which entailed considerably lower manning levels than in the case prepared by the NUM. Moreover, the colliery was not kept open on the merits of the case prepared for the Union, because the pit was not taken to the IRB. Since Darfield was known to be a strong case, the NCB preferred to avoid the embarrassment of another case like Bates at the IRB, and to maintain production at the pit on management's terms.

Increasingly, pits have been closed without opposition as management persuaded miners to vote for immediate redundancy rather than the uncertainty and delay of the IRB. By December 1987, seventy-seven pits had closed since the start of the 1984–85 strike, and the 103 collieries in operation nationally included a number already destined for closure. In Yorkshire, Wheldale, Wath, Redbrook and Woolley were closed with agreement, Treeton was cut back and earmarked for closure and Manvers was put into the review procedure.

7.5 Coalfield Communities Campaign

The initiative to establish the Coalfield Communities Campaign (CCC) was taken while the miners' strike was still in progress. Local authorities were concerned at the detrimental effect upon local economies of the NCB pit closure programme. Jobs were being lost in areas where employment prospects were bleak and local authorities recognised that the Government had to be pressurised either into halting the pit closure programme or into generating alternative employment. Barnsley MBC circulated authorities in coal-mining areas to discuss the possibility of a co-ordinated campaign. The NUM dispute and the CCC were each prompted by NCB policy, but the independence of the CCC from the strike was established from the outset.

Local authority studies

Several authorities had undertaken studies on the economic importance of coal-mining to their localities. The report by Barnsley MBC, published a few months before the first CCC meeting, outlined changes in coal demand, the markets for the grades of coal produced in Barnsley, and the potential loss of jobs through the application of new technology.[199] In 1981, twenty-two per cent of those employed in the Borough worked in coal-mining, compared with thirty per cent in 1971 and a 1981 national figure of three per cent. Future job losses were expected to have a more serious impact because of high unemployment and changes in regional policy. As a primary industry, coal-mining was shown to have few secondary employment effects, perhaps fourteen jobs for every 100 miners. Expenditure multiplier effects were found to be more significant, representing twenty-one further jobs for every 100 miners. Despite the reconstruction of the Barnsley Area, coal-mining employment was expected to decline in importance, adding to serious regional unemployment and presenting challenges for the authority in attracting new employers and retraining miners.

Reports on the local effects of coal-mining decline had also been published by West Yorkshire MCC and Easington DC in December 1984. The report of WYMCC charted the decline of coal-mining in the county and predicted the loss of a further 2,500 jobs through effective exhaustion and 2,000 through closures on economic grounds, while other jobs could be lost through new technology. The report emphasised that industries in which employment nationally had grown since 1981 were underrepresented in the West Yorkshire coalfield. The environmental effects of coal-mining reduced the attractiveness of the area to prospective investors and there were 'no signs of generating new firms by indigenous initiatives in anything like sufficient numbers to make any impression on the problem'.[200] The WYMCC report called for more support for the coal industry to slow the decline but, accepting that decline was inevitable, emphasised the need for measures designed to promote alternative employment in the area. The major policy demands were for the restoration of Intermediate Area Status under UK regional economic policy (which the area had enjoyed from 1970 to 1982), the establishment of an EC Community Programme for coal-mining areas, and for support for vocational retraining from the European Social Fund.

The Easington DC report assessed the economic and social costs of further pit closures in Easington. Coal-mining decline was shown to have contributed most to the high unemployment in the district but the local labour market was still sensitive to further losses of mining jobs. The mobility of redundant mineworkers was limited by their coal-specific skills and the age distribution of the workforce. The Easington report estimated direct costs of £65m to be borne by local and central government and the EC, arising from the coal-mining job losses anticipated in the District Plan.[201] Indirect costs were estimated to be between at least £13·1m and £32·8m, representing a total quantifiable cost of between £78·1m and £97·8m.[202] Hudson and colleagues also drew attention to the social costs of pit closures:

closures strike deep into the social fabric of mining communities. There has been an intimate connection over a long period between the growth of coal mining and the development of communities and villages in Easington District . . . the loss of coal mining jobs and colliery closures removes the material basis around which the social relations of coal mining communities have been constructed and to which the culture and way of life in these villages crucially relate. Thus colliery closures . . . represent . . . most fundamentally a threat to a way of life in coal mining communities that was highly valued by many living in them.[203]

At the January meeting some representatives feared that co-ordinated action might have a detrimental effect upon local campaigns but the majority view was that the CCC could encourage individual initiatives and be enhanced by them. There was evidently more awareness among representatives of coalfield authorities of the effects of coal-mining job losses than of their causes. Hence the three reports produced independently were vague about predicted job losses and depended upon information released by NCB areas, and WERG's projections of job losses due to new technology.[204] The same was also true of reports produced in February by Derbyshire CC and in July by South Yorkshire CC since the background research was largely complete before the CCC came into existence.

Objects of the campaign

One of the present authors attended the January meeting and explained that the crises experienced by coal communities, while differing in degree and detail, derived from the NCB restructuring programme which introduced new capacity in the context of a shortfall of demand (see Chapter one). Only a national analysis of the restructuring, including detailed assessments of individual collieries, could produce realistic estimates of future coal-mining employment in the areas; such a study was volunteered and became the first CCC Working Paper.[205] The source of the crisis demanded a nationally-co-ordinated campaign to confront the issues in place of individual authority responses. While agreement was reached on strategic necessity for a national campaign, differences over the campaign's tactics and objectives were never fully resolved.

After agreeing to launch an eight-week campaign, tactical differences became apparent. Representatives from Chesterfield, Easington and Lothian argued that the campaign should be launched immediately to inject new energy into the miners' strike, since if the strike was lost there would be little prospect of reversing coal-mining decline. Others were not prepared to participate in a campaign before the resolution of the dispute and the prevailing view was that if the campaign began while the strike was in progress: 'Few of the target groups would be able to separate our arguments from those of the dispute. Many would pay no attention whatsoever to "yet another coal dispute story". We could expect little press coverage'.[206] The meeting agreed that the campaign would be formally launched at Westminster 'with a seminar opened by a senior statesman not directly associated with the current dispute'.

The meeting established a Steering Group which met the following month and noted that several authorities had reiterated their wish to see the CCC express support for the miners: 'There is . . . no chance the coalfield communities will unite in a campaign around the issue of the dispute. The NUM has itself been split geographically. It would be fatuous in the extreme to expect that the local authorities would be less so'.[207] Correspondence from the authorities involved demonstrated that tactical differences reflected underlying differences over the fundamental objectives of the campaign. These differences were apparent in the differing emphasis of the three reports published in 1984, and represented a source of continuing tension within the CCC.

Those authorities wishing to support the miners' dispute emphasised the need to make the economic case for maintaining the UK coal industry. Those wishing to distance the campaign from the miners argued that the NUM had adequately stated this case and that the campaign should 'concentrate on the knock-on effects of the decline in coal on the rest of the economy rather than on the coal industry's decline itself'.[208] Combining 'the participatory forces into a united group with a policy that all support' was the major challenge of the CCC, and it was recognised that 'any factional split will inevitably damage the campaign'.[209] To ensure the broadest possible participation, the CCC proposed short-term and long-term campaigns running simultaneously. Immediate publication of local studies, such as that undertaken for Easington, would be possible without requiring the participation of other member councils. However, the 'main purpose' of the CCC would be a long-term campaign to call upon the Government, the EEC and the NCB:

* to increase substantially national and local initiatives to maintain existing employment and create new jobs in the coalfield communities;
* to inject the extra resources necessary to overcome the environmental problems caused by coal mining;
* to recognise social as well as economic priorities in deciding government policies relating to the coal industry;
* to ensure a more just distribution of the EEC budget, giving rise to increased EEC resources devoted to coalfield communities;
* to increase marketing activities and assistance for coal use where technological advances now make it the most clean and economic fuel;
* to prepare a new energy White Paper which gives full consideration to the role of coal;
* to commit the EEC to an integrated energy policy.[210]

The campaign was directed primarily at the Government, so it was launched at Westminster with the aid of lobbying consultants on 15 May 1985. The first CCC Working Paper was presented at the national launch since its analysis of job losses both demonstrated the scale of the crisis facing the coalfield communities and emphasised the need to develop a co-ordinated strategy to ameliorate the economic and social consequences.[211] A large number of representatives from member authorities attended the media conference, but only seven MPs and a dozen media representatives were present, so media coverage was minimal. Simultaneously, six regional launches in Manchester, Matlock, Durham, Sterling, Barnsley and Merthyr Tydfil were held, which attracted considerable local media interest. The parliamentary presentation in the afternoon was attended by forty-five MPs of all parties and one member of the Lords.

Three further working papers were released during the eight-week campaign, each serving to highlight specific aspects of the coal crisis as it affected coalfield communities. Fothergill and Gudgin examined the structure and trends of employent in the coalfield areas to assess alternative job prospects in the light of the impending loss of coal-mining jobs. They concluded:

> other sectors – manufacturing and services – are no longer providing new jobs to offset those lost in the coal industry. Indeed, they are reinforcing the decline. In the 1980s we estimate that for every job lost in mining in coalfield areas an additional job has been lost in other sectors of the local economy. Manufacturing industry has been the main source of these other job losses. Looking to the future, there appears to be little prospect of a spontaneous upturn in the economy in coalfield areas to offset the job losses that are anticipated.[212]

Moreover, the scope for local economic initiatives was thought to be small because of the ineffectiveness of regional policy during a recession. The problems could only be tackled by a government prepared to implement a coal-based energy policy and to reflate the national economy, which would both stimulate coal demand and create the conditions under which regional policy could have an impact.

Wright argued that national energy policy would have a greater impact upon the future of coalfield communities than any local initiatives and criticised the failure of successive governments to introduce energy policies designed to optimise the use of existing resources in the long-term.[213] Short-term price movements had been allowed to dictate the contraction of production of coal, the primary fuel with most abundant reserves, in order to deplete more rapidly smaller reserves of oil and gas. In the fourth Working Paper, Hudson and colleagues outlined the socio-economic implications of coal-mining decline, which they had detailed earlier in the Easington report,[214] and argued that the NCB should take these broader issues into account when determining policy. In recognising that job losses in the coalfields have not been confined to coal-mining, they stressed the need for reindustrialisation to proceed with a 'coherent energy policy and a policy for communities'.

The CCC agreed to conclude the eight-week campaign with a visit of fifty representatives to the European Parliament on 11 July 1985. The British

Labour Group of MEPs had obtained a travel subsidy to fund the cost of specialists who were to present papers. The EEC proposed to remove all coal subsidies from 1 January 1986, so it was necessary for the CCC to brief MEPs 'to ensure that coal's place in an integrated European energy policy remains paramount'.[215] The visit to Luxembourg and meeting with Nik Mosar, EEC Energy Commissioner, demonstrated the gulf between the attitudes of the Commission and the CCC towards the coal industry.[216] Nevertheless, it was felt that the visit contributed to the coal subsidies debate and the eventual decision that these were to be maintained.

A long-term programme

In June the Steering Group recommended that the campaign be translated into a longer-term programme, since it was feared that 'the awareness of coalfield issues may not long survive if the campaign does not keep it well bolstered'.[217] The CCC was expected to facilitate communication between local authorities and to carry more weight than individual authorities with the Government, media and European institutions. The major purpose of the long-term campaign was to influence the policy of the present Government and future governments to give greater consideration to the economic and social issues facing the coalfield communities.

The Steering Group acknowledged that the working papers 'have had a definite impact' and anticipated that 'their usefulness will increase as the campaign progresses', so recommended the publication of the first four working papers in a single volume as part of the long-term campaign. All four papers had outlined aspects of the severity of the coal crisis and warned of the dangers of seeking palliatives in place of genuine solutions. To avoid the criticism that the CCC had been negative, the Steering Group asked academic advisers to prepare further papers addressing means of alleviating the problems. The second volume of working papers was therefore qualitatively different in focus from the first. Parkinson recommended that Coalfield Community Action Zones be established, within which strategies of change would be developed to improve the environment, encourage new enterprise and foster new skills.[218] Young argued that practical schemes could be introduced within the framework of existing policies but that because of the scale of the problem: 'piecemeal, ad hoc responses will be totally inadequate and seen as tokenism. What is needed is Government commitment in the form of extra financial resources, particularly over the next three financial years, and a changed policy framework'.[219] The other two papers in the second volume assessed the potential for expanding coal demand in manufacturing industry and public services. Gibson reviewed new and traditional industrial market prospects in the light of new coal-burning technologies which improve efficiency and reduce pollution.[220] Christian illustrated the economic advantages of a solid fuel policy for Doncaster MBC and argued for the more widespread adoption of coal-firing by local authorities.[221]

Representatives of the CCC had met Walker at Thatcher's invitation during the eight-week campaign and found him willing to maintain a dialogue, although he shared few of the views expressed by the CCC.[222] Young believed that Walker would, nevertheless, be receptive to suggestions relating to NCB (Enterprise) Ltd. Six months into the strike, before the CCC was organised, the Government had established NCB (Enterprise) Ltd in order to demonstrate its concern to provide alternative employment in the coalfield areas.[223] Although there was little evidence that NCB(E) could compensate for the massive job losses planned in coal-mining, the CCC was obliged not to be dismissive of any initiative, particularly if they wished to inflence the Energy Secretary.

Merrick Spanton Chairman of NCB(E) was invited to address a general meeting and explained to CCC members that the role of NCB(E) was:

1. to create alternative employment for men put out of work;
2. to create alternative employment for young men who ought to have come into the coal industry;
3. to create jobs and prosperity for the whole community.[224]

These objectives had been pursued through developing MSC retraining programmes, assisting established enterprise agencies, and providing loans of up to twenty-five per cent of the total finance required for ex-miners and others in coalfield areas for investment to create jobs. According to Spanton, between October 1984 and September 1985, NCB(E) had contributed £3.3m to 153 projects creating 2,379 jobs; the cost of each job was estimated to be £8,500. The aim of NCB(E) was to provide 500 new jobs per month over five years to compensate for job losses in the coal industry, Spanton told the CCC. In the debate which followed Spanton's address, many members expressed grave doubts that NCB(E) could provide sufficient jobs to compensate for the decline of coal. Spanton explained that NCB(E) was not aimed specifically at areas facing most closures and emphasised that it was no more than a 'partial response to regional problems'.

The changing emphasis of the CCC was apparent in the first national conference held in Wakefield on 19–21 February 1986. National CCC chairman Councillor Hedley Salt reiterated that the campaign was not an attempt to rerun the recent dispute, and emphasised the 'possibilities' for the coalfield communities. Dr Karlheinz Reichert, head of the European Coal Directorate, outlined EEC energy objectives for 1995 which included a 'twenty-five per cent improvement in energy efficiency', the continued 'restructuring of . . . the coal industry', and the production of forty per cent of electricity from nuclear power.[225] Coal Minister David Hunt pursued two themes: performance since the strike and the record of NCB(E). In Hunt's view, the 'outstanding improvements in attitude and productivity throughout the British coalfields' indicated that miners wanted to put the 'tragic strike' behind them and help promote 'an expanding competitive coal industry'.[226] Hunt also claimed that NCB(E) was attaining its objective of creating 500 jobs from investing £750,000 per month. Professor Brian McCormick argued the case for a regional employment premium, taxing employment in the South-East to subsidise employment in the North. McCormick also condemned the conservation of mining, on the grounds that coal-mining had damaged the ability of coal areas to attract capital *and* kept wages artifically low.[227] Orme identified three priorities for a future coal industry: its continued public ownership, investment in coal and manufacturing industries, and developing new markets for coal. A new *Plan for Coal* would be an essential part of a national energy policy:

> It is vital that our resources are used in a sensible way, looking at long-term sufficiency rather than short-term gain. It is crucial that exploration and use of our energy resources are planned on a national basis so that their potential is maximized and availability sustained. They should never be left to the uninhibited market forces which can operate with little concern for Britain's long-term national interest.[228]

Fothergill supported Orme's approach, and accused the Coal Minister of misleading the conference with his claim that there would be a high coal burn in the power stations: this was inconsistent with the fact that five nuclear power stations will be on-stream within the next four years, each reducing CEGB coal requirements by about 3m tonnes.[229] Moreover, Hunt had exaggerated the number of jobs created by NCB(E) which had: 'probably only assisted marginally in the creation of those jobs, if at all. The sort of figures that the Minister was quoting of £2,700 per job are simply unbelievable . . . the costs of creating jobs are more like £30,000 or £40,000 per job'.[230] Fothergill argued that local economic initiatives were ineffective; the problems of the coalfield communities could only be tackled through national policy. The essential elements of such a policy would be the reflation of the economy, a ban on using oil for generating electricity and on importing electricity from France, the removal of nuclear power stations from base-load operation, and a moratorium on new nuclear power stations.

In its first years the CCC had come full circle. The authorities had originally been brought together because of the similar local problems experienced in the coalfield communities. These local problems were shown to be the result of a coherent national strategy pursued by the Government and the NCB. Local initiatives to ameliorate the effects of national policy were found to be inadequate, so the solution to the coal crisis must entail a national energy policy. While the CCC continued to advocate local economic measures.[231] in addition to supporting a coal-based national energy policy, its emphasis altered from publicising the issues to contributing to policy formulation.[232] During its first year the CCC had addressed eight Parliamentary Back-Bench and Select Committees, but the major contributions which the CCC made to policy formation were through evidence to the Energy Committee and through publicising Fothergill's arguments which influenced LP energy policy (see below).

In November 1985 the CCC was invited to submit evidence to the Energy Committee inquiry into the coal industry. The first CCC Memorandum outlined the responsibilities of the Government, the NCB and local authorities to the coalfield communities, and argued that the social costs of coal-mining decline had been inadequately met by both the NCB and the Government.[233] The Memorandum called upon the Government to formulate 'a sensible energy policy which gives appropriate emphasis to the use of coal' and to give a higher priority to measures to alleviate the effects of job losses in the coalfield areas. When the CCC representatives presented their oral evidence, the Energy Committee sought clarification on the social cost-benefits of pit closures, the difference between colliery and factory closures, and the priorities of an energy policy.[234] The supplementary questions provided the CCC with an opportunity to make the social and economic case against pit closures and to reiterate the sterilisation arguments made in our Memorandum.[235] The CCC was also able to outline the major considerations of an energy policy, but did not use the opportunity to present the logic of a coal-based energy strategy for the UK.[236]

7.6 Prospects for coal

The Coalfield Communities Campaign demonstrated that the future of the coal industry could not be secured through local authority campaigning alone. The authorities could promote coal locally through coal-firing in public facilities, but, as Wright made clear, these moves are likely to be 'frustrated if they run counter to national trends in energy use'.[237]

The NUM's manifesto for the 1945 general election had proclaimed that 'the future of the mining industry is one of the main issues in this election'.[238] The same could have been said of the 1987 election, which represented the last opportunity for voters to choose between nuclear power and coal. It is ironic for the miners that Labour lost the 1987 election when LP policy was more supportive of the coal industry than at any time since 1945. In this concluding section we contrast Labour's energy policy with the prospects for coal under the Conservative Government.

Labour's energy policy

The failure of the labour movement to mobilise behind the miners contributed to their defeat in the 1984–85 strike, yet, paradoxically, the debate prompted by the strike altered labour movement energy policies in favour of coal. In contrast with the lack of support shown by the PLP during the strike, and despite the powerful pro-nuclear lobby within the TUC, before the 1987 general election the LP had formulated a coal-based energy policy entailing the expansion of the nationalised industry. At the 1984 LP Conference the NEC reaffirmed support for a 'comprehensive, integrated energy policy,based on an expanding and healthy coal industry'.[239] Composite Motion 64 (moved by Scargill) also included a declaration that 'the manifesto of the next Labour Government will reaffirm whole-hearted commitment to the *Plan for Coal* and introduce an integrated energy policy based on an expanding and healthy coal industry'.[240] Both the NEC Statement and Composite Motion 64 were carried, but Composite Motion 65 (moved by Jim Slater of NUS) calling for a freeze on further nuclear power stations on environmental grounds, was lost by 2,967,000 to 3,483,000 votes.[241]

At the first NUM Annual Conference after the strike, Orme invited the mining unions to meet LP leaders to develop a *Plan for Coal* which a Labour Government 'could start to implement from day one'.[242] Although Orme warned the miners that they had to work *with* the party, the proposal had Kinnock's 'full enthusiastic support' and was aimed at guaranteeing a 'bright expanding future' for the coal industry. At the 1985 LP Conference, a motion calling for nuclear power to be phased out was lost, despite the impressive performance of the SERA Chairman from Islington CLP.[243] Discussions between the LP and NUM leaders took place over the next twelve months, during which time the NUM stressed the need for a new *Plan for Coal* to be 'located within the context of a Socialist energy policy' where competition between fuels is ended.[244] The conflict within the labour movement over nuclear power resurfaced at the 1986 TUC, where a resolution from the FBU, NUS and NUM, calling for the closure of existing nuclear power stations was narrowly defeated by 4,581,000 votes to 4,641,000.[245] The EETPU and other unions with a significant membership employed in nuclear power stations naturally sought to protect their members' jobs. Within the LP there was the analogous problem of five marginal Labour constituencies which contained nuclear power stations. Despite these difficulties, the 1986 LP Conference passed a motion calling for a run-down of nuclear power in five years, but this did not attain the two-thirds majority necessary to make it a manifesto commitment. A motion from the Executive, however, was overwhelmingly supported and committed the party to halting the nuclear power programme and to phasing out all existing nuclear plants before 2000.

Following the 1986 LP Conference, the party's energy policy was developed for publication in 1987.[246] The energy sector was seen to be crucial for LP economic strategy, less for the employment within the energy sector than for its role in 'relaxing the major constraints on the level of activity and employment in the economy as a whole'.[247] Cheap energy was to have been an essential part of stimulating the recovery of manufacturing industry and of providing for peoples' needs. Labour's policy reaffirmed that coal, gas and electricity must be socially owned and that the LP intended to 'take British Gas back into full social ownership' and to 'reacquire a strategic stake in the oil industry'.[248] The energy policy was also designed to improve the balance of payments deficit by restraining energy imports; this implied the use of coal wherever possible in place of oil, as well as a depletion policy for the nation's gas and oil reserves. The NUM argued for a depletion policy for coal because BC strategy was sterilising vast reserves; the LP therefore promised a review of the industry's extraction policy: 'Labour . . . will not allow the rapid mining of the best seams if it leads to the sterilisation of other extractable reserves of coal. Labour understands that a non-nuclear energy policy requires us to ensure that coal is available as a resource for future generations.[249]

The most important objective for the NUM was to secure a non-nuclear energy policy to restore demand for coal from electricity generation. The commitment to phase out nuclear power by the end of the century represented a major change of LP policy that would have had a dramatic impact on the coal industry. Moreover, the non-nuclear stance was expected to become a major election issue following the Government's approval of Sizewell B.[250] The Magnox stations (which do not comply with modern safety standards) were to have been closed during Labour's first full term, but the seven new AGRs could not be closed until replacement capacity in the form of three large coal-fired stations was completed between 1995 and 2000.

Until the nuclear power stations were closed, new technology and other factors increasing the productivity of coal-mining would have continued to create surplus capacity. Total UK coal demand was expected to reach a low point of 105m tonnes in 1988 but a Labour government would have avoided the 1960s approach of pit closures and sterilisation during the period of stagnant coal demand. During the 1990s, demand deficiency for coal would have given way to supply deficiency as a result of the policies pursued by earlier Conservative governments. To avoid excessive demands being imposed upon the coal industry during the 1990s, energy conservation was viewed as essential.[251] The success of conservation measures would have determined the rate of closure of the Magnox stations, which would be 'kept under review in the light of emerging electricity demand and supplies from other plant'.[252] The NUM suggested that nuclear power station operation should be restricted to peak winter demand and left idle in the summer.[253]

Continuation of Conservative policies

The miners' defeat allowed Conservative policies to continue to erode the British coal-mining industry. The rate of colliery closures increased, regardless of the new Independent Review Body (see above). Within a year of the end of the strike, forty-seven of the 180 collieries which were operating in March 1984 were closed, while employment had fallen from 181,400 to 138,500 in the same period.[254] In May 1986 the CEGB announced that electricity price increases would be reduced as a result of an agreement with BC to obtain cheaper coal.[255] The pincer movement of CEGB monopsony and improvements in miners' productivity therefore continued to create surplus capacity and lead to more collieries being defined as 'uneconomic'.

Within two years of the end of the strike, Conservative policies were endorsed by the Energy Committee in its report on *The Coal Industry*, published on 24 February 1987. As a result of their inquiry, the Committee believed: 'it would be unrealistic to ignore the possibility that the rate of capacity reduction could be in excess of 2m tonnes each year . . . Together with the steady further improvement in productivity as best practice techniques are diffused more widely throughout the industry, these circumstances could result in substantial job losses throughout the rest of the 1980s and into the 1990s.'[256] The Energy Committee accepted our employment projections undertaken for the CCC and for WYMCC, although they were reluctant to put an exact figure on the likely job losses, agreeing that 'a continuing haemorrhage of jobs from British coalmining is more than likely into the 1990s'.[257] While the Select Committee accepted that ownership of coal reserves should continue to be vested in the crown and BC should remain in the public sector 'for the immediate future', they also recommended providing greater opportunities for private coal undertakings by removing the size limitations for licensed mines.[258]

The more serious long-term threat of the continued expansion of nuclear power was confirmed shortly after the second anniversary of the end of the strike. Sir Frank Layfield's report following the Sizewell Inquiry recommended that the CEGB should be allowed to build a PWR on the Suffolk site. The Government endorsed the Sizewell recommendation, and the CEGB made a planning application in relation to Hinkley Point for the second of a family of six PWRs planned to be built by 2000.[259] The Government's intention to privatise the electricity supply industry may halt the nuclear programme, given the dubious record of nuclear power, but imported coal is expected to displace between 30m tonnes and 48·6m tonnes of IDMC, representing job losses of between 47,000 and 69,000.[260] Robinson and Sykes recommend the simultaneous privatisation of electricity and coal, claiming that between 50,000 and 75,000 more jobs must be shed from coal-mining by 1992 to make it internationally competitive.[261]

The Conservatives' 1987 election victory guaranteed that British coal-mining will continue to decline in accordance with Scenario III outlined in Chapter one. By 1990, we estimate that there will be sixty-nine operating collieries (or their equivalent in capacity terms), and 50,119 employees (see Table 10). Similar projections of job losses made during the strike were dismissed as exaggeration by the NCB. Miners may recall part of MacGregor's first letter to strikers:

> This is a strike which should never have happened. It is based on very serious misrepresentation and distortion of the facts. At great financial cost miners have supported the strike for fourteen weeks because your leaders have told you this:
>
> That the Coal Board is out to butcher the coal industry.
>
> That we plan to do away with 70,000 jobs.
>
> That we plan to close around 86 pits, leaving only 100 working collieries.

IF THESE THINGS WERE TRUE I WOULD NOT BLAME MINERS FOR GETTING ANGRY OR FOR BEING DEEPLY WORRIED. *BUT THESE THINGS ARE ABSOLUTELY UNTRUE.* I STATE THAT CATEGORICALLY AND SOLEMNLY. YOU HAVE BEEN DELIBERATELY MISLED.[262]

Three years later, it is clear that it was MacGregor who was misleading the miners. By the end of March 1987 the workforce had declined from a pre-strike level of 181,400 to 109,900, a loss of 71,500 jobs.[263]

By interpolating between the March 1984 figure and our 1990 projection, we anticipated 115,760 employees would be left in March 1987, whereas there were only 109,900. Far from exaggerating the scale of the job losses, therefore, our projections have slightly underestimated the rate of job contraction which has been achieved to date. The reduction of forty per cent of mining jobs in three years represents an annual contraction of sixteen per cent; a higher rate than experienced in the great contraction of the 1960s. Unless there is a change of government, or a dramatic change in Conservative policies, by the end of the century there will be fewer than 25,000 miners in the country, and virtually no coal-mining outside the central coalfield (see Table 12). The miners' defeat in the 1984–85 strike will not only have done irreparable damage to the labour movement, but will also have allowed the effective destruction of the British coal-mining industry.

Notes

Introduction

1 For a brief review of the books and pamphlets written to date, see J. Winterton, 'The miners' strike: lessons from the literature', *Industrial Tutor*, Vol. 4, Nos. 4–5, 1987, pp. 92–106.

2 E. T. Hiller, *The Strike: A Study in Collective Action*, Chicago, Ill., Chicago University Press, 1928.

3 J. Steuben, *Strike Strategy*, New York, Gaer, 1950.

4 J. McIlroy, *Strike!, How to Fight. How to Win*, London, Pluto, 1984.

5 N. H. Nie *et al.*, *Statistical Package for the Social Sciences*, New York, McGraw-Hill, 1970, Version 6, 1975.

6 J. Hartley, J. Kelly and N. Nicholson, *Steel Strike*, London, Batsford, 1983.

7 See A. Taylor, *The Politics of the Yorkshire Miners*, London,Croom Helm, 1984, pp. 166–70 for an account of the role of the Yorkshire Area NUM panels.

8 M. Bulmer, 'Sociological models of the mining community', *Sociological Review*, Vol. 23, No. 1, 1975, p. 71.

Chapter 1

1 Confidential Report to Ezra from Wilfred Miron, 6 December 1973.

2 S.Bradshaw, 'The impact of high technology on the politics of mining', *Listener*, 25 April 1985, pp. 2–4.

3 BBC TV, 13 December 1973.

4 *Sunday Express*, 6 January 1974.

5 Taylor, *op.cit.*, p. 249.

6 Conservative Research Department, *Economic and Industrial Situation*, Notes on Current Politics No. 1, London, Conservative Party, 21 January 1974, p. 10.

7 *The Times*, 29 January 1974.

8 V. L. Allen, *The Militancy of British Miners*, Shipley, Moor Press, 1981, pp. 238–9.

9 NCB, *Plan for Coal*, London, NCB, 1974.

10 Department of Energy, *Coal Industry Examination: Final Report 1974*, London, HMSO, 1974.

11 D. Ezra, *Prospects for Expansion in Coal*, Cadman Memorial Lecture, London, Royal Society, 17 November 1976; National Coal Board, *Plan 2000*, London, NCB, 1976.

12 Department of Energy, *Coal for the Future: Progress with 'Plan for Coal' and Prospects to the Year 2000*, London, HMSO, 1977.

13 Interview with Rt Hon Tony Benn MP, Westminster, 17 July 1985.

14 Department of Energy, *Energy Policy: A Consultative Document*, Green Paper, Cmnd. 7101, London, HMSO, 1978, p. 31.

15 R. Ormerod, 'Corporate planning and its use of O.R. in the NCB: a personal view', *Journal of the Operational Research Society*, Vol. 34, No. 6, 1983, p. 461.

16 A. D. W. Jones, J. J. Quinn and K. M. Woolley, *Evaluation of Coal-Winning*

Technology, London, Technical Change Centre, 1984, pp. 65–9.

17 A. Burns, M. Newby and J. Winterton, 'The restructuring of the British coal industry', *Cambridge Journal of Economics*, Vol. 9, No. 1, 1985 pp. 93–110.

18 D. Feickert, 'Of men and MINOS', *Computing Europe*, November 1979, pp. 30–1.

19 A. Burns *et al.*, 'The miners and new technology', *Industrial Relations Journal*, Vol. 14, No. 4, 1983, pp. 7–20.

20 K. W. Chandler, 'MINOS – a computer system for central control at collieries', Second International Conference on Centralised Control Systems, London, 20–3 March 1978.

21 E. Horton, 'Mining techniques in the 1980s', *Mining Engineer*, February 1983, pp. 451–5.

22 C. C. Cooper, 'Improving machine utilisation and reliability', COMMIT 82, Computer-based Mine Management Information Technology Exhibition and Symposium, Harrogate, 8–10 December 1982.

23 J. Winterton, 'Computerised coal: new technology in the mines', *Digging Deeper: Issues in the Miners' Strike*, ed. H. Beynon, London, Verso, 1985, p. 233.

24 J. Winterton, 'The politics of new technology in British coal mining', Conference of Socialist Economists, Sheffield, 10–12 July 1987.

25 Letter to Lawrence Daly from Owen Briscoe, 6 November 1981.

26 A. D. W. Jones *et al.*, *op. cit.*, pp. 47–9.

27 Sharlston face-worker.

28 J. Cleary, 'FIDO (Face Information Digested On-line) at Bold Colliery', *Mining Engineer*, November 1981, pp. 281–9.

29 A. Burns, M. Newby and J. Winterton, 'New technology and the restructuring of work in British coal mining', British Sociological Association Conference, University of Bradford, 2–5 April 1984.

30 C. Goodrich, *The Miner's Freedom*, Boston, Marshall Jones, 1925, p. 57.

31 A. Scargill, *Miners in the Eighties*, Barnsley, NUM (Yorkshire Area), 1981.

32 A. Burns *et al.*, *An Interim Assessment of MINOS*, Report No. 4, Working Environment Research Group, University of Bradford, 1982.

33 NUM, *The Miners and the Battle for Britain*, London, NUM, 1980, p. 41.

34 *The Times*, 23 August 1982.

35 Energy Committee, *Pit Closures*, Minutes of Evidence – National Union of Mineworkers, HC 80, London, HMSO, 25 November 1982, p. 25.

36 *Ibid*, p. 62.

37 Energy Committee, *Pit Closures*, Minutes of Evidence, National Coal Board, HC 80–i, London, HMSO, 1 December 1982, p. 123.

38 Energy Committee, *Pit Closures*, Report, HC 135, London, HMSO, 21 December 1982, pp. x, xv.

39 *Ibid.*, p. xx.

40 Monopolies and Mergers Commission, *National Coal Board: A report on the efficiency and costs in the development, production and supply of coal by the NCB*, Cmnd. 8920, London, HMSO, 1983, p. 54.

41 *Ibid.*, p. 367.

42 *Ibid.*, p. 168.

43 P. Minford and P. Kung, *The Costs and Benefits of Pit Closures*, Watford, Public Service Research Centre, 1984, p. 19.

44 P. Minford, 'Labour market equilibration in an open economy', *Oxford Economic*

Papers, November 1983, pp. 207–44.

45 P. Minford, *et al.*, 'The Liverpool macroeconomic model of the United Kingdom', *Economic Modelling*, January 1984, pp. 24–62.

46 B. Robinson, 'Economic background to the coal dispute', *Economic Outlook*, Vol. 9, No. 1, October 1984, p. 30.

47 *Ibid.*, p. 31.

48 B. Robinson, 'How large a coal industry?', *Economic Outlook*, Vol. 9, No. 3, December 1984, p. 4.

49 *Ibid.*, p. 2.

50 G. Davies and D. Metcalf, *Pit Closures: Some Economics*, Weekend World, London Weekend Television, 1984.

51 J. Lloyd, *Understanding the Miners' Strike*, London, Fabian Society, 1985, p. 31.

52 Davies and Metcalf, *op. cit.*, p. 7.

53 A. Glyn, *The Economic Case Against Pit Closures*, Sheffield, NUM, 1984.

54 Energy Committee, *Pit Closures*, Minutes of Evidence – National Union of Mineworkers, *op. cit.*, p. 61.

55 *Ibid.*, p. 109.

56 *Ibid.*, p. 110.

57 Glyn, *op. cit.*, 1984, p. 2.

58 *Ibid*, p. 5.

59 *Ibid.*, p. 7.

60 T. Berry *et al.*, 'NCB accounts – a mine of mis-information?', *Accountancy*, Vol. 96, No. 1097, January 1985, p. 10.

61 D. Brewer, 'The role of the accountant in the mining industry', *Mining Engineer*, August 1985, pp. 53–7.

62 P. Wright, 'Bottomless pits?' *New Statesman*, 30 March 1984, pp. 12–13.

63 D. Cooper and T. Hopper (eds.), *Debating Coal Closures, Economic Calculation in the Coal Dispute 1984–5*, Cambridge, CUP, 1988.

64 K. O'Donnell, 'Brought to account: the NCB and the case for coal', *Capital and Class*, No. 26, 1985, pp. 105–23.

65 B. Towers, 'Posing larger questions: the British miners' strike of 1984–85', *Industrial Relations Journal*, Vol. 16, No. 2, 1985, p. 12.

66 Lloyd, *op. cit.*, p. 30.

67 O'Donnell, *op. cit.*, p. 110.

68 Towers, *op. cit.*, p. 12.

69 A. Glyn, *One Million Jobs a Year*, London, Verso, 1985.

70 O'Donnell, *op. cit.*, p. 109.

71 A.Burns, M. Newby and J. Winterton, *Second Report on MINOS*, Report No. 6, Working Environment Research Group, University of Bradford, 1984.

72 *Diverse Reports*, Channel 4 TV, 28 November 1984.

73 E. F. Schumacher, 'Coal – the next fifty years', NUM Study Conference, Britain's Coal, 25–6 March 1960, London, NUM, pp. 6–17.

74 Political and Economic Planning, *op. cit.*, 1966. p. 139.

75 *Ibid.*, p. 140.

76 M. Barratt Brown, *What Really Happened to the Coal Industry?*, Nottingham, Institute for Workers' Control, 1972, p. 12.

77 H. Aage, 'Economic arguments on the self-sufficiency of natural resources', *Cambridge Journal of Economics*, Vol. 8, 1984, p. 105.

78 Monopolies and Mergers Commission, *op. cit.*, p. 167.

79 Glyn, *op. cit.*, 1984, p. 12

80 T. Cutler *et al.*, *The Aberystwyth Report on Coal*, Economic History Department, University College of Wales, 1985, p. 50.

81 A. Goodlad, 'The appropriate energy mix for Britain', Town and Country Planning Association Conference, A Responsible Energy Policy for Britain – The Choices, Bristol, 15–16 October 1985.

82 NCB, *op. cit.*, 1974.

83 CEGB, *Corporate Plan*, London, CEGB, 1978.

84 Department of Energy, *op. cit.*, 1977.

85 Department of Energy, *Proof of Evidence for Sizewell 'B' Enquiry*, London, HMSO, 1982.

86 Trades Union Congress, *Review of Energy Policy*, London, TUC, 1981.

87 Minutes of 13th Meeting of Ministerial Committee on Economic Strategy, Cabinet E(79), Tuesday 23 October 1979.

88 *New Statesman*, 6 August 1982, 7 January 1983.

89 C. Sweet, *The Price of Nuclear Power*, London, Heinemann, 1983.

90 P. Bunyard, *Nuclear Britain*, London, New English Library, 1981; R. Curtis, E. Hogan and S. Horowitz, *Nuclear Lessons*, Harrisburg, Pa., Stackpole Books, 1980.

91 S. Fothergill, 'Nuclear power and the coal industry', Coalfield Communities Campaign Working Paper No. 10, Vol. 3, Barnsley, CCC, 1985, p. 30.

92 CEGB, *Annual Report and Accounts 1984/85*, London, CEGB, 1985, p. 44.

93 G. Manners, 'The demand for British coal', Town and Country Planning Association Conference, Economic Prospects for the Coalfields: A Case Study of the Yorkshire Coalfield, Sheffield, 22 November 1985.

94 Electricity Council, *Medium term development plan 1985–92*, London, HMSO, 1985.

95 *Guardian*, 28 April 1986.

96 K. Howells, 'British coal in the late 1980s', Study Conference on the British Coalfields, Northern College, 31 January–2 February 1986.

97 *First Tuesday*, ITV, 3 July 1984.

98 H. Radice, 'The national economy: a Keynesian myth?', *Capital and Class*, No. 22, 1984, p. 127.

99 P. James, *The Future of Coal*, London, Macmillan, 1982, p. 11.

100 P. Wright and I. D. Rutledge, 'Coal worldwide: the international context of the British miners' strike', *Cambridge Journal of Economics*, Vol. 9, No. 4, 1985, pp. 303–26.

101 P. Wright and I. D. Rutledge, 'The oil companies, world coal and the British miners' strike', *International Labour Reports*, No. 7, January–February 1985, pp. 7–8.

102 MMC, *op. cit.*, p. 233.

103 R. J. Loarridge, 'Contemporary technology in South African coal production and usage', B. Sc. dissertation, University of Bradford, 1986.

104 J. M. Griffin and V. J. Calarco, 'The global shift to lower rank, surface mineable coal', *The Coal Situation*, Chase Manhattan Bank, Vol. 5, No. 1, December 1984.

105 L. Turner, *Coal's Contribution to UK Self-Sufficiency*, London, Heinemann, 1984.

106 J. S. Jeans, 'The coal crisis and the paralysis of British industry', *Nineteenth Century*, Vol. 34, No. 201, November 1893, pp. 791–801.

107 MMC, *op. cit.*

108 G. McCloskey, 'The international coal market: comparison with British production costs and prices', Memorandum No. 82, Energy Committee, *The Coal Industry*, HC 165–I, London, HMSO, 1986, p. 381.

109 C. Robinson and E. Marshall, *Can Coal be Saved?*, Hobart Paper 105, London, IEA, 1985, p. 34.

110 Howells, *op. cit.*

111 G. Manners, *Coal in Britain: An Uncertain Future*, London, Allen and Unwin, 1981, p. 83.

112 Department of the Environment *et al.*, *Coal and the Environment*, Commission on Energy and the Environment [Ch. Flowers], London, HMSO, 1981.

113 MMC, *op. cit.*, p. 257.

114 Robinson and Marshall, *op. cit.*, 1985, p. 41.

115 *World in Action*, ITV, 29 October 1984.

116 D. Whitfield, 'Coal: a privatisation postponed?', *Capital and Class*, No. 20, Spring 1985 p. 11.

117 Department of the Environment *et al*, *op. cit.*

118 C. Sweet, 'Why coal is under attack: nuclear powers in the energy establishment', *Digging Deeper: Issues in the Miners' Strike*, ed. H. Beynon, London, Verso, 1985, pp. 207–9.

119 J. Winterton, 'The source of the crisis', Coalfield Communities Campaign, Working Paper No. 1, Vol. 1, Barnsley, CCC, 1985, pp. 1–7.

120 D. Ezra, 'A review of Plan for Coal', *Mining Engineer*, January 1981, pp. 479–87.

121 K. Moses, 'New mines – the way forward now', *Mining Engineer*, May 1982, pp. 641–5.

122 Manners, *op. cit.*, 1981.

123 Burns *et al.*, *Interim Report on MINOS*, *op. cit.*

124 P.Hills and D. Cope, 'Planning for an expanding coal industry', *The Planner*, Vol. 48, 1979, p. 173.

125 *Guardian*, 14 June 1982.

126 Moses, *op. cit.*, p. 642.

127 C. Shepherd, 'The reconstruction of the Barnsley Area of the NCB', Midlands Institute of Mining Engineers, January 1984.

128 *Barnsley Chronicle*, 30 December 1983.

129 Burns *et al.*, *op. cit.*, 1982.

130 Mining Department, *Production and Productivity Bulletin*, No. 16, March 1984, London, NCB.

131 *The Times*, 19 January 1981.

132 *Financial Times*, 28 July 1981.

133 *The Times*, 2 August 1982.

134 *Guardian*, 3 November 1982.

135 *Morning Star*, 16 June 1983.

136 NCB, 'NCB reply to the NUM claim for increased wages from November 1983', London, NCB.

137 NCB, *Report and Accounts 1983/84*, London, NCB, 1984.

138 Glyn, *op. cit.*, 1984.

139 Burns *et al.*, *Second Report on MINOS*, *op. cit.*, p. 24.

140 J. Winterton, 'The source of the crisis', *op. cit.*

141 NCB, 'New strategy for coal', Joint Policy Advisory Committee, Memorandum, 15 October 1985.

142 J. Winterton and R. Winterton, *Implications of NCB Restructuring Policies for West Yorkshire*, Wakefield, West Yorkshire Metropolitan County Council, 1985.

143 *First Tuesday*, ITV, 3 July 1984.

144 For 1984 the number of collieries and employment statistics are the latest available immediately before the strike, while the output figures are estimates of production in the absence of the overtime ban. Investment relates to the total estimated cost of major capital projects in progress in 1983–84. The number of collieries and output for 1990 is based on the lowest-cost, long-life collieries necessary to meet budgeted output following reorganisation. The maximum 1990 employment levels (and minimum job losses) include the effects of pit closures *only*, while the minimum 1990 employment levels (and maximum job losses) also include the effects of a further application of new technology. Original statistics from NCB *Report and Accounts 1983/84*, London, NCB, 1984.

145 Robinson and Marshall, *Can Coal be Saved?*, *op. cit.*, pp. 19–20.

146 Sweet, *op. cit.*, 1985, p. 205.

147 Whitfield, *op. cit.*, p. 6.

148 J. Moore, *Why Privatise?*, London, Conservative Political Centre, 1983.

149 NUM, *Campaign for Coal*, Sheffield, NUM, 1983.

150 *The Times*, 5 April 1984.

151 *The Times*, 5 March 1984.

152 *Ibid.*

153 Adam Smith Institute, *Energy Policy*, London, ASI, 1984.

154 MMC, *op. cit.*, p. 371.

155 *Financial Times*, 30 March 1984.

156 *Financial Times*, 6 December 1984.

157 *The Times*, editorial, 5 March 1985.

158 P. Rost (MP) and M. Pargeter, 'Pits, privatisation and politics', *Economic Affairs*, Vol. 15, No. 4, July–September 1985, p. 39.

159 Robinson and Marshall, *Can Coal be Saved?*, *op. cit.*, pp. 7–8.

160 *Ibid.*, p. 45.

161 *Ibid.*, p. 44.

162 *Financial Times*, 17 September 1984.

Chapter two

1 NEC Report, June 1980, p. 60.

2 Allen, *op. cit.*, 1981. p. 307.

3 NUM, *op. cit.*, 1980.

4 NEC Report, May 1981, p. 12.

5 *Ibid.*, p. 55.

6 Taylor, *op. cit.*, p. 282.

7 The wording of the Yorkshire ballot held in January 1981 was as follows:

Voting form – pit closures

As a direct consequence of the Government's economic policy, the National Coal Board are to implement a programme to close pits and make thousands of miners redundant. Already a substantial number of collieries in Yorkshire are threatened.

Over the past few years many pits have been closed, thousands of jobs lost, and job opportunities are now being denied to the sons of miners.

We urge every member to vote to give authority to the Union to take industrial action (including strike action, if necessary) to oppose the closure of any pit other than on grounds of exhaustion. Remember, the next pit scheduled for closure could be yours.

Are you in favour of giving the NUM Yorkshire Area authority to take various forms of Industrial action (including strike action, if necessary) to stop the closure of any pit, unless on the grounds of exhaustion?

8 'Statement to the Unions', from George Hayes, South Yorkshire Area Director, 18 February 1981.

9 J. Gormley, *Battered Cherub*, London, Hamish Hamilton, 1982, p. 177.

10 *Ibid.*, p. 181.

11 Interview with Allen McKay MP, Westminster, 15 July 1985.

12 Allen, *op. cit.*, 1981, p. 313.

13 South Yorkshire Area Colliery Review Meeting, Minutes, 23 April 1981.

14 *Daily Express*, 13 January 1982.

15 NEC Report, May 1982, p. 43.

16 The wording of the October 1982 ballot was as follows:

National ballot – pit closures–wages claim 28/29 October 1982

Are you in favour of the Special Conference recommendation that the National Executive Committee be given authority to take Industrial Action (if necessary) to prevent the closure or partial closure of any pit, plant or unit (other than on grounds of exhaustion) and at the same time bring about a satisfactory settlement of our wages claim?

17 *Observer*, 31 October 1982.

18 *Morning Star*, 8 November 1982.

19 Taylor, *op. cit.*, p. 285.

20 The status of collieries showed no significant correlation with the ballot results in 1981 (R = 0·01) or 1982 (R = 0·19). In Yorkshire the thirty per cent fall in the margin of support for action between the ballots of January 1981 and October 1982 was very uneven across the coalfield so there is no significant correlation (R = 0·14) between the 1981 and 1982 results on a pit-by-pit basis. The drop in support ranged form seventy per cent at Ricall (where only ninety-eight miners voted in 1982) to nine per cent at Rossington, while at Dodworth the margin actually increased by five per cent. At the majority of pits the drop was within the range twenty to forty per cent irrespective of their status, and four of the nine collieries where the vote fell less than this were known to be secure. However, five of the six pits where support fell by over forty-five per cent, were classed as secure: the four Selby pits and Denby Grange in Calderdale. The sixth colliery, Bullcliffe Wood, neighbouring Denby Grange, was expected to close but miners would transfer to the new Calder Drift. In addition to assumed job security, miners at these six collieries were on the periphery of the Yorkshire coalfield. By contrast, the vote held up best in the heart of the coalfield at the collieries in the southern half of the Barnsley Area and at the older western Doncaster pits. About half of the collieries in the heartland are as secure (or at least were regarded as being so before the strike) as those where support for action deteriorated most.

21 *Morning Star*, 12 January 1983.

22 C. Jones, 'Miners' strike called off', *New Statesman*, 14 January 1983.

23 *Guardian*, 24 February 1983.

24 *The Times*, 23 August 1982.

25 C. Jones, 'New year strike by miners', *New Statesman*, 17/24 December 1982.

26 *Scotsman*, 7 March 1983.

27 Triple Alliance, *What is the Future?*, London, Triple Alliance, 1980.

28 Triple Alliance, *Coal, Rail and Steel Fighting for the Future*, London, Triple Alliance, 1983.

29 MMC, *op. cit.*, p. 168.

30 *The Times*, 21 September 1983.

31 NUM, '1983 wage claim', Sheffield, NUM.

32 NCB, 'NCB reply to the NUM claim', *op. cit.*

33 *The Times*, 14 October 1983.

34 NUM, *Campaign for Coal*, *op. cit.*

35 *Financial Times*, 2 November 1983.

36 *Financial Times*, 8 November 1983.

37 *The Times*, 20, 21 November 1983.

38 *Daily Telegraph*, 28 November 1983.

39 *The Times*, 17 November 1983.

40 *The Times*, 21 December 1983.

41 *The Times*, 29 November 1983.

42 *Sunday Times*, 4 December 1983.

43 *The Times*, 21 December 1983.

44 *The Times*, 22 December 1983.

45 R. Ottey, *The Strike: An Insider's Story*, London, Sidgwick and Jackson, 1985, p. 55.

46 *The Times*, 24 January 1984.

47 *Sunday Times*, 15 January 1984; *The Times*, 16 January 1984.

48 *The Times*, 14 January 1984.

49 *Daily Telegraph*, 31 January 1984.

50 *Daily Telegraph*, 3 February 1984.

51 *The Times*, 24 February 1984.

52 J. Winterton and J. Rentoule, 'Confronting the new technology', *New Statesman*, 20 January 1984.

53 *Newsnight*, BBC TV, 16 February 1984.

54 NEC Minutes, 9 February 1984.

55 G. Goodman, *The Miners' Strike*, London, Pluto, 1985, p. 68.

56 R. Hyman, *Strikes*, London, Fontana, 1972, p. 115.

57 C. T. B. Smith *et al.*, *Strikes in Britain*, Department of Employment Manpower Paper No. 15, London, HMSO, 1978, p. 42.

58 The ranking of MacGregor as a cause correlated with the ranking of opposition to pit closures (R = 0·18), and revenge for 1974 correlated with avoidance of job losses (R = 0·14). The correlation between the objective of a new technology agreement and the importance of reorganisation (R = 0·34) and new technology (R = 0·40) as strike issues was the most significant.

59 The reaction of the branch meeting, ranging from unanimous support to threats of strike-breaking, correlated significantly with the percentage in favour of strike action in the 1982 ballot (R = 0·37), the drop in support between the 1981 and 1982 ballots (R = 0·22), but not with the status of the colliery (R = 0·16).

60 Rule 41 of the National Union prescribed that: 'In the event of a dispute arising in any Area or applying to the workers in any Branch likely or possible to lead to a stoppage of work or any other industrial action short of a strike the questions involved must be immediately reported by the appropriate official of the Area in question to the National Executive Committee which shall deal with the matter forthwith, and in no case shall a cessation of work or other form of industrial action short of a strike take place by the workers without the previous sanction of the National Executive Committee', *Rules, Model Rules, Standing Orders*, London, NUM, 1978.

61 BBC TV, 7 March 1984.

62 Rule 43 of the National Union prescribed that:
'In the event of national action being proposed by the Union in pursuance of any of the objects of the Union, the following provision shall apply:

That a national strike shall only be entered upon as the result of a ballot vote of the members taken in pursuance of a resolution of Conference, and a strike shall not be declared unless 55 per cent of those voting in the ballot vote in favour of such a strike. If a ballot vote be taken during the time a strike is in progress, a vote of 55 per cent of those taking part in the ballot shall be necessary to continue the strike', *Rules, Model Rules, Standing Orders*, London, NUM, 1978.

63 Ottey, *op. cit.*, pp. 60–9.

64 BBC TV, 10 March 1984.

65 H. Beynon, 'Introduction', *Digging Deeper: Issues in the Miners' Strike*, ed. H. Beynon, London, Verso, 1985, p. 7.

66 *Weekend World*, ITV, 11 March 1984; BBC TV, 31 March 1984.

67 ITV, 15 April 1984; *Sunday Times*, 15 April 1984.

68 Department of Employment, *Democracy in Trade Unions*, Green Paper, Cmnd. 8778, London, HMSO, 1983.

69 *Look North*, BBC TV, 20 March 1984; BBC TV, 21 March 1984.

70 *Newsnight*, BBC TV, 11 April 1984.

71 *The Times*, 1 April 1984.

72 ITV, 27 March 1984.

73 Ottey, *op. cit.*, p. 81.

74 *TV Eye*, 29 March 1984.

75 NEC Minutes, 12 April 1984, p. 106.

76 Ottey, *op. cit.*, p. 63.

77 P. Wilsher, D. Macintyre and M. Jones, *Strike: Thatcher, Scargill and the Miners*, London, Coronet, 1985, p. 78.

78 *Panorama*, BBC TV, 30 April 1984.

79 *Newsnight*, 14 May 1984.

80 NEC Minutes, 10 May 1984, p. 141.

81 *Ibid.*, pp. 177–83.

82 *Ibid.*, p. 141.

83 NEC Minutes, 14 June 1984, p. 195.

84 NEC Minutes, 10 July 1984, p. 231.

85 NEC Minutes, 26 July 1984, pp. 557–8.

86 ITV, 8 August 1984.

87 Goodman, *op. cit.*, p. 75.

88 Wilsher *et al.*, *op. cit.*, p. 107.

89 B. Fryer, 'Trade unionism, public policy and the law', *Freedom and Fairness: Empowering People at Work*, ed. K. Coates, Nottingham, Spokesman, 1986, p. 87.

90 R. J. Waller, *The Dukeries Transformed*, Oxford, Clarendon, 1983.

91 F. Richards, *The Miners' Next Step*, London, Junius, 1984, p. 25.

92 Taylor, *op. cit.*, p. 195.

93 *Financial Times*, 17 April 1984.

94 MMC, *op. cit.*, p. 278.

95 J. Winterton, 'How the bonus scheme has cheated the miners', *Tribune*, 6 April 1984.

96 Incomes Data Services, 'Miners' pay: Monopolies Commission Report reveals major changes', Report 405, July 1983, p. 25.

97 Royal Commmission on Trade Unions and Employers' Associations 1965–1968, *Report* Ch. Donovan, Cmnd. 3623, London, HMSO, 1968; National Board for Prices and Incomes, *Payment by Results Systems*, Report No. 65, Cmnd. 3627, London, HMSO, 1968.

98 EAC Report, 11–12 July 1984, p. 311.

99 NUM (Nottingham Area), *An Appraisal of the British Coalmining Industry and an Examination of the Possible Consequences following Pit Closures in the NUM (Nottinghamshire Area)*, Mansfield, NUM (Nottinghamshire Area), 1983.

100 J. Winterton, 'Security blankets', *New Statesman*, 6 April 1984.

101 *Today*, BBC Radio 4, 1 February 1985.

102 J. Winterton, 'The source of the crisis', *op. cit.*

103 Nottinghamshire County Council, Economic Development Committee, *Report on Employment Structure and Prospects in Nottinghamshire*, 24 July 1985.

104 A. Callinicos and M. Simons, *The Great Miners' Strike of 1984–5 and its Lessons*, London, SWP, 1985, pp. 74–6.

105 D. Thomas, 'The unexpected miners', *New Society*, 22 March 1984.

106 M. Freeman, *Our Day Will Come*, London, Junius, 1985, p. 37.

Chapter three

1 E. T. Hiller, *The Strike: A Study in Collective Action*, Chicago, Ill., Chicago University Press, 1928, pp. 115–6.

2 *Ibid.*, p. 69.

3 Hartley *et al.*, *op. cit.*, p. 16.

4 Callinicos and Simons, op. cit., p. 14.

5 Fryston picket interviewed on television, BBC TV, 14 March 1984.

6 M. Jones, *Killed on the Picket Line 1984: The Story of David Gareth Jones*, London, New Park, 1985.

7 SCC Minutes, 2 April 1984.

8 SCC Minutes, 30 April 1984.

9 SCC Minutes, 13 April 1984.

10 SCC Minutes, 27 April, 1984.

11 SCC Minutes, 4 May 1984.

12 SCC Minutes, 30 May 1984.

13 SCC Minutes, 18 April 1984.

14 SCC Minutes, 30 April 1984.

15 SCC Minutes, 25 May 1984.

16 SCC Minutes, 24 April 1984.

17 SCC Minutes, 25 April 1984.

18 SCC Minutes, 27 April 1984.

19 SCC Minutes, 30 May 1984.

20 SCC Minutes, 1 May 1984.

21 SCC Minutes, 4 May 1984.

22 NEC Minutes, 14 June 1984, p. 195.

23 NCC Minutes, 19 June 1984, p. 221.
24 NCC Minutes, 29 August 1984, p. 630.
25 *Look North*, BBC TV, 20 March 1984.
26 Hartley *et al*, *op. cit.*
27 M. Upham, 'British Steel: retrospect and prospect', *Industrial Relations Journal*, Vol. 11, No. 2, 1980, pp. 5–21.
28 BBC TV, 30 March 1984.
29 BBC TV, 2 April 1984.
30 BBC TV, 4 April 1984.
31 BBC TV, 5 April, 1984.
32 BBC TV, 18 April 1984.
33 ITV, 4 May 1984.
34 BBC TV, 11 May 1984.
35 *Newsnight*, BBC TV, 15 June 1984.
36 BBC TV, 16 June 1984.
37 SCC Minutes, 25 May 1984.
38 BBC TV, 27 May 1984.
39 SCC Minutes, 28 May 1984.
40 BBC TV, 30 May 1984.
41 SCC Minutes, 1 June 1984.
42 SCC Minutes, 4 June 1984.
43 SCC Minutes, 6 June 1984.
44 *Ibid.*
45 SCC Minutes, 8 June 1984.
46 NEC Minutes, 14 June 1984, p. 195.
47 B. Jackson and T. Wardle, *The Battle for Orgreave*, London, Canary, 1986.
48 SCC Minutes, 18 June 1984.
49 NCC Minutes, 19 June 1984, p. 220.
50 *Ibid.*
51 BBC TV, 27 June 1984.
52 NCC Minutes, 27 June 1984, p. 222.
53 NEC Minutes, 28 June 1984, p. 219.
54 NEC Minutes, 29 June 1984. p. 226.
55 *Ibid.*, p. 227.
56 BBC TV, 2 July 1984.
57 NCC Minutes, 4 July 1984, p. 581.
58 Letter to steel-workers from Bob Haslam, BS Chairman, 'Our business: Steel jobs in danger', 4 July 1984.
59 *Newsnight*, BBC TV, 11 July 1984.
60 BBC TV, 13 July 1984.
61 BBC TV, 23 August 1984.
62 BBC TV, 25 August 1984.
63 BBC TV, 19 September 1984.
64 NCC Minutes, 26 September 1984, p. 654.
65 NEC Minutes, 28 June 1984, p. 217.
66 NCC Minutes, 4 July 1984, p. 581.
67 NCC Minutes 27 June 1984, p. 222.
68 NCC Minutes 4 July 1984, p. 581.

69 BBC TV, 20 September 1984.
70 NCC Minutes, 26 September 1984, p. 654.
71 NCC Minutes, 10 October 1984, p. 661.
72 NCC Minutes, 24 October 1984, p. 679.
73 NCC Minutes, 14 November 1984, p. 716.
74 NCC Minutes, 21 November 1984, p. 730.
75 NCC Minutes, 28 November 1984, p. 731.
76 NCC Minutes, 5 December 1984, p. 767.
77 NCC Minutes, 12 December 1984, p. 768.
78 SCC Minutes, 27 April 1984.
79 SCC Minutes, 18 April 1984.
80 SCC Minutes, 19 April 1984.
81 SCC Minutes, 29 May 1984.
82 SCC Minutes, 1 June 1984; 25 June 1984.
83 SCC Minutes, 4 July 1984.
84 NCC Minutes, 19 June 1984, p. 220.
85 NCC Minutes, 1 August 1984, p. 595.
86 NCC Minutes, 24 October 1984, p. 679.
87 NCC Minutes, 31 October 1984, p. 681.
88 NCC Minutes, 19 December 1984, p. 5.
89 NCC Minutes, 19 June 1984, p. 221.
90 NEC Minutes, 1 November 1984, p. 676.
91 NCC Minutes, 19 June 1984, p. 220.
92 NEC Minutes, 1 November 1984, p. 676.
93 J. Steuben, *Strike Strategy*, New York, Gaer, 1950, p. 100.
94 Callinicos and Simons, *op. cit.*, pp. 92–3.
95 McIlroy, *op. cit.*, 1984, p. 110.
96 *Ibid*, p. 106.
97 SCC Minutes, 30 April 1984.
98 SCC Minutes, 4 June 1984.
99 SCC Minutes, 22 May 1984.
100 Sharlston miner.
101 SCC Minutes, 19 April 1984.
102 SCC Minutes, 22 May 1984.
103 SCC Minutes, 18 May 1984.
104 Steuben, *op. cit.*, pp. 117, 141.
105 McIlroy, *op. cit.*, 1984, p. 125.
106 Hartley *et al.*, *op. cit.*, p. 44.
107 Steuben, *op. cit.*, p. 141.
108 *Ibid*, p. 146.
109 McIlroy, *op. cit.*, 1984, p. 125.
110 H. Beynon, 'Authority and change in the coalfields', *Journal of Law and Society*, Vol. 12, No. 3, 1985, p. 397.
111 Steuben, *op. cit.*, pp. 130, 140.
112 Callinicos and Simons, *op. cit.*, p. 101.
113 SCC Minutes, 6 June 1984.
114 A significant negative correlation ($R = -0{\cdot}29$) was obtained between the percentage of members involved in local picketing and the degree of support for the strike at the initial branch meetings.

115 Hiller, *op. cit.*, p. 195.

116 *Ibid.*, p. 108.

117 C. Kerr and A. Siegel, 'The interindustry propensity to strike: an international comparison', *Industrial Conflict*, eds. A. Kornhauser, R. Dubin and A. M. Ross, New York, McGraw-Hill, 1954, pp. 189–212.

118 W. H. Scott *et al.*, *Coal and Conflict*, Liverpool, Liverpool UP, 1963, p. 151.

119 Allen, *op. cit.*, 1981, p. 84.

120 Kerr and Siegel, *op. cit.*

Chapter four

1 Interview with Max Madden MP, Westminster, 15 July 1985.

2 ITV, 5 July 1985.

3 See, for example, D. Coates, *The Labour Party and the Struggle for Socialism*, Cambridge, CUP, 1975; L. Panitch, *Social Democracy and Industrial Militancy*, Cambridge, CUP, 1976.

4 H C Deb., 17 May 1984, cols. 508, 573, 594.

5 H C Deb., 7 June 1984, col. 496.

6 ITV, 19 January 1985.

7 Interview with Rt Hon Stan Orme MP, Westminster, 16 July 1985.

8 ITV, 16 May 1984, 12 August 1984; BBC TV, 6 December 1984.

9 ITV, 13 November 1984.

10 H C Deb., 7 June 1984, col. 458.

11 Interview with Rt Hon Stan Orme MP, Westminster, 16 July 1985.

12 H C Deb., 7 June 1984, col. 456.

13 *Ibid.*, cols. 456–7.

14 H C Deb., 17 May 1984, cols. 491–2.

15 BBC TV, 2 October 1984.

16 BBC TV, 3 October 1984.

17 BBC TV, 14 October 1984.

18 H C Deb., 7 June 1984, col. 460.

19 *Ibid.*, col. 468.

20 *Ibid.*, col. 460.

21 *Ibid.*, col. 462.

22 *Ibid.*, col. 480.

23 *Ibid.*, col. 494.

24 *Ibid.*, col. 477.

25 Interview with Bryan Davies, Westminster, 15 July 1985.

26 ITV, 28, 30 October 1984.

27 BBC TV, 1, 5 November 1984.

28 NEC Minutes, 13 December 1984, p. 764.

29 NEC Minutes, 24 January 1985, p. 10.

30 Interview with Bryan Davies, Westminster, 15 July 1985.

31 Interview with Derek Fatchett MP, Westminster, 16 July 1985.

32 Interview with Max Madden MP, Westminster, 15 July 1985.

33 Interview with Martin Redmond MP, Westminster, 16 July 1985.

34 Interview with Rt Hon Tony Benn MP, Westminster, 17 July 1985.

35 Campaign Group of Labour MPs, *Annual Report, June 1983 – October 1984*, London, Campaign Group, 1984, p. 11.

36 Interview with Bryan Davies, Westminster, 15 July 1985.

37 Interview with Max Madden MP, Westminster, 15 July 1985.

38 Interview with Rt Hon Tony Benn MP, Westminster, 17 July 1985.

39 D. Massey and H. Wainwright, 'Beyond the coalfields: the work of the Miners' Support Groups', *Digging Deeper: Issues in the Miners' Strike*, ed. H. Beynon, London, Verso, 1985 p. 151.

40 Labour Research Department, *Solidarity with the Miners*, London, LRD, 1985, p. 18.

41 *Ibid*, p. 14.

42 BBC TV, 15 April 1984.

43 *Weekend World*, ITV, 1 April 1984.

44 BBC TV, 8 July 1984.

45 McIlroy, *op. cit.*, 1984, p. 12.

46 *Weekend World*, ITV, 1 April 1984.

47 ITV, 9 April 1984.

48 ITV, 19 May 1984.

49 BBC TV, 11 May 1984.

50 ITV, 11 June 1984.

51 ITV, 12 June 1984.

52 BBC TV, 25 June 1984.

53 BBC TV, 5 April 1984.

54 BBC TV, 9, 10 May 1984.

55 BBC TV, 20, 21 May 1984.

56 ITV, 14, 17, 21 May 1984.

57 BBC TV, 27 June 1984.

58 BBC TV, 30 July 1984.

59 BBC TV, 26 August 1984.

60 *Sunday Times*, 26 August 1984.

61 TUC *Report 1984*, p. 653; NEC Minutes, 18 October 1984, p. 674.

62 ITV, 1 September 1984.

63 BBC TV, 31 August 1984.

64 TUC *Report 1984*, pp. 398–9.

65 *Ibid.*, pp. 400–1.

66 *Ibid.*, p. 402.

67 *Ibid.*, p. 403.

68 *Ibid.*, p. 405.

69 BBC TV, 21 September 1984.

70 ITV, 16 October 1984.

71 *Union World*, Channel 4 TV, 20 October 1984.

72 NEC Minutes, 15 November 1984, p. 712.

73 BBC TV, 13 November 1984.

74 NEC Minutes, 13 December 1984, p. 764.

75 SDC Report, 3 March 1985, p. 102.

76 TGWU *Record*, December 1986.

77 Worsborough Community Group, *The Heart and Soul of it*, Barnsley, Bannerworks, 1985.

78 LRD, *op. cit.*, p. 11.

79 M. Adeney and J. Lloyd, *The Miners' Strike 1984–5, Loss without Limit*, London, RKP, 1986, p. 228.

80 LRD, *op. cit.*, p. 5.

81 NCC Minutes, 18 July 1984, p. 583.

82 Barnsley Women Against Pit Closures, *Women Against Pit Closures*, Barnsley, BWAPC, 1984, p. 7.

83 Jean Miller, 'Relations with the NUM: Barnsley', *The Cutting Edge: Women and the Pit Strike*, ed. V. Seddon, London, Lawrence and Wishart, 1986, p. 227.

84 K. MacKey, 'Women against pit closures', *The Cutting Edge: Women and the Pit Strike*, ed. V. Seddon, London, Lawrence and Wishart, 1986, p. 52.

85 North Yorkshire Women Against Pit Closures, *Strike 84–85*, Leeds, NYWAPC, 1985, p. 7.

86 BWAPC member.

87 Fryston miner's wife.

88 South Kirkby miner's wife.

89 NYWAPC, *op cit.*, p. 8.

90 Jean Miller, *op. cit.*, p. 232.

91 Callinicos and Simons, *op. cit.*, p. 180.

92 P. Gibbon and D. Steyne, *Thurcroft: A Village and the Miners' Strike*, Nottingham, Spokesman, 1986, p. 215.

93 NYWAPC, *op. cit.*, p. 12.

94 Gascoigne Wood miner.

95 Gibbon and Steyne, *op. cit.*, p. 214.

96 Savile miner.

97 Barnsley Women Against Pit Closures, *Women Against Pit Closures, Volume 2*, Barnsley, BWAPC, 1985, p. 21.

98 NYWAPC, *op. cit.*, p. 12.

99 Jean Miller, *op. cit.*, pp. 228–9.

100 *Ibid.*, p. 232.

101 Leeds Miners' Support Group member.

102 Callinicos and Simons, *op. cit.*, p. 178.

103 Jean Miller, *op. cit.*, p. 237.

104 *Striking Back*, Cardiff, WCCPL and NUM (South Wales Area), 1985, p. 37.

105 L. Loach, 'We'll be here right to the end . . . and after: women in the miners' strike', *Digging Deeper: Issues in the Miners' Strike*, ed. H. Beynon, London, Verso, 1985, p. 173.

106 Linda Jenkins, quoted in Worsborough Community Group, *op. cit.*, np.

107 Loach, *op cit.*, p. 174.

108 Pete McCulloch, quoted in Worsborough Community Group, *op. cit.*, np.

109 NCC Minutes, 25 July 1984, p. 585.

110 NCC Minutes, 18 July 1984, p. 583.

111 Jean Miller, *op. cit.*, p. 228.

112 *Ibid.*

113 Fryston branch official's wife.

114 BWAPC, *op. cit.*, 1985, p. 60.

115 MacKey, *op. cit.*, p. 59.

116 NCC Minutes, 1 August 1984, p. 595.

117 NCC Minutes, 15 August 1984, p. 627.

118 NCC Minutes, 12 September 1984, p. 627.

119 Jean Miller, *op. cit.*, p. 236.

120 *Ibid.*

121 H C Deb., 7 June 1984, col. 483.

122 Callinicos and Simons, *op. cit.*, p. 127.

123 Interview with Richard Caborn MP, Westminster, 16 July 1985.

124 Interview with William Michie MP, Westminster, 17 July 1985.

125 Interview with Richard Caborn MP, Westminster, 16 July 1985.

126 NCC Minutes, 18 July 1984, p. 583.

127 NCC Minutes, 25 July 1984, p. 585.

128 NCC Minutes, 9 August 1984, p. 596.

129 NCC Minutes, 12 September 1984, p. 642.

130 NCC Minutes, 19 December 1984, p. 5.

131 Barry Seal MEP (Lab., Yorkshire West), 'Report' (to constituents), 1984 (*mimeo*).

132 NCC Minutes, 26 September 1984, p. 653.

133 NCC Minutes, 10 October 1984, p. 661.

134 SCC Minutes, 8 May 1984.

135 SCC Minutes, 15 May 1984.

136 SCC Minutes, 4 July 1984.

137 SCC Minutes, 6 June 1984.

138 *Ibid.*

139 SCC Minutes, 7 June 1984.

140 L. Sutcliffe and B. Hill, *Let Them Eat Coal: the Political use of Social Security during the Miners' Strike*, London, Canary, 1985, pp. 6, 42.

141 Adeney and Lloyd, *op. cit.*, p. 227.

142 Sutcliffe and Hill, *op. cit.*, p. 37.

143 Gibbon and Steyne, *op. cit.*, p. 218.

144 *Ibid.*, p. 211.

145 *Ibid.*, p. 209.

146 *Striking Back*, *op. cit.*, p. 45.

147 Worsborough Community Group, *op. cit.*, np.

148 Trade Union and Community Resource and Information Centre, *Benefits in Dispute*, Leeds, TUCRIC, 1982.

149 Claimants' Union, *Strikers' Handbook*, London, Federation of Claimants' Unions, 1984; Merseyside Welfare Rights Resource Centre, *Aid for Strikers*, Liverpool, MWRRC, 1984.

150 A. Dunn, *The Strikers' Moneyguide*, London, Campaign for Alternative Industrial and Technical Systems, 1985.

151 National Association of Citizens' Advice Bureaux, *A Report on the 1984/55 Coal Dispute and the CAB Service*, London, NACAB, 1985, p. 39.

152 TUCRIC *Bulletin*, Summer 1984.

153 NACAB, *op. cit.*, pp. 62–3.

154 BWAPC, *op. cit.*, 1984, p. 10; 1985, p. 10.

155 Manton branch official.

156 Ferrymoor Riddings branch official.

157 Selby miner.

158 Adeney and Lloyd, *op. cit.*, p. 229.

159 BWAPC, *op. cit.*, 1984, p. 10.

160 Wilsher *et al.*, *op. cit.*, p. 180.

161 Gibbon and Steyne op. cit., p. 207.

162 C. Salt and J. Layzell, *Here We Go: Women's Memories of the 1984/85 Miners' Strike*, London, London Political Committee, Co-operative Retail Services, 1985, p. 69.

163 BWAPC, *op. cit.*, 1985, p. 56.

164 Allerton Bywater miner.

165 Interview with Richard Caborn MP, Westminster, 16 July 1985.

166 NCC Minutes, 21 November 1984, p. 730.

167 *Miners' Strike News*, 8 February 1985 (MCG).

168 *Miners' Strike News*, 16 February 1985 (MCG).

169 *Miners' Strike News*, 23 February 1985 (MCG).

170 Tony Littlewood, quoted in Worsborough Community Group, *op. cit.*, np.

Chapter five

1 R. Harrison. 'The labour movement and the strike', Coal and Community Conference, University of Sheffield, 7–9 March 1986.

2 Goodman, *op. cit.*, pp. 69–70; I. MacGregor, *The Enemies Within: the Story of the Miners' Strike 1984–5*, London, Collins, 1986, p. 116.

3 K. Jeffrey and P. Hennessy, *States of Emergency: British Governments and Strikebreaking since 1919*, London, RKP, 1983.

4 *Ibid.*, p. 195.

5 R. Maudling, *Memoirs*, London, Sidgwick and Jackson, 1978, p. 161.

6 Jeffrey and Hennessy, *op. cit.*, p. 255.

7 'Appomattox or civil war?', *Economist*, 27 May 1978.

8 *Ibid.*

9 *Daily Mirror*, 6 June 1984.

10 Iron and Steel Trades Confederation, *New Deal for Steel*, London, ISTC, 1980.

11 P. S. Bagwell, *End of the Line*, London, Verso 1984.

12 M. Crick, *Scargill and the Miners*, Harmondsworth, Penguin, 1985.

13 Wilsher *et al.*, *op. cit.*, p. 26.

14 L. Turner, *op. cit.*

15 *Guardian*, 16 January 1985.

16 Sweet, *op. cit.*, 1985, p. 202.

17 H. Beynon and P. McMylor, 'Decisive power: the new Tory state against the miners', *Digging Deeper: Issues in the Miners' Strike*, ed. H. Beynon, London, Verso, 1985, p. 37.

18 CEGB, *op. cit.*, 1985, p. 17.

19 *Economist*, 3 June 1972.

20 J. Gennard and R. Lasko, 'Supplementary benefit and strikers', *British Journal of Industrial Relations*, Vol. 12, No. 1, 1974, p. 10.

21 C. Jones and T. Novak, 'Welfare against the workers: benefits as a political weapon', *Digging Deeper: Issues in the Miners' Strike*, ed. H. Beynon, London, Verso, 1985, p. 93.

22 Conservative Political Centre, *Financing Strikes*, London, Con. Party, 1974.

23 Gennard and Lasko, *op. cit.*, 1974; 'The individual and the strike', *British Journal of Industrial Relations*, Vol. 13, No. 3, 1975, pp. 346–70; J. W. Durcan and W. E. J. McCarthy, 'The state subsidy theory of strikes: an examination of statistical data for the period 1956–1970', *British Journal of Industrial Relations*, Vol. 12, No. 1, 1974, pp. 26–47.

24 H C Deb., 7 June 1984, col. 458.

25 BBC TV, 21 November 1985.

26 BBC TV, 22 November 1985.

27 Sutcliffe and Hill, *op. cit.*, p. 11.

28 *Ibid.*, p. 7.

29 M. Kettle and L Hodges, *Uprising!, The police, the people and the riots in Britain's cities*, London, Pan, 1982.

30 BSSRS Technology of Political Control Group, *TechnoCop, New Police Technologies*, London, Free Association, 1985.

31 B. Schwarz, 'Let them eat coal: the Conservative Party and the strike', *Digging Deeper: Issues in the Miners' Strike*, ed. H. Beynon, London, Verso, 1985, p. 56.

32 A. Flanders, *Management and Unions*, London, Faber, 1970.

33 K. W. Wedderburn, *The Worker and the Law*, Harmondsworth, Penguin, 1986, p. 33.

34 R. Miliband, *The State in Capitalist Society*, London, Weidenfeld and Nicolson, 1969, p. 127.

35 Donovan *Report*, *op. cit.*

36 Department of Employment and Productivity, *In Place of Strife: A Policy for Industrial Relations*, White Paper, Cmnd. 3888, London, HMSO, 1969.

37 Panitch, *op. cit.*

38 Committee of Members of the Inns of Court Conservative and Unionist Society, *A Giant's Strength*, London, ICCUS, 1958; Conservative Political Centre, *Fair Deal at Work*, London, Con. Party, 1968.

39 Weekes, B. *et al.*, *Industrial Relations and the Limits of Law*, Oxford, Blackwell, 1975.

40 Clegg, *op. cit.*, pp. 319–28.

41 J. McMullen, *Employment Law Under the Tories*, London, Pluto, 1981, pp. 4–16.

42 Trades Union Congress, *Guide on the Conduct of Disputes*, London, TUC, 1979.

43 Department of Employment, *Employment Code of Practice (Picketing) Order 1980*, SI 1980 No. 1757, London, HMSO, 1980.

44 BBC Radio 4, 7 October 1984.

45 *Financial Times*, 30 March 1984.

46 *Today*, BBC Radio 4, 14 October 1984.

47 *Read (Transport) Ltd* v. *NUM (S. Wales Area)* [1985] IRLR 67.

48 *Wight Contractors* v. *NUM (S. Wales Area)*, unreported, 19 June 1984.

49 J. McIlroy, 'The law struck dumb? Labour law and the miners' strike', *Policing the Miners' Strike*, eds. B. Fine and R. Millar, London, Lawrence and Wishart, 1985, p. 88.

50 *The Times*, 6 June 1984.

51 K.D. Ewing, 'The strike, the courts and the rule books', *Industrial Law Journal*, Vol. 14, 1985, p. 160.

52 J. McIlroy, 'Police and pickets: the law against the miners', *Digging Deeper: Issues in the Miners' Strike*, ed. H. Beynon, London, Verso, 1985, p. 117.

53 *The Times*, 27 June 1984.

54 Wedderburn, *op. cit.*, p. 731.

55 *Clarke* v. *Chadburn* [1984] IRLR 350.

56 Ottey, *op. cit.*, pp. 142–3.

57 Crick, *op. cit.*, pp. 111–12.

58 *Taylor (R.) and Foulstone* v. *NUM (Yorkshire Area)* [1984] IRLR 445.

59 Ewing, *op. cit.*

60 *Taylor* v. *NUM (Derbyshire Area)* [1984] IRLR 440.

61 *Fettes* v. *NUM*, unreported, 9 November 1984 (Scotland).

62 *Thomas* v. *NUM (South Wales Area)* [1985] 2 WLR 1081; [1985] IRLR 136.

63 Channel 4 News, 28 September 1984.

64 Scargill addressing SDC, SDC Report, 3 December 1984, p. 749.

65 *Taylor* v. *NUM (Derbyshire Area)* (No. 2) [1985] IRLR 65, 99.

66 Adeney and Lloyd, *op. cit.*, pp. 174–5.

67 Ottey, *op. cit.*, p. 136.

68 Section 16 Employment Act 1980 restricts the immunity afforded to peaceful picketing by Section 13 Trade Union and Labour Relations Act 1974 to picketing at or near one's own place of work.

69 *Thomas* v. *NUM (South Wales Area)* [1985] 2 WLR 1081; [1985] IRLR 136.

70 Writ of summons issued against Scargill and others by Sharp and others in the High Court, 14 January 1985, CH1985 S No. 157.

71 *Financial Times*, 1 September 1984.

72 *Diverse Reports*, Channel 4 TV, 4 October 1984.

73 *Lyons* v. *Wilkins* [1896] 1 Ch 811 (CA); [1899] 1 Ch 255 (CA).

74 B. Bercusson, 'One hundred years of conspiracy and protection of property, time for a change', *Modern Law Review*, Vol. 40, 1977, p. 268.

75 *The Times*, 19 March 1984.

76 M. Kettle, '1926 and now: if the miners lose again', *New Socialist*, February 1985, p. 25.

77 T. Bunyan, *The Political Police in Britain*, London, Quartet, 1977, pp. 181–3.

78 *Guardian*, 1 July 1980.

79 *Tribune*, 20 April 1984.

80 *Striking Back*, *op. cit.*, pp. 140–2.

81 M. Abdel-Rahim, *Strike Breaking in Essex*, London, Canary Press, 1985, pp. 50–3.

82 BSSRS, *op. cit.*, p. 89.

83 *New Scientist*, 12 January 1984.

84 Bunyan, *op. cit.*, p. 270.

85 Abdel-Rahim, *op. cit.*, pp. 17–18.

86 *Moss* v. *McLachlan* [1985] IRLR 76.

87 *R.* v. *Mansfield JJ, ex parte Sharkey* [1984] IRLR 496.

88 L. Christian, 'Restriction without conviction: the role of the courts in legitimising police control in Nottinghamshire', *Policing the Miners' Strike*, eds. B. Fine and R. Millar, London, Lawrence and Wishart, 1985, p. 131.

89 BSSRS, *op. cit.*, pp. 14, 88.

90 Letter to the Prime Minister from Tony Benn, 7 May 1984.

91 Letter to Tony Benn from the Prime Minister, 23 May 1984.

92 *Broome* v. *DPP (Hunt* v. *Broome)* [1974] ICR 84 (HL).

93 J. Rozenberg, 'The legacy of legal changes', *Listener*, 7 March 1985, pp. 5–6.

94 *Kavanagh* v. *Hiscock* [1974] ICR 282.

95 R. Kidner, *Trade Union Law*, London, Stevens, 1979, p. 204.

96 *Broome* v. *DPP (Hunt* v. *Broome)* [1974] ICR 84 (HL).

97 *Kavanagh* v. *Hiscock* [1974] ICR 282.

98 H. Beynon, 'Introduction', *op. cit.*, p. 2.

99 J. Kay and K. Millidene, 'Intruders, not peacekeepers', *The Cutting Edge: Women and the Pit Strike*, ed. V. Seddon, London, Lawrence and Wishart, 1986, p. 187.

100 J. Winterton, 'Police activities', *New Statesman*, 27 April 1984.

101 BSSRS, *op. cit.*, p. 90.

102 Hartley *et al.*, *op. cit.*, pp. 54–6.

103 T. Lane and K. Roberts, *Strike at Pilkingtons*, London, Fontana, 1971, p. 171.

104 J. Field, 'Police monitoring: the Sheffield experience', *Policing the Miners' Strike*, eds. B. Fine and R. Millar, London, Lawrence and Wishart, 1985, p. 212.

105 *Striking Back*, *op. cit.*, pp. 93–5.

106 Jackson and Wardle, *op. cit.*

107 MacGregor, *op. cit.*, p. 205.

108 T. Bunyan, 'From Saltley to Orgreave via Brixton', *Journal of Law and Society*, Vol. 12, No. 3, 1985, pp. 301–2; R. East, H. Power and P. A. Thomas, 'The death of mass picketing', *Journal of Law and Society*, Vol. 12, No. 3, 1985, p. 312.

109 J. Coulter, S. Miller and M. Walker, *State of Siege*, London, Canary, 1984, pp. 86–90; J. Rose 'Intimidating behaviour', *New Statesman*, 17 August 1984.

110 P. Gordon, 'If they come in the morning . . . the police, the miners and black peoples', *Policing the Miners' Strike*, eds. B. Fine and R. Millar, London, Lawrence and Wishart, 1985, p. 165.

111 D. Waddington, 'Maltby: a study in community resistance', Coal and Community Conference, University of Sheffield, 7–9 March 1986.

112 T. Kaye, 'A village at war: a report to the Yorkshire Area NUM concerning the policing of Fitzwilliam, West Yorkshire, on Monday 9th July 1984', University of Warwick, 1984, (*mimeo*).

113 Callinicos and Simons, *op. cit.*, pp. 168–9.

114 BWAPC, *op. cit.*, 1985, p. 56.

115 J. Percy-Smith and P. Hillyard, 'Miners in the arms of the law: a statistical analysis', *Journal of Law and Society*, Vol. 12 No. 3, 1985, p. 345.

116 Abdel-Rahim, *op. cit.*, pp. 56–8.

117 Percy-Smith and Hillyard, *op. cit.*, p. 350.

118 *Ibid.*, p. 351.

119 N. Blake, 'Picketing, justice and the law', *Policing the Miners' Strike*, eds. B. Fine and R. Millar, London, Lawrence and Wishart, 1985, p. 114.

120 H C Deb., 31 July 1984, col. 276.

121 L. Christian, *op. cit.*, p. 135.

122 M. Walker, *A Turn of the Screw*, London, Canary, 1985.

123 ITV, 16 May 1985.

124 BBC TV, 8 November 1985, 22 November 1985.

125 BBC TV, 17 July 1985.

126 BBC TV, 26 September 1985, 1 October 1985.

127 BBC TV, 11 October 1985.

128 South Yorkshire Police Committee, *Policing Policy during the Strike of the National Union of Mineworkers (Yorkshire Area), 1984*, Report of an Inquiry by the Special Sub-Committee of the South Yorkshire Police Committee, 4 February 1985, pp. 13, 23.

129 *Ibid.*, pp. 9, 24.

130 P. Wright, *Policing the Coal Industry Dispute in South Yorkshire*, Report to South Yorkshire Police Committee for the Chief Constable, Sheffield, South Yorkshire Police, 1985.

131 P. Wright, *Enquiry into Incidents at Stainforth and Armthorpe on 21 August 1984*, Report to South Yorkshire Police Committee from the Chief Constable, Sheffield, South Yorkshire Police, 25 September 1985; P. Wright, *Charges of Riot and Unlawful Assembly*,

Report to South Yorkshire Police Committee from the Chief Constable, Sheffield, South Yorkshire Police, 25 September 1985.

132 S. Spencer, *Police Authorities during the Miners' Strike*, London, Cobden Trust, 1985.

133 Sheffield Policewatch, *Taking Liberties: Policing during the Miners' Strike, April–Oct 1984*, Sheffield, Sheffield Policewatch, 1984.

134 National Council for Civil Liberties, *Civil Liberties and the Miners' Dispute*, London, NCCL, 1984, p. 6.

135 *Ibid.*, p. 36.

136 A. Jones and R. Thompson, *After the Miners' Strike – What Next?* London, International Marxist Group, 1972, pp. 29–30.

137 See the criticisms of mass-picketing by: K. Howells, 'Stopping out: the birth of a new kind of politics', *Digging Deeper: Issues in the Miners' Strike*, ed. H. Beynon, London, Verso, 1985, p. 145; P. Carter, 'Striking the right note', *Marxism Today*, Vol. 29, No. 3, 1985, pp. 28–31; George Bolton in 'A strike to the finish: a roundtable discussion', *Marxism Today*, Vol. 28, No. 9, pp. 9–15.

138 Callinicos and Simons, *op. cit.*, p. 119.

139 Glasgow Media Group, *Bad News*, London, RKP, 1976; Glasgow Media Group, *More Bad News*, London, RKP, 1980; P. Beharrell and G. Philo, *Trade Unions and the Media*, London, Macmillan, 1977.

140 R. Michels, *Political Parties*, Glencoe, Ill., Free Press, 1949.

141 D. Douglass, *Tell us Lies about the Miners*, Doncaster, Direct Action Movement, 1985, p. 11.

142 Adeney and Lloyd, *op. cit.*, pp. 28–52.

143 *Sunday Express*, 10 October 1982.

144 J. Reid, 'What Scargill means', *New Society*, 17 January 1985 pp. 91–3.

145 V. L. Allen, 'Miners' man', *New Society*, 24 January 1985, pp. 136–7.

146 D. Warren, *The Key to my Cell*, London, New Park, 1982 p. 17; R. Clutterbuck, *Britain in Agony, the Growth of Political Violence*, London, Faber, 1978, pp. 82–8.

147 Hartley, *et al*, *op. cit.*, p. 53.

148 S. C. Dixon, 'A critical analysis of press coverage of the firemen's dispute', B. Sc. dissertation, University of Bradford, 1980.

149 J. Neale, *Memoirs of a Callous Picket*, London, Pluto, 1983.

150 Kettle and Hodges, *op. cit.*

151 *The Times*, 9 May 1984.

152 H C Deb., 4 February 1985, col. 649.

153 *Labour Research*, January 1986.

154 *The Times*, 26 July 1985.

155 Home Office, *Review of Public Order Law*, White Paper, Cmnd. 9510, London, HMSO, 1985.

156 Douglas, *op. cit.*, pp. 19–20.

157 *Mail on Sunday*, 5 August 1984.

158 *The Times*, 6 August 1984.

159 *Daily Star*, 2 October 1984.

160 *Daily Express*, 24 November 1984.

161 Douglas, *op. cit.*, p. 17.

162 *Sun*, 20 November 1984.

163 *Daily Mail*, 5 February 1985.

164 *Sunday Times*, 26 August 1984.

165 A. Hetherington, 'Is TV news impartial? Miners and the media', *Listener*, 17 January 1985, pp. 5–6.

166 *Sunday Times*, 26 August 1984; D. Jones *et al.*, *Media Hits the Pits*, London, Campaign for Press and Broadcasting Freedom, 1985, pp. 23–5.

167 Adeney and Lloyd, *op. cit.*, p. 256.

168 *Sun*, 10 January 1984.

169 *Sun*, 13 January 1984.

170 *Sunday Times*, 26 February 1984.

171 *Sunday Times*, 11 March 1984.

172 *Daily Express*, 15 May 1984.

173 *Sun*, 15 May 1984.

174 M. Hollingsworth, 'Scargill's Sun support', *New Stateman*, 6 July 1984.

175 P. Flather, 'Academics at the coalface', *Times Higher Education Supplement*, 7 December 1984, p. 12.

176 Adeney and Lloyd, *op. cit.*, p. 254.

177 A. Gerlis, 'Knuckles to briefcases', *New Statesman*, 5 November 1982, pp. 20–21; J. Cunningham, 'The covert acceleration of union busting', *Guardian*, 2 March 1984.

178 J. Field, 'Union-free management: experiences and resources from North America', *Trade Union Studies Journal*, Summer 1983, pp. 14–15.

179 *Labour Research*, February 1984.

180 P. Gillman, 'Missing facts in Ian McGregor's history', *New Statesman*, 18 March 1983, p. 4.

181 *World in Action*, ITV, 15 October 1984.

182 MacGregor, *op. cit.*, pp. 51–3.

183 J. W. Hevener, *Which Side Are You On? The Harlan County Coal Miners, 1931–39*, Urbana, University of Illinois Press, 1978.

184 MacGregor, *op. cit.*, p. 38.

185 *Ibid.*, p. 36.

186 Steuben, *op. cit.*, pp. 232–62.

187 *Ibid.*, pp. 245–6.

188 Adeney and Lloyd, *op. cit.*, p. 184.

189 R. Stephenson, 'The UDM and the miners' strike: democratic breakaway or union busting?', BA dissertation, University of Bradford, 1986, p. 65.

190 M. Hollingsworth, 'Who is paying the piper?', *New Statesman*, 21 December 1984.

191 D. Hart, *The Soul Politic*, London, Adam Smith Institute, 1985.

192 Wilsher *et al.*, *op. cit.*, p. 183; Adeney and Lloyd, *op. cit.*, p. 163.

193 MacGregor, *op. cit.*, p. 224.

194 *Ibid.*, p. 216.

195 *Ibid.*, pp. 58, 229, 234.

196 M. Hollingsworth, 'Using miners to bust the Union', *New Statesman*, 14 December 1984; Adeney and Lloyd, *op. cit.*, p. 164.

197 *Daily Mirror*, 18 October 1984.

198 M. Hollingsworth, 'Who is paying the piper?' *op. cit.*

199 *The Times*, 13 September 1984.

200 MacGregor, *op. cit.*, pp. 258, 323.

201 *Ibid.*, p. 322.

202 Adeney and Lloyd, *op. cit.*, p. 200.

203 Letter to NUM strikers from Ian MacGregor, 'Your Future in Danger', nd (posted 13 June 1984).

204 Adeney and Lloyd, *op. cit.*, p. 187.

205 MacGregor, *op. cit.*, p. 230.

206 *Ibid.*, p. 222.

207 Letter to Kiveton Park strikers from R. Taylor, Colliery Manager, 2 July 1984.

208 Letter to Doncaster strikers from Albert Tuke, Area Director, 5 July 1984.

209 SCC Minutes, 5 July 1984.

210 SCC Minutes, 9 July 1984.

211 Letter to NUM strikers from Ian MacGregor, July 1984. The NCB proposed discussion document of 18 July was also circulated with another letter in August 1984.

212 *Financial Times*, 13 August 1984.

213 NCB leaflet to strikers, 'You and who else?', November 1984.

214 Letter to NUM members in Doncaster from working miners, December 1984.

215 Letter to Treeton miners from G. E. Thomas, Colliery Manager, 3 December 1984.

216 Letter to Kiveton Park miners from R. Taylor, Colliery Manager, 3 December 1984.

217 NCB leaflet to strikers, 'Earn it or lose it', December 1984.

218 K. Coates and T. Topham, *Trade Unions and Politics*, Oxford, Blackwell, 1986, pp. 147–69.

Chapter six

1 O. Ashenfelter and G.E. Johnson, 'Bargaining theory, trade unions and industrial strike activity', *American Economic Review*, Vol. 59, 1969, pp. 35–49.

2 Adeney and Lloyd, *op. cit.*, p. 24.

3 G. Blazyca, J. Harrison and B. Morgan, 'Figures behind the strike, Nigel's sums', *New Statesman*, 10 August 1984, pp.11–12.

4 ITV, 30 August 1984.

5 *Newsnight*, BBC TV, 19 September 1984.

6 *The Times*, 2 January 1985.

7 BBC TV, 4 March 1985.

8 BBC TV, 19 June 1985.

9 ITV, 17 July 1985.

10 NCB *Report and Accounts 1984/85*, London, NCB, 1985, p. 2.

11 CEGB, *op. cit.*, 1985, p. 8.

12 BBC TV, 30 July 1985.

13 SDC Report, 3 March 1985, p. 102.

14 *Sunday Times*, 3 March 1985.

15 *The Times*, 4 February 1985.

16 Henley Centre for Forecasting, *The Cost of the 1984 Coal Industry Dispute*, Newsnight, London, BBC TV, 17 September 1984, p. 22.

17 *The Times*, 20 March 1985.

18 *The Times*, 21 March 1985.

19 *Sunday Times*, 24 March 1985.

20 *Newsnight*, BBC TV, 8 March 1985.

21 ITV, 1 March 1985.

22 Henley Centre for Forecasting, *The Cost of the 1984 Coal Industry Dispute*, Newsnight, London, BBC TV, 17 September 1984.

23 A. M. Ross, *Trade Union Wage Policy*, Berkeley, Cal., University of California Press, 1948, p. 74.

24 *The Times*, 4 February 1985.

25 *The Miner*, February 1984.

26 *Sunday Times*, 6 May 1984.

27 *Labour Research*, June 1984.

28 *Weekend World*, ITV, 22 July 1984.

29 BBC TV, 23 July 1984.

30 P. Wright, 'Coal stocks: glaring contradictions', *New Statesman*, 27 July 1984.

31 *Guardian*, 18 September 1984.

32 BBC TV, 19 September 1984.

33 P. Wright, 'Interim projections of coal stocks at power stations in the UK', University of Sheffield Division of Continuing Education, 1984.

34 CEGB, *op. cit.*, pp., 18, 20.

35 Wright, 'Interim projections of coal stocks', *op. cit.*

36 Wilsher *et al.*, *op. cit.*, p. 224.

37 *Ibid.*, p. 220.

38 NCC Minutes, 10 October 1984, p. 661.

39 NCC Minutes, 19 December 1984, p. 5.

40 TUC, *Report to General Council on Fuel Stocks at Power Stations*, London, TUC Industrial Relations Department, 1984 (*mimeo*).

41 Wilsher *et al.*, pp. 227–8.

42 Interview with Rt Hon Stan Orme MP, Westminster, 16 July 1985.

43 BBC TV, 25 April 1984.

44 BBC TV, 11 May 1984.

45 NEC Minutes, 23 May 1984, p. 192.

46 BBC TV, 23 May 1984.

47 NEC Minutes, 14 June 1984, p. 194.

48 *Ibid.*

49 H C Deb., 7 June 1984, col. 457.

50 BBC TV, 13 June 1984.

51 *Morning Telegraph*, 8 June 1984.

52 NEC Minutes, 14 June 1984, p. 194.

53 *Guardian*, 28 June 1984.

54 NEC Minutes, 28 June 1984, p. 219.

55 Interview with Rt Hon Stan Orme MP, Westminster, 16 July 1985.

56 NEC Minutes, 10 July 1984, p. 229.

57 NEC Minutes, 26 July 1984, p. 554.

58 NCB Draft Agreement, 18 July 1984.

59 NEC Minutes, 30 August 1984, p. 626.

60 ITV, 22 August 1984.

61 *Newsnight*, BBC TV, 23 August 1984.

62 Channel 4 TV, 31 August 1984.

63 *Guardian*, 6 September 1984.

64 Adeney and Lloyd, *op. cit.*, p. 195.

65 Letter from Kevan Hunt, NCB Industrial Relations, to Peter Heathfield, 5 September 1984.

66 Letter from Peter Heathfield to Kevan Hunt, 5 September 1984.
67 Letter from Ned Smith to Peter Heathfield, 6 September 1984.
68 NEC Minutes, 13 September 1984, p. 640.
69 NEC Minutes, 16 September 1984, p. 647.
70 ITV, 15 September 1984.
71 Interview with Rt Hon Stan Orme MP, Westminster, 16 July 1985.
72 NEC Minutes, 5 September 1984, p. 637.
73 NCC Minutes, 22 August 1984, p. 629.
74 ITV, 19 September 1984.
75 NEC Minutes, 13 September 1984, p. 640.
76 BBC TV, 15 October 1984.
77 BBC TV, 30 October 1984.
78 BBC TV, 27 November 1984.
79 NEC Minutes, 29 November 1984, p. 728.
80 Wilsher *et al.*, *op. cit.*, p. 229; Adeney and Lloyd, *op. cit.*, p. 213.
81 BBC TV, 14, 18 December 1984.
82 ITV, 7 January 1985.
83 NEC Minutes, 10 January 1985, p. 3.
84 NEC Minutes, 24 January 1985, p. 10.
85 ITV, 24 January 1985.
86 NEC Minutes, 30 January 1985, p. 17.
87 NEC Minutes, 1 February 1985, pp. 21–2.
88 BBC TV, 4 February 1985.
89 *Newsnight*, BBC TV, 6 February 1985.
90 Interview with Rt Hon Stan Orme MP, Westminster, 16 July 1985.
91 NEC Minutes, 15 February 1985, p. 33.
92 *Ibid*, p. 34.
93 NEC Minutes, 16 February 1985, p. 37.
94 NEC Minutes, 17 February 1985, p. 41.
95 BBC TV, 20 February 1985, p. 44.
96 SDC Report, 21 February 1985, pp. 56–7.
97 *Ibid.*, p. 69.
98 NEC Minutes, 28 February 1985, p. 72.
99 *Ibid.*
100 Interview with Rt Hon Stan Orme MP, Westminster, 16 July 1985.
101 Adeney and Lloyd, *op. cit.*, p. 215.
102 Hiller, *op. cit.*, p. 93.
103 Letter to Brodsworth strikers form J. M. Smith, Colliery Manager, 4 January 1985.
104 NCB leaflet to strikers, 'You and who else?', 6 November 1984.
105 ITV, 17 February 1985.
106 BBC TV, 24 February 1985.
107 Hiller, *op. cit.*, p. 210.
108 *Ibid.*, p. 100.
109 *Ibid.*, p. 211.
110 NEC Minutes, 3 March 1985, p. 83.
111 SDC Report, 3 March 1985, p. 88.
112 *Ibid.*
113 NEC Minutes, 3 March 1985, p. 109.

114 SDC Report, 3 March 1985, pp. 90–1.
115 *Ibid.*, p. 94.
116 *Ibid.*, p. 98.
117 *Ibid.*, pp. 94–5.
118 *Ibid.*, pp. 95–6.
119 *Ibid.*, p. 97.
120 *Ibid.*, p. 100.
121 *Ibid.*, p. 99.
122 *Ibid.*, p. 91.
123 *Ibid.*, p. 100.
124 *Ibid.*, p. 102.
125 *Ibid.*, p. 103.
126 BBC TV, 4 March 1985.
127 ITV, 6 March 1985.
128 Hiller, *op. cit.*, p. 212.
129 BBC TV, 5 March 1985.
130 Wheldale miner.
131 Gascoigne Wood branch official.
132 Ricall miner.
133 Letter to all NUM members from Ian MacGregor, March 1985.
134 BBC TV, 4 March 1985.
135 Letter to Kiveton Park miners from R. Taylor, Colliery Manager, 4 March 1985.
136 Letter to Brodsworth miners from J. M. Smith, Colliery Manager, nd (March 1985).
137 Steuben, *op. cit.*, p. 297.
138 NEC Minutes, 14 June 1984, p. 197.
139 NEC Minutes, 15 November 1984, p. 727.
140 NEC Minutes, 13 December 1984, p. 763.
141 NEC Minutes, 1 February 1985, p. 26.
142 SDC Report, 21 February 1985, p. 60.
143 *Ibid.*, p. 69.
144 SDC Report, 3 March 1985, p. 105.
145 NEC Minutes, 7 March 1985, p. 111.
146 BBC TV, 4 March 1985.
147 BBC TV, 26 March 1985.
148 BBC TV, 9 May 1985.
149 NEC Minutes, 28 March 1985, p. 116.
150 Gibbon and Steyne, *op. cit.*, p. 241.
151 Nostell branch official.
152 NEC Minutes, 7 March 1985, p. 112.
153 *Ibid.*
154 Employment Committee, *The Dismissal of National Coal Board Employees*, HC 416, London, HMSO, 19 June 1985, p. 2.
155 *Ibid.*, p. 3.
156 *Ibid.*, p. 4.
157 *Ibid.*
158 BBC TV, 20 June 1985.
159 BBC TV, 18 July 1985.

160 H C Deb., 15 July 1985, col. 12.
161 BBC TV, 20 June 1985.
162 Interview with Rt Hon Gerald Kaufmann MP, Westminster, 16 July 1985.
163 Press release from Tony Benn, 6 May 1985.
164 Miners' Amnesty (General Pardon) Bill, H C Bill 1984–5 (172), 28 June 1985.
165 Interview with Rt Hon Tony Benn MP, Westminster, 17 July 1985.
166 BBC TV, 3 September 1985.
167 H C Deb., 15 July 1985, col. 7.
168 BBC TV, 2 October 1985.
169 Interview with Rt Hon Tony Benn MP, Westminster, 17 July 1985.
170 BBC TV, 30 September 1985.
171 Campaign Group of Labour MPs, *Justice: The Miners' Strike 1984–85*, London, Verso, 1986, pp. 53–4.
172 R. Samuel, B. Bloomfield and G. Boanas, *The Enemy Within: Pit Villages and the Miners' Strike of 1984–5*, London, RKP, 1986, p. 238.
173 Gibbon and Steyne, *op. cit.*, p. 241.
174 BBC TV, 13 January 1986.

Chapter seven

1 Steuben, *op. cit.*, p. 265.
2 *The Times*, 21 May 1985.
3 Goodman, *op. cit.*, pp. 101–2.
4 G. Kirk, *Schumacher on Energy*, London, Cape, 1982.
5 Wilsher *et al.*, *op. cit.*, pp. 181–2.
6 MacGregor, *op. cit.*, pp. 186, 290.
7 *Panorama*, BBC TV, 4 March 1985.
8 *Sunday Telegraph*, 10 March 1985.
9 BBC TV, 7 April 1985.
10 NEC Minutes, 7 March 1985, pp. 112–3.
11 NEC Minutes, 28 March 1985, p. 116.
12 BBC TV, 15 March 1985.
13 ITV, 8 September 1985.
14 BBC TV, 12 March 1985.
15 ITV, 13 April 1985.
16 *TV Eye*, ITV, 23 May 1985.
17 Wheldale miner.
18 Jack Taylor, seminar, Northern College, 24 January 1986.
19 Channel 4 TV, 28 August 1986.
20 BBC TV, 8 September 1986.
21 ITV, 1 April 1987.
22 *Newsline*, 18 September 1985.
23 Barnsley Main branch official (Interview by Peter Clarney, Northern College, 29 January 1986).
24 Frickley miner.
25 NCB Memorandum, 'New strategy for coal', September 1985.
26 *Guardian*, 27 November 1986.
27 NUM Research, 'Hours of work: 1908 Coal Mines Regulation Act', December 1986.

28 ITV, 8 February 1987.

29 ITV, 7 March 1987.

30 *Newsnight*, BBC TV, 3 July 1987.

31 BBC TV, 28 June 1987.

32 ITV, 7 July 1987.

33 *Miners' Strike News*, 16 February 1985 (MCG).

34 Samuel *et al.*, *op. cit.*, p. 231.

35 Woolley miner.

36 *Miners' Strike News*, 16 February 1985 (MCG).

37 P. Kahn, 'A new politics for coal', *New Socialist*, September 1985, pp. 26–30.

38 H C Deb., 15 July 1985, col. 5.

39 Wheldale miner.

40 Frickley miner.

41 Letter to Ted Scott, NUM Branch Secretary, Stillingfleet, from J. Henderson, Colliery General Manager, 5 February 1986.

42 Wheldale miner.

43 BC, *Code of Conduct*, London, BC, 1987.

44 BBC TV, 14 July 1987.

45 BBC TV, 15 July 1987.

46 BBC TV, 16 July 1987.

47 Letter to Frickley strikers from Colliery Manager, 16 July 1987.

48 BBC TV, 18 July 1987.

49 BBC TV, 19 July 1987.

50 BBC TV, 20 July 1987.

51 Letter to all North Yorkshire miners from Albert Tuke, Area Director, 17 July 1987.

52 Letter to all North Yorkshire miners from Albert Tuke, Area Director, 20 August 1987. The collieries named were: Woolley, South Kirkby, Nostell, Redbrook, Prince of Wales, Markham Main and Hatfield Main.

53 See, for example, *Yorkshire Post*, 22 August 1987.

54 The ballot asked: 'Are you in favour of giving authority to the National Executive Committee to call various forms of industrial action in the event of British Coal refusing to withdraw their new vicious disciplinary code and discuss the questions involved with the NUM?'

55 BBC TV, 12 August 1987.

56 BBC TV, 24 August 1987; ITV, 26 August 1987.

57 BBC TV, 3 September 1987.

58 BBC TV, 18 September 1987.

59 Letter to all NUM members from Bob Haslam, 20 September 1987.

60 BBC TV, 8, 12 October 1987.

61 Letter to Bentley miners from A. Davies, Colliery Manager, 6 January 1988.

62 Letter to South Yorkshire miners from T. D. Holmes, Area Industrial Relations Officer, 8 January 1988.

63 *Yorkshire Post*, 22 January 1988.

64 NEC Minutes, 26 July 1984, p. 593.

65 NEC Minutes, 30 August 1984, p. 626.

66 Adeney and Lloyd, *op. cit.*, p. 267.

67 MacGregor, *op. cit.*, p. 229.

68 NEC Minutes, 27 September 1984, p. 652.

69 NEC Minutes, 18 October 1984, p. 660.

70 *Ibid.*, p. 661.

71 NEC Minutes, 1 November 1984, p. 676; 15 November 1984, p. 712.

72 Rule 30 of the NUM (Nottingham Area) *Rules and Standing Orders*, provides that: In all matters in which the Rules of this Union and those of the National Union conflict, the Rules of the National Union shall apply, and in all cases of doubt or dispute the matter shall be decided by the National Executive Committee of the National Union and such decision shall be subject to the right of appeal set out in Rule 30 of the Rules of the National Union (NUM Nottingham Area, *Rules and Standing Orders*, 1980, p. 41).

73 NEC Minutes, 13 December 1984, p. 765.

74 NEC Minutes, 10 January 1985, p. 1.

75 Adeney and Lloyd, *op. cit.*, p. 268.

76 NEC Minutes, 10 January 1985, p. 3.

77 BBC TV, 12 January 1985.

78 Adeney and Lloyd, *op. cit.*, p. 261.

79 NEC Minutes, 28 March 1985, p. 117.

80 *Newsnight*, BBC TV, 18 January 1985.

81 BBC TV, 14, 15 January 1985.

82 BBC TV, 23 January 1985.

83 BBC TV, 18 February 1985.

84 SDC Report, 3 March 1985, p. 92.

85 ITV, 13 April 1985.

86 BBC TV, 11 May 1985.

87 ITV, 15 May 1985.

88 'Conference July 1985, NUM Rules Revision', Nottingham Area NUM (*mimeo*).

89 ITV, 14 May 1985.

90 BBC TV, 3 July 1985.

91 BBC TV, 6 July 1985.

92 ITV, 7 July 1985.

93 BBC TV, 6 July 1985.

94 BBC TV, 8 July 1985.

95 ITV, 9 July 1985.

96 *Financial Times*, 15 August 1985.

97 Rule 14(c) includes the passage: 'No person shall be present at a branch general meeting other than members of the branch convening the meeting except that the branch committe may, with the approval of the Area official who is agent for their branch, invite such person or persons to be present to give advice or assistance to the members present on any matter to be discussed at the meeting'. (NUM Nottingham Area, *Rules and Standing Orders*, 1980, p. 21).

98 BBC TV, 12 July 1985.

99 ITV, 15 July 1985.

100 BBC TV, 27 July 1985.

101 BBC TV, 7 July 1985.

102 ITV, 9 July 1985.

103 BBC TV, 8 July 1985.

104 ITV, 14 July 1985.

105 *Newsnight*, BBC TV, 21 October 1985.

106 BBC TV, 16 June 1986.
107 BBC TV, 21 October 1985.
108 *Newsnight*, BBC TV, 21 October 1985.
109 Adeney and Lloyd, *op. cit.*, p. 276.
110 *The Miner*, 25 November 1985.
111 Channel 4 News, 4 October 1985.
112 BBC TV, 15 October 1985.
113 BBC TV, 19 October 1985.
114 ITV, 20 October 1985.
115 ITV, 21 October 1985.
116 BBC TV, 5, 11 November 1985; ITV, 4 December 1985.
117 BBC TV, 29 October 1985.
118 BBC TV, 26 November 1985.
119 BBC TV, 26 October 1985; ITV, 27 October 1985.
120 BBC TV, 1 November 1985.
121 ITV, 28 November 1985.
122 BBC TV, 5 December 1985.
123 BBC TV, 17 June 1986.
124 BBC TV, 8 September 1986.
125 *Yorkshire Miner*, June 1987.
126 BBC TV, 22 June 1987.
127 *Newsnight*, BBC TV, 3 July 1987.
128 BBC TV, 20 July 1987.
129 BBC TV, 22 November 1985.
130 BBC TV, 6 December 1985.
131 BBC TV, 20 June 1986.
132 *New Society*, 25 October 1985.
133 Stephenson, *op. cit.*
134 Letter from Nigel Ashfield, former Secretary, Lea Hall strike committee, 10 July 1986, cited by Stephenson, *op. cit.*
135 Adeney and Lloyd, *op. cit.*, p. 366.
136 *Yorkshire Miner*, June 1987.
137 Adeney and Lloyd, *op. cit.*, p. 277.
138 *New Society*, 13 December 1985.
139 BBC TV, 10 January 1986.
140 Hiller, *op. cit.*, p. 213.
141 Former Forum activist.
142 Jack Taylor, seminar, Northern College, 24 January 1986.
143 BBC TV, 30 March 1987.
144 *Calendar*, ITV, 18 January 1988.
145 *Calendar*, ITV, 22 January 1988.
146 Jack Taylor, Seminar, Northern College, 24 January 1986.
147 S. Miller, '"The best thing that ever happened to us": women's role in the coal dispute', *Journal of Law and Society*, Vol. 12, No. 3, 1985, p. 363.
148 Loach, *op. cit.*, p. 178.
149 *Striking Back*, *op. cit.*, p. 33.
150 S. Taylor, 'Grub up for the miners!', *The Cutting Edge: Women and the Pit Strike*, ed. V. Seddon, London, Lawrence and Wishart, 1986, p. 83.

151 Salt and Layzell, *op. cit.*, p. 83.

152 Samuel *et al.*, *op. cit.*, p. 240.

153 Salt and Layzell, *op. cit.*, p. 24.

154 Gibbon and Steyne, *op. cit.*, p. 286.

155 J. Hamilton, 'Give us a future, give us our jobs!', *The Cutting Edge: Women and the Pit Strike*, ed. V. Seddon, London, Lawrence and Wishart, 1986, p. 221.

156 S. Corrigan, C. Cunningham and M. Thorburn, 'Fife women stand firm', *The Cutting Edge: Women and the Pit Strike*, ed. V. Seddon, London, Lawrence and Wishart, 1986, p. 46.

157 Gibbon and Steyne, *op. cit.*, p. 267.

158 *Coalfield Woman*, October 1985.

159 *Coalfield Woman*, January 1986.

160 NYWAPC, *op. cit.*, p. 60.

161 *Coalfield Woman*, July 1985.

162 D. Phillips, 'Warm welcome, sharp realities: Gwent', *The Cutting Edge: Women and the Pit Strike*, ed. V. Seddon, London, Lawrence and Wishart, 1986, p. 247.

163 *Coalfield Woman*, July 1985.

164 Salt and Layzell, *op. cit.*, p. 82.

165 *Socialist Viewpoint*, September 1985.

166 Adeney and Lloyd, *op. cit.*, p. 223.

167 'Miners' Campaign Group', April 1985 (MCG leaflet).

168 *Financial Times*, 30 December 1985.

169 G. B. Baldwin, *Beyond Nationalization, the Labour Problems of British Coal*, Cambridge, Mass, Harvard UP, 1955, pp. 303–8.

170 NUM circular to Area Secretaries, AS 13/73, 10 January 1973.

171 Letter to Syd Schofield from Denby Grange Branch Secretary, 20 January 1973.

172 Letter to Michael Eaton from Arthur Scargill, 19 February 1973.

173 Letter to Michael Eaton from Jack Leigh, 10 July 1973.

174 NACODS Agreement, 23 October 1984.

175 BBC TV, 25 March 1985.

176 BBC TV, 16 April, 1985.

177 BBC TV, 7 May 1985.

178 BBC TV, 11 April 1985.

179 BBC TV, 3 June 1985.

180 BBC TV, 5 June 1985.

181 ITV, 16 May 1985.

182 BBC TV, 24 May 1985.

183 BBC TV, 3 June 1985.

184 ITV, 7–12 June 1985.

185 BBC TV, 12 July 1985.

186 ITV, 21 August 1985, 10 September 1985.

187 BBC TV, 8, 9 October 1985.

188 BBC TV, 9, 17 October 1985.

189 BBC TV, 13, 28 June 1985.

190 BBC TV, 30 October 1985.

191 BBC TV, 8 October 1985.

192 BBC TV, 8 July 1985.

193 BBC TV, 26 May 1985, 13 June 1985, 1 July 1985.

194 MacGregor, *op. cit.*, p. 296.

195 *The Times*, 8 November 1986.

196 *The Times*, 29 June 1986.

197 *The Times*, 14 October 1986.

198 *The Times*, 7 December 1985.

199 Barnsley Metropolitan Borough Council, *Coal Mining and Barnsley, A Study of Employment Prospects*, Barnsley, BMBC, 1984.

200 West Yorkshire Metropolitan County Council, *Prospects for the West Yorkshire Coalfield*, Wakefield, WYMCC, 1984, p. 25.

201 R. Hudson, F. Peck and D. Sadler, *Undermining Easington: Who'll pay the price of pit closures?*, Easington, Easington District Council, 1984, p. 204.

202 *Ibid.*, p. 209.

203 *Ibid.*, pp. 216–28.

204 Burns, A., *et al.*, *An Interim Assessment of MINOS*, Report No. 4, Working Environment Research Group, University of Bradford, 1982; Burns *et al.*, *Second Report on MINOS*, *op. cit.*

205 Winterton, 'The source of the crisis', *op. cit.*

206 CCC Minutes, 11 January 1985.

207 CCC Minutes, 8 February 1985.

208 *Ibid.*

209 CCC Minutes, 11 June 1985.

210 Cllr. Hedley Salt, Chairman CCC, address at national launch, 15 May 1985.

211 Winterton, 'The source of the crisis', *op. cit.*

212 S. Fothergill and G. Gudgin, 'Job prospects in the coalfields', Coalfield Communities Campaign, Working Paper No. 2, Vol. 1, Barnsley: CCC, 1985, pp. 16–17.

213 P. Wright, 'Coal, communities and energy policy', Coalfield Communities Campaign Working Paper No. 3, Vol. 1, Barnsley, CCC, 1985, pp. 19–24.

214 R. Hudson, D. Sadler and E. Wade, 'Social costs of coal-mining decline: a position statement', Coalfield Communities Campaign Working Paper No. 4, Vol. 1, Barnsley, CCC, 1985, p. 27.

215 CCC Minutes, 14 June 1985.

216 CCC Minutes, 19 July 1985.

217 CCC Minutes, 14 June 1985.

218 E. Parkinson, 'Coalfield communities and their environment', Coalfield Communities Campaign Working Paper No. 5, Vol. 2, Barnsley, CCC, 1985, pp. 1–10.

219 S. Young, 'The scope for locally-based social and economic initiatives in the coalfield communities', Coalfield Communities Campaign Working Paper No. 7, Vol. 2, Barnsley, CCC, 1985, p. 22.

220 J. Gibson, 'Coal-prospects and new industrial opportunities', Coalfield Communities Campaign, Working Paper No. 6, Vol. 2. Barnsley, CCC, 1985, pp. 11–16.

221 G. S. Christian, 'The heating of public buildings, the case for coal', Working Paper No. 8, Vol. 2, Barnsley, Coalfield Communities Campaign, 1985, pp. 25–9.

222 CCC Minutes, 19 July 1985.

223 *Employment Gazette*, September 1985.

224 CCC Minutes, 11 October 1985.

225 K. Reichert, 'Coal in the Community', Coalfield Communities Campaign National Conference, Coalfield Communities Looking Forward, Wakefield, 19–21 February 1986, p. 13.

226 D. Hunt (MP), 'The opportunities for coal', Coalfield Communities Campaign National Conference, Coalfield Communities Looking Forward, Wakefield, 19–21 February 1986, p. 23.

227 B. J. McCormick, address to the Coalfield Communities Campaign National Conference, Coalfield Communities Looking Forward, Wakefield, 19–21 February 1986, p. 35.

228 S. Orme (Rt Hon), 'Energy and employment: the case for coal', Coalfield Communities Campaign National Conference, Coalfield Communities Looking Forward, Wakefield, 19–21 February 1986, p. 74.

229 S. Fothergill, address to Coalfield Communities Campaign National Conference, Coalfield Communities Looking Forward, Wakefield, 19–21 February 1986, Barnsley, CCC, p. 80.

230 *Ibid.*, p. 87.

231 S. Monk, *Retraining Opportunities for Miners who have been made Redundant*, Coalfield Communities Campaign Special Report No. 1, Barnsley, CCC, 1986.

232 CCC Minutes, 10 January 1986.

233 Energy Committee, *The Coal Industry*, Vol. I, *Memoranda*, HC 165–I, London, HMSO, 19 January, 18 June, 5 November 1986, pp. 186–205, Memorandum 35 (submitted by Coalfield Communities Campaign).

234 Energy Committee, *The Coal Industry*, Vol. II, *Minutes of Evidence*, HC 165–II, London, HMSO, 5 November 1986, pp. 105–18, Minutes of Evidence of the Coalfield Communities Campaign.

235 Energy Committee, *The Coal Industry*, Vol. I, *Memoranda*, HC 165–I, London, HMSO, 19 January, 18 June, 5 November 1986, pp. 314–20, Memorandum 60 (submitted by J. Winterton and R. Winterton).

236 *Ibid*, pp. 343–9, Memorandum 75 (submitted by Coalfield Communities Campaign).

237 P. Wright, 'Coal communities and energy policy', *op. cit.*, p. 19.

238 NUM General Election Manifesto, 1945 (cited in R. Page Arnot, *The Miners: One Union, One Industry*, London, Allen and Unwin, 1979, p. 98.)

239 LP, *Report of the Annual Conference of the Labour Party 1984*, London, LP, 1985, p. 285.

240 *Ibid.*

241 *Ibid.*, p. 44.

242 ITV, 5 July 1985.

243 BBC TV, 2 October 1985.

244 NUM Research, 'Plan for Coal: Labour Party discussions', Sheffield, NUM, 1985.

245 BBC TV, 4 September 1986.

246 LP, 'Labour's energy policy', London, LP, 1986.

247 *Ibid.*, p. 2.

248 *Ibid.*, p. 39.

249 *Ibid.*, p. 24.

250 BBC TV, 13 March 1987.

251 LP, *op. cit.*, 1987, p. 5.

252 *Ibid.*, p. 16.

253 NUM Research, 'Plan for Coal', *op. cit.*

254 NCB, *Report and Accounts 1985/1986*, London, NCB, 1986, p. 19.

255 BBC TV 28 May 1986.

256 Energy Committee, *The Coal Industry*, First Report, HC 165, London, HMSO, 24 February 1987, p. xlvi.

257 *Ibid*., p. xlvii.

258 *Ibid*., p. lxvii.

259 BBC TV, 12 March 1987.

260 NUM Research, 'The sale of the century: privatising power', December 1987.

261 C. Robinson and A. Sykes, *Privatise Coal*, Policy Study No. 85, London, Centre for Policy Studies, 1987, p. 66.

262 Letter to NUM strikers from Ian MacGregor, 'Your Future in Danger', nd. (posted 13 June 1984).

263 BC, *Report and Accounts, 1986/1987*, London, BC, 1987, p. 19.

Strike chronology

1 March 1984 Cortonwood closure announced at South Yorkshire Area Colliery Review Meeting.

5 March 1984 Yorkshire Area Council votes to implement 1981 ballot decision and instructs branches to stop work from the last shift on Friday 9 March.

6 March 1984 Reduction of 4m tonnes capacity entailing the loss of 20,000 jobs announced by NCB at CINCC meeting; Scottish Area Executive call strike from 12 March in response to proposed cut-backs in Scotland; spontaneous strikes at several Yorkshire collieries.

8 March 1984 NEC endorses Yorkshire and Scottish strikes. Government announces increases in redundancy payments for miners aged betweeen twenty-one and fifty years.

9 March 1984 Durham and Kent Area Executives agree to support strike. Nottinghamshire Executive calls for a pit-head ballot.

12 March 1984 Strike solid in Yorkshire, Kent, and most of South Wales and Scotland; flying pickets deployed to other areas.

13 March 1984 Harworth, Bevercotes and Thoresby pits picketed out; NRC begins monitoring the strike.

14 March 1984 High Court grants NCB injunction to prevent Yorkshire miners from picketing other areas; NRC drafts 8,000 police into Nottinghamshire from over twenty forces.

15 March 1984 David Jones killed while picketing at Ollerton. Notts NUM leaders call strike pending the outcome of an area ballot; Yorkshire Area Executive agrees to defy the court injunction on picketing.

16 March 1984 seventy-three per cent of Notts miners vote against strike. DHSS staff instructed to deduct all 'notional resources' (eg. food parcels) received by striker households from benefit entitlements.

17 March 1984 Area ballots in Midlands, North-East and North-West coalfields all heavily against strike.

18 March 1984 Kent police intercept miners at Dartford tunnel; BWAPC holds first meeting.

19 March 1984 Police operations maintain production at pits in Midlands. Yorkshire Area NUM found to be in contempt of court (but action postponed indefinitely); Taylor announces £100,000 Industrial Action Fund virtually exhausted.

20 March 1984 Kent miners lose court case against police over intercept policy; pickets prevent coal entering Scunthorpe steelworks from private wharf.

21 March 1984 Chadburn and Richardson urge Notts miners to cross picket lines.

26 March 1984 NUM leaders ask TUC not to become involved in the dispute.

27 March 1984 Caucus meeting of seven right-wing NEC members to co-ordinate campaign for a ballot; Triple Alliance agree to picket ports and steelworks, and to plan a series of sympathy stoppages.

28 March 1984 Miners blockade M18 and A38.

29 March 1984 Transport unions offer to black the movement of all coal by rail, sea, road and on the docks.

30 March 1984 ASLEF Executive instructs members not to cross NUM picket lines and not to move coal.

31 March 1984 NOP poll suggest fifty-one per cent of miners would vote for a strike, thirty-four per cent against.

2 April 1984 Yorkshire Strike Co-ordinating Committee (SCC) holds first meeting.

3 April 1984 Walker presses for a national ballot; NUS implements decision to black all coal imports; Federation of Rail Unions repeats call for members not to move coal or cross NUM picket lines.

4 April 1984 Ravenscraig steel-workers agree to use coal carried by private hauliers.

5 April 1984 Steel-workers vote to work normally.

6 April 1984 Operation Policewatch launched in Sheffield.

9 April 1984 NUS provide offices for NUM to co-ordinate support of other unions; NUM food kitchen opened in Barnsley.

10 April 1984 Two coke hauliers begin legal action against South Wales Area's secondary picketing of Port Talbot steelworks; emergency Commons debate on policing the miners' strike.

11 April 1984 NACODS vote 7,638 to 6,661 in favour of strike action over pit closures, less than the two-thirds majority required.

12 April 1984 NEC proposes to reduce majority required for strike from fifty-five per cent to a simple majority, and to convene an SDC; Kinnock calls for a national ballot.

17 April 1984 High Court grants injunction to prevent South Wales strikers picketing Port Talbot steelworks.

19 April 1984 SDC reduces majority required in strike ballot, votes against a national ballot, and agrees to establish a National Co-ordinating Committee (NCC).

20 April 1984 Scargill claims CEGB coal stocks will last nine weeks; TGWU threaten national docks strike if dockers are sacked for supporting miners.

24 April 1984 Scargill refuses to meet NCB to discuss rescheduling pit closures.

25 April 1984 NCB launches back-to-work campaign advertisements.

26 April 1984 Scargill calls for a national demonstration in Nottinghamshire.

30 April 1984 Scargill declares strike can be won without the support of Nottinghamshire miners.

1 May 1984 NCB grants Nottinghamshire miners a day off with pay, and 7,000 demonstrate in Mansfield against the strike.

3 May 1984 BSC intensifies use of scab road haulage to maintain coal deliveries to steelworks.

6 May 1984 Power-station coal stocks estimated at 20m tonnes.

8 May 1984 NUM guarantees coke supplies to Llanwern steelworks; NUR is asked to ban iron-ore deliveries to BSC plants.

10 May 1984 NEC discusses a new disciplinary rule and postpones branch elections for the duration of the strike.

11 May 1984 Thatcher declares the Government will not intervene in the coal dispute.

12 May 1984 Steelworks receive coal by rail following inter-union agreement; women's rally in Barnsley.

14 May 1984 Mansfield rally attended by 15,000; serious charges brought against over sixty participants; Scargill claims CEGB will be in desperate trouble in a few weeks; print-workers prevent *Sun* publishing defamatory picture of Scargill.

16 May 1984 Anne Scargill arrrested on picket line for 'wilful obstruction'.

17 May 1984 Police siege of Blidworth.

18 May 1984 Moses launches pilot back-to-work campaign in North Derbyshire with a letter to strikers.

19 May 1984 Scargill meets leaders of overseas unions in Paris: Australian, French and Polish unions agree to halt coal shipments to Britain.

21 May 1984 Murray condemns Yorkshire and Humberside Regional TUC for organising day of action in support of miners.

23 May 1984 First meeting between NUM leaders and MacGregor since the start of the strike ends after one hour; Megarry instructs Nottinghamshire Area to hold Union elections, ruling the NEC postponement void.

25 May 1984 High Court grants injunctions upholding the Notts miners' 'right to work' in defiance of Union instructions; coke convoys from Orgreave to Scunthorpe begin.

26 May 1984 Megarry rules that there is no basis for calling the strike 'official'.

28 May 1984 Over 2,000 pickets assemble at Orgreave.

29 May 1984 NCB claim fifteen pits at risk from geological problems caused by strike. Over 5,000 pickets at Orgreave; eighty-two arrested, sixty-nine injured.

30 May 1984 Scargill charged with obstruction at Orgreave.

4 June 1984 Government agrees to accept NCB losses.

5 June 1984 Daily Mirror reveals Government agreed to rail-workers' pay claims to divorce them from the miners' dispute.

7 June 1984 First major parliamentary debate on the strike; miners lobby Parliament; over 100 arrested.

8 June 1984 NCB and NUM negotiators meet in Edinburgh.

11 June 1984 Pro-strike left heavily defeated in Notts branch elections.

12 June 1984 Wales Regional TUC day of action in support of miners.

13 June 1984 NUM-NCB negotiations collapse; Scargill predicts strike will last until winter; MacGregor sends letter 'Your future in danger' to every striker.

14 June 1984 MacGregor suggests NCB may organise a strike ballot; NEC constitutes NCC.

15 June 1984 Joe Green killed by a lorry while picketing Ferrybridge power station.

17 June 1984 Maltby police station attacked.

18 June 1984 5,000 pickets at Orgreave; ninety arrested, over one hundred injured, including Scargill.

19 June 1984 First NCC meeting resolves to negotiate a national agreement on coke and coal deliveries to steelworks.

21 June 1984 MacGregor writes to all NUM members, predicitng that they will lose the strike.

25 June 1984 Pickets blockade iron ore at Llanwern.

26 June 1984 1,000 pickets converge on Coal House (Doncaster), in defiance of SCC instructions.

27 June 1984 ISTC leaders agree to accept coal from any source; SE Regional TUC demonstration in support of miners, including one-day stoppage by rail-workers and Fleet Street printers.

28 June 1984 NEC proposes disciplining members for actions 'detrimental to the interests of the Union'.

29 June 1984 Sirs tells NUM that ISTC support will not extend to industrial action.

2 July 1984 Working miners take control in Notts Area Council elections; Kiveton Park Manager writes to colliery workforce.

4 July 1984 Robert Haslam writes to steelworkers advising them not to support the miners.

5 July 1984 NCB resumes negotiations with NUM; Albert Tuke writes to Doncaster miners urging them to return to work.

7 July 1984 BS organises convoy of iron ore from Immingham to Scunthorpe, bringing Immingham and Grimsby dockers into the dispute; women's national office established in St. James's House.

8 July 1984 Notts strikers occupy committee rooms to prevent Area Council meeting to mandate representatives to oppose the new disciplinary rule.

9 July 1984 National docks strike called over BS use of non-registered labour; apparent progress in NCB-NUM negotiations; police riot in Fitzwilliam.

10 July 1984 Notts working miners granted High Court injunction to prevent EAC unless Area Council is allowed to meet; docks strike begins.

11 July 1984 SDC pass disciplinary rule change with two-thirds majority in defiance of court injunction; TGWU votes to extend action to ports outside the dock labour scheme.

12 July 1984 MacGregor announces scabs will not lose their jobs if expelled from the NUM, thereby terminating closed-shop agreement.

15 July 1984 ACAS intervenes in docks strike.

18 July 1984 Sir Robert Megarry rules new disciplinary rule unlawful; NCB-NUM negotiations end with NCB continuing to insist that pits must be closed unless they can be 'beneficially developed'; MacGregor writes to strikers with details of the settlement proposed by the Board.

19 July 1984 Thatcher describes striking miners as Britain's 'enemy within'; court grants injunction to opencast operator against South Wales Area NUM.

21 July 1984 Docks strike called off.

22 July 1984 First national conference of Women Against Pit Closures held at Northern College.

28 July 1984 First discussions between NUM and TUC leaders.

30 July 1984 South Wales Area NUM fined £50,000 for contempt of court; assets seized.

31 July 1984 Lawson claims £400m spent on stirke is 'a worthwhile investment'.

6 August 1984 Foulstone and Taylor begin action against Yorkshire Area NUM for failure to hold a ballot.

10 August 1984 SDC, boycotted by delegates from Nottinghamshire, Leicestershire and South Derbyshire, reiterates disciplinary rule change.

11 August 1984 National Women Against Pit Closures demonstration in London.

13 August 1984 NCCL begins inquiry into policing the dispute.

15 August 1984 Twelve collieries have closed since the strike began; Welsh miners occupy Birmingham offices of Price Waterhouse, sequestrators; Spanton tells NACODS Executive that members will not be paid unless they cross picket lines; first scab in Yorkshire Area (at Gascoigne Wood).

16 August 1984 Sequestrators seize £707,000 from South Wales Area NUM.

17 August 1984 NCB claims strike will cost more than 20,000 jobs originally in dispute.

20 August 1984 Solitary miner is escorted to work by 1,000 police at Gascoigne Wood.

21 August 1984 First formal discussion of strike by TUC General Council; MacGregor suggests Scargill should be prosecuted for 'criminal conspiracy' to intimidate working miners; police attack Armthorpe community.

23 August 1984 Second docks dispute provoked by steel-workers unloading *Ostia* which dockers had blacked at Hunterston.

24 August 1984 NCB rejects Orme's formula to settle dispute; Cowan declares negotiations will only resume when NUM abandons policy of opposing closures on grounds other than exhaustion; TGWU dockers' representatives vote 78–11 to strike from weekend.

28 August 1984 Cabinet meeting discusses docks strike and coal stocks.

30 August 1984 South Wales miners occupy transporter bridge in Newport and BS jetty at Port Talbot; NUM leaders meet TUC General Council in Brighton to discuss extent of support that could be generated at Congress.

2 September 1984 Robert Maxwell arranges for NCB-NUM negotiations to be resumed.

3 September 1984 TUC votes 10–1 to increase financial and industrial support to the miners.

4 September 1984 At TUC, Kinnock condemns picket-line violence; negotiations called off.

6 September 1984 TGWU flying pickets deployed to ports still working.

7 September 1984 NCB-NUM negotiations resumed; NUM agrees to consult TUC.

8 September 1984 Paul Womersley, fourteen years old, killed in accident whilst digging for coal.

9 September 1984 NCB-NUM negotiations begin in Edinburgh, moving to Selby, Doncaster and London to avoid reporters.

11 September 1984 Nat WMC formally constituted; MacGregor and Smith return to London during adjournment of Edinburgh negotiations.

12 September 1984 NACODS leaders agree to national ballot with a recommendation of strike action.

13 September 1984 Striking miners refused funeral grants by DHSS.

14 September 1984 Negotiations end in stalemate after a week.

18 September 1984 Second docks strike abandoned; 3,500 pickets at Kiveton Park.

21 September 1984 Steel-workers reject request to halt production.

24 September 1984 Scottish judge refuses to rule that the strike is unlawful without a national ballot; police attack Maltby community.

25 September 1984 Scargill and Kinnock hold 'constructive' discussions.

26 September 1984 NACODS offered compromise to avert strike.

28 September 1984 NACODS vote 82·5% in favour of strike; High Court rules strike 'unlawful' in Derbyshire, 'unofficial' in Yorkshire.

1 October 1984 NCB-NACODS negotiations resumed; Scargill and others served with writ at LP Conference for continuing to describe strike as 'official'.

2 October 1984 Kinnock condemns police and picket violence, but LP Conference gives overwhelming support to NUM.

3 October 1984 NCB-NACODS negotiations involve ACAS; Leslie Curtis, Police Federation Chairman, claims certain officers would be unable to work with a Labour administration.

6 October 1984 NCB-NUM negotiations involve ACAS.

9 October 1984 Leon Brittan guarantees that the Government will provide the money for policing the strike; Ottey resigns from NEC; mass picket at Cortonwood.

10 October1984 NUM fined £200,000, Scargill £1,000 for contempt of court.

11 October 1984 NCB-NACODS-NUM negotiations at ACAS.

14 October1984 Twenty-two arrested for coal-picking in Grimethorpe.

15 October 1984 ACAS negotiations collapse.

16 October 1984 NACODS calls national strike from 25 October; TUC agrees to coordinate further meetings to increase the effectiveness of the strike.

17 October 1984 Scargill's fine paid anonymously.

19 October 1984 EETPU power workers vote by eighty-four per cent not to support miners.

20 October 1984 Eaton appointed NCB spokesman.

22 October 1984 NCB-NACODS negotiations at ACAS.

24 October 1984 NACODS call off strike in return for modifications to colliery review procedure and agreement to pay deputies.

25 October 1984 High Court orders sequestration of NUM assets after failure to pay £200,000 contempt fine; NCB-NUM negotiations at ACAS.

28 October 1984 News breaks that NUM Chief Executive Officer, Windsor, has met Gadaffi in Libya.

29 October 1984 Eaton removed as NCB spokesman, Kirk suspended as Director of Public Relations.

30 October 1984 Eccles, TUC Chairman, says NUM should compromise; BR warns employees of job losses if they support miners.

31 October 1984 ACAS negotiations collapse.

1 November 1984 NUM launch series of rallies to boost morale; working miners begin action to make NEC members individually liable for the £200,000 fine.

2 November 1984 NCB offers Christmas bonus and holiday pay to miners who return by 19 November.

3 November 1984 Kinnock 'too busy' to attend NUM rallies; High Court injunction forbids North Derbyshire Area from spending any funds on the strike, Price Waterhouse seize NUM assets; SDC affirms continuation of strike.

5 November 1984 Taylor and Foulstone begin action to prevent Yorkshire Area officials controlling Union funds.

6 November 1984 Kirk resigns over Tory involvement in NCB public relations; 3,000 attend rally in Edinburgh.

7 November 1984 Dublin Court rules that £2·75m NUM funds be frozen; TUC report on fuel stocks concludes little likelihood of power cuts.

9 November 1984 Mass-picket at Cortonwood in response to solitary strike-breaker; petrol bomb attack on police station in South Yorkshire; Eccles claims one hundred per cent victory for miners no longer possible.

12 November 1984 Barricades and petrol bombs in several battles between police and strikers in Yorkshire.

13 November 1984 Hostile reception for Willis for his condemnation of picket violence at a South Wales rally; Scargill notes that strike would have been won already if TUC had given miners the same support that CBI had given NCB.

18 November 1984 Two brothers die coal-picking.

21 November 1984 Government deducts further £1 per week notional strike pay from benefits to strikers' dependants.

24 November 1984 NCB claim 5,952 miners returned this week.

27 November 1984 Getty donates £100,000 to hardship fund.

28 November 1984 NUM funds traced to Zurich and Luxembourg.

29 November 1984 Public verbal attack on Scargill from McKay (North Wales).

30 November 1984 High Court appoints Receiver to administer NUM funds; Wilkie killed driving a working miner to the colliery, three strikers charged with murder.

1 December 1984 Brewer appointed Receiver.

3 December 1984 SDC votes to continue Union's defiance of court, despite sequestration and receivership.

4 December 1984 Brewer fails to obtain NUM funds in Luxembourg.

5 December 1984 TUC Liaison Committee meets Walker and King to discuss strike, without knowledge of NUM, NEC or TUC General Council.

6 December 1984 TUC Liaison Committee reiterates support for NUM and discusses ways of assisting the Union.

7 December 1984 Brewer resigns, replaced by Arnold.

9 December 1984 TUC implores Government and NCB to reopen negotiations; NCCL demands formal inquiry into dispute.

10 December 1984 Government agrees to meet costs if Price Waterhouse fail to obtain NUM funds.

11 December 1984 Notts miners vote heavily in favour of new constitution giving greater autonomy from National Union.

13 December 1984 LP and NUM leadership hold first formal meeting.

14 December 1984 Scargill fined £250 for obstruction; TUC Liaison Committee present their proposals for a settlement to Walker.

20 December 1984 Notts Area Council endorses area rule change.

21 December 1984 High Court instructs TUC to furnish details of its assistance to NUM during strike.

27 December 1984 North Derbyshire miners campaign to remove strike supporters from NEC.

30 December 1984 NCB claim miners who return could earn £1,000.

1 January 1985 Strike costs estimated at £85m per week; Kent miners given jail sentences of three to five years.

3 January 1985 Kinnock joins picket line in his constituency for the first time since the strike began.

10 January 1985 NEC agrees to widen negotiating team and threatens to expel dissident Notts Area.

11 January 1985 Local authorities agree to promote Coalfield Communities Campaign, but maintain its independence from the NUM dispute.

12 January 1985 Richardson suspended by Notts Area.

16 January 1985 Richardson wins right in court to retain limited role as area official.

17 January 1985 NUR-ASLEF regional strike in support of rail-workers sacked at Coalville for refusing to transport coal.

21 January 1985 Heathfield and Smith have informal talks.

26 January 1985 Amnesty for 500 sacked strikers becomes an issue.

29 January 1985 Preliminary negotiations collapse.

30 January 1985 Sequestrators recover 5m NUM funds.

31 January 1985 Willis assures NEC that TUC will not be a party to negotiations, but will seek a 'procedural route' to negotiations.

2 February 1985 Yorkshire Area WAPC conference at Northern College.

4 February 1985 NCB claims further 2,318 miners return to work; ACAS involved in attempting to restart negotiations; adverse effect of oil burn on balance of payments estimated at £2,800m.

7 February 1985 NCB refuses to negotiate.

11 February 1985 Injunction restrians South Wales Area from organising other than peaceful picketing in numbers not exceeding six; TUC regional day of action.

12 February 1985 Willis meets MacGregor; injunction restrains Yorkshire Area from organising at eleven pits, picketing other than peaceful picketing in numbers not exceeding six at a miner's own place of work.

15 February 1985 NEC rejects NCB proposals presented by Willis.

19 February 1985 TUC Liaison Committee meets Thatcher, then drafts further proposals for settlement.

20 February 1985 NEC rejects TUC document.

21 February 1985 SDC rejects TUC document.

22 February 1985 NCB plans 'final push' to encourage sufficient miners to return to bring the total working to fifty per cent.

24 February 1985 London rally in support of miners, 101 arrested after disturbances near Whitehall.

25 February 1985 NCB claim record number returning to work; TUC Finance and General Purposes Committee denies input to the NCB settlement proposal.

26 February 1985 South Wales moves to end the strike by an organised return without agreement, supported by Durham.

27 February 1985 NCB claims fifty per cent now working, including twenty per cent in Yorkshire.

28 February 1985 NEC calls SDC.

1 March 1985 Durham, Lancashire, South Wales and COSA Areas vote for organised return; Yorkshire Area Council delegates vote 42–22 to continue strike, but to reconvene following branch meetings.

2 March 1985 Reconvened Yorkshire Area Council votes 38–31 to return, but reversed on a card vote to 557–561.

3 March 1985 SDC votes 98–91 to return on 5 March; deterioration of pits expected to cost £600m.

4 March 1985 Yorkshire Area Council votes to return unconditionally; Scotland and Kent vote to stay on strike as amnesty for 718 sacked miners unresolved.

5 March 1985 Miners return in Yorkshire and South Wales; half of Yorkshire and Scotland still out with Kent.

10 March 1985 Kent and Scotland vote to return; Richardson dismissed by Notts Area Executive.

Appendix A

Interview schedule

Pages 304–15 are a facsimile reproduction of the interview schedule used for data collection.

1

MINERS' STRIKE INTERVIEW SCHEDULE

Case: [| |] (1 2 3) Colliery: ________________

How many people work here? [| | |] (4 5 6 7)

RESPONDENT'S DETAILS

Name: ________________ Job:(card 3) [ ] (8) Age: [|] (9 10)

Have any relatives of yours ever worked in coal mining ?
- grandfather/grandmother ? [ ] 11
- father/mother ? [ ] 12
- son/daughter ? [ ] 13
- brother/sister ? [ ] 14
- other relative ? [ ] 15

How far do you live from your place of work ? [|] (16 17) miles

Is the community where you live dependent upon your pit:
absolutely(2) partly(1) not at all(0) ? [ ] 18

Do you hold a position on your local NUM branch ?
no (0) yes (1) DK (2) [ ] 19
(if yes, specify)

Do you hold a position on the area panel or executive ?
no (0) yes (1) DK (2) [ ] 20
(if yes, specify)

Before the strike, how often did you attend NUM branch meetings ?
regularly(1) occasionally(2) never(0) ? [ ] 21

During the strike how many union meetings (including rallies) did you attend (approximately) ? [|] (22 23)

Which issues do you think were most important in bringing about the NCB's pit closure programme:
(card M rank)
- new capacity like Selby [ ] 24
- reorganization of existing mines [ ] 25
- new technology (MINOS) [ ] 26
- lack of demand for coal [ ] 27
- appointment of Ian MacGregor [ ] 28
- defeat of Tory Government in 1974 [ ] 29

Did you take an active part in the following strikes:
- 1969 unofficial Yorkshire strike ? [ ] 30
- 1970 unofficial Yorkshire strike ? [ ] 31
- 1972 national strike ? [ ] 32
- 1974 national strike ? [ ] 33

How many times do you estimate you have taken local strike action since working for the NCB ? [| |] (34 35 36)

During the strike which activities have taken up most of your time ? (card C rank):

- collecting/fund raising ☐ 37
- speaking at meetings ☐ 38
- manning strike centre ☐ 39
- local picketing ☐ 40
- flying pickets ☐ 41
- sabotage against NCB ☐ 42
- retribution against scabs ☐ 43

On average, how many days per week did you spend on the above activities ? ☐ 44

Did your involvement in the strike:
increase(1) stay the same(2) decrease(0) ? ☐ 45

Were you arrested during the dispute ? yes(1)/no(0) ☐ 46

<u>if yes</u>, what was the charge ? ______________ ☐ 47

How did you find out what was happening during the strike? (card A rank)

- television/radio ☐ 48
- newspapers ☐ 49
- <u>Miner</u>/<u>Yorkshire Miner</u> ☐ 50
- local strike bulletins ☐ 51
- area executive ☐ 52
- panel meetings ☐ 53
- branch meetings ☐ 54
- strike centre ☐ 55
- friends ☐ 56
- picket line ☐ 57

When did you first think that there would be a strike:

- before the overtime ban ? 1
- during the overtime ban ? 2
- not until the strike began ? 3 ☐ 58

How many weeks did you think it would last ? ☐☐ 59 60

What did you hope the strike would achieve ? (card B rank)

- no pit closures ☐ 61
- no job losses ☐ 62
- 4 day week ☐ 63
- new technology agreement ☐ 64
- consolidation of bonus ☐ 65
- new plan for coal ☐ 66

Are you an active member of any political party ? yes(1)/no(0) ☐ 67

2 STRIKE MOBILIZATION

Before the strike, was the colliery ever described as:

uneconomic	3
experiencing geological difficulties	2
nearing exhaustion	1 ☐ 1

Before the strike was the colliery considered:

under immediate closure threat	3
a short-life colliery	2
a long-life colliery	1 ☐ 2

Are any of the following MINOS facilities installed at, or planned for, this colliery ?

MINOS control room	☐ 3
coal preparation plant	☐ 4
coal clearance	☐ 5
IMPACT (condition monitoring)	☐ 6
MIDAS (system 70 000 shearer)	☐ 7
horizontal shearer guidance	☐ 8
FACE (chock advance)	☐ 9
FIDO (face delay reports)	☐ 10

What date did the strike begin at this colliery ? ☐☐☐☐☐☐ 11 12 13 14 15 16

How was the action called ?

branch meeting	1
decision of panel	2
instructions from area	3
flying pickets	4 ☐ 17

If picketed out, where were the pickets from ?

this colliery	1
within panel	2
within area	3
another area	4 ☐ 18

If from another area, which ? ________________ ☐ 19

What was the majority membership reaction ?

unanimous support	1
majority support	2
objections to picketing	3
objections to strike	4
calls for national ballot	5
threats of strike breaking	6 ☐ 20

3 STRIKE INVOLVEMENT

Can you tell me roughly how many people were involved in the different activities essential to the strike (card C):

collecting/fund raising	1-3
speaking at meetings	4-6
manning strike centre	7-9
local picketing	10-12
flying pickets	13-15
sabotage against NCB	16-18
retribution against scabs	19-21

During which periods (card 1) were the numbers involved in each activity (card C) highest ?

collecting/fund raising	22
speaking at meetings	23
manning strike centre	24
local picketing	25
flying pickets	26
sabotage against NCB	27
retribution against scabs	28

I want to ask some questions about the types of people involved in each of these activities (card C).

Firstly, which age groups (card 2) were most involved in each activity (card C) ?

collecting/fund raising	29
speaking at meetings	30
manning strike centre	31
local picketing	32
flying pickets	33
sabotage against NCB	34
retribution against scabs	35

Secondly, which occupations (card 3) were most involved in each activity (card C) ?

collecting/fund raising	36
speaking at meetings	37
manning strike centre	38
local picketing	39
flying pickets	40
sabotage against NCB	41
retribution against scabs	42

Thirdly, how did kinship (card 4) affect people's involvement (card C) ?

collecting/fund raising	43
speaking at meetings	44
manning strike centre	45
local picketing	46
flying pickets	47
sabotage against NCB	48
retribution against scabs	49

Fourthly, did the community (card 5) in which people were living affect their involvement (card C) ?

- collecting/fund raising [ ] 50
- speaking at meetings [ ] 51
- manning strike centre [ ] 52
- local picketing [ ] 53
- flying pickets [ ] 54
- sabotage against NCB [ ] 55
- retribution against scabs [ ] 56

Fifthly, did previous strike experience (card 6) play a part in determining people's involvement (card C) ?

- collecting/fund raising [ ] 57
- speaking at meetings [ ] 58
- manning strike centre [ ] 59
- local picketing [ ] 60
- flying pickets [ ] 61
- sabotage against NCB [ ] 62
- retribution against scabs [ ] 63

Finally, did a person's politics (card 7) affect their involvement (card C) ?

- collecting/fund raising [ ] 64
- speaking at meetings [ ] 65
- manning strike centre [ ] 66
- local picketing [ ] 67
- flying pickets [ ] 68
- sabotage against NCB [ ] 69
- retribution against scabs [ ] 70

4 STRIKE BREAKING

I now want to ask some questions about strike breaking and strike breakers.

When did the first scabs go in ? [| | | | |] 1–6

When was the first organized return to work ? [| | | | |] 7–12

How many were involved at that time ? [| |] 13–15

Can you tell me roughly how many people were involved in different strike-breaking activities (card D) ?

- against picketing [| |] 16–18
- against the strike [| |] 19–21
- returned after Xmas [| |] 22–24
- returned before Xmas [| |] 25–27
- back to work campaign [| |] 28–30

During which periods (card 1) were the numbers involved in each activity highest (card D) ?

- against picketing [ ] 31
- against the strike [ ] 32
- returned after Xmas [ ] 33
- returned before Xmas [ ] 34
- back to work campaign [ ] 35

5

As with those active in the strike, I want to ask some questions about the types of people involved in these strike breaking activities (card D).

Firstly, which <u>age groups</u> (card 2) were most involved in each activity (card D) ?

- against picketing ☐ 36
- against the strike ☐ 37
- returned after Xmas ☐ 38
- returned before Xmas ☐ 39
- back to work campaign ☐ 40

Secondly, which <u>occupations</u> (card 3) were most involved in each activity (card D) ?

- against picketing ☐ 41
- against the strike ☐ 42
- returned after Xmas ☐ 43
- returned before Xmas ☐ 44
- back to work campaign ☐ 45

Thirdly, how did <u>kinship</u> (card 4) affect people's involvement (card D) ?

- against picketing ☐ 46
- against the strike ☐ 47
- returned after Xmas ☐ 48
- returned before Xmas ☐ 49
- back to work campaign ☐ 50

Fourthly, did the <u>community</u> (card 5) in which people were living affect their involvement (card D) ?

- against picketing ☐ 51
- against the strike ☐ 52
- returned after Xmas ☐ 53
- returned before Xmas ☐ 54
- back to work campaign ☐ 55

Fifthly, did previous strike <u>experience</u> (card 6) play a part in determining people's involvement (card D) ?

- against picketing ☐ 56
- against the strike ☐ 57
- returned after Xmas ☐ 58
- returned before Xmas ☐ 59
- back to work campaign ☐ 60

Finally, did a person's <u>politics</u> (card 7) affect their involvement (card D) ?

- against picketing ☐ 61
- against the strike ☐ 62
- returned after Xmas ☐ 63
- returned before Xmas ☐ 64
- back to work campaign ☐ 65

5 STRIKE ORGANIZATION

I want to ask some questions about how the strike was organized locally.

How were picketing targets communicated to pickets ?

Were targets ever discussed by branch committee ?
no (0) yes (1) DK (2) ☐ 1

Were targets ever discussed at branch meetings ?
no (0) yes (1) DK (2) ☐ 2

Did pickets provide intelligence reports for future targets ?
no (0) yes (1) DK (2) ☐ 3

Were pickets organized into cells ?
no (0) yes (1) DK (2) ☐ 4

Did pickets ever decide their own targets ?
no (0) yes (1) DK (2) ☐ 5

Would you characterize strike organization as (card E):
- centralized (organized by NUM) 1
- hierarchical (area-panel) 2
- two-way process (information) 3
- autonomous (branch-pickets) 4 ☐

Did the character of strike organization (card E) alter at diferent phases of the strike (card 1) ?
- mobilization ☐ 7
- Orgreave era ☐ 8
- TUC era ☐ 9
- NCB Xmas campaign ☐ 10
- fragmentation ☐ 11

Were people appointed with responsibility for the picket ?
no (0) yes (1) DK (2) ☐ 12

What sort of people took charge of the picket ?
- age groups (card 2) ☐ 13
- occupations (card 3) ☐ 14
- kinship (card 4) ☐ 15
- community (card 5) ☐ 16
- experience (card 6) ☐ 17
- politics (card 7) ☐ 18

What instructions were issued (and by whom) concerning conduct of the picket ? (ask for copy)

What arrangements were made for the eventuality of arrests ? (ask for documents)

Roughly how many people were arrested from this colliery during the strike ? ☐☐☐ 19-21

Do you have a record of the charges against those arrested and the results of cases ?

no (0) yes (1) DK (2) ☐ 22

(<u>if yes</u>, ask to see records)

Of the pickets you attended, what percentage would you describe as (card F):

peaceful	☐☐	23-24
ritual push	☐☐	25-26
violent	☐☐	27-28

During which <u>periods</u> (card 1) could picketing be most nearly characterized in these terms (card F):

peaceful	☐	29
ritual push	☐	30
violent	☐	31

What <u>factors</u> (card G) influenced the <u>character</u> of picketing (card F) most ? (rank)

area of target	☐	32
importance of target	☐	33
police numbers	☐	34
police activity at picket	☐	35
police forces involved	☐	36
police activity in community	☐	37
picket organizers	☐	38
number of pickets	☐	39
who was picketing	☐	40
phase of strike	☐	41
presence of media	☐	42
other	☐	43

(Prompt for elaboration on major items). How did 'factor' influence the character of the picket ?

On what proportion of pickets would you describe the <u>attitude</u> of the police as (card H):

friendly, joking	☐☐	44-45
quiet, courteous	☐☐	46-47
tense	☐☐	48-49
belligerent	☐☐	50-51
violent	☐☐	52-53

Was there a <u>pattern</u> to police-picket relations over the duration of a particular picket ?

no (0) yes (1) DK (2) ☐ 54

(<u>if yes</u>, elaborate)

Did it happen that the same pickets confronted the same police regularly ?

no (0) yes (1) DK (2) ☐ 55

(<u>if yes</u>, elaborate)

6 STRIKE MAINTENANCE

Which were the most important sources of income to the branch overall (card J rank):

- public donations ☐ 1
- street collections ☐ 2
- collections at meetings ☐ 3
- Labour Party ☐ 4
- trades councils ☐ 5
- other trade unions ☐ 6
- Miners' Support Group ☐ 7
- area/panel ☐ 8
- other (specify) ☐ 9

Was the branch adopted by (twinned with) any institutions ?

no (0) yes (1) DK (2) ☐ 10

(specify) ____________________

What was the approximate total weekly income to the branch from all sources during the different phases of the strike (card B) ?

- mobilization ☐☐☐ 11–13
- Orgreave era ☐☐☐ 14–16
- TUC era ☐☐☐ 17–19
- NCB Xmas campaign ☐☐☐ 20–22
- fragmentation ☐☐☐ 23–25

Which were the most important sources of food given to the branch overall (card J):

- public donations ☐ 26
- street collections ☐ 27
- collections at meetings ☐ 28
- Labour Party ☐ 29
- trades councils ☐ 30
- other trade unions ☐ 31
- Miners' Support Group ☐ 32
- area/panel ☐ 33
- other (specify) ☐ 34

Name any companies or shops that regularly supplied food for food parcels.

During which periods (card 1) did the branch receive most food from all sources ? (rank)

- mobilization ☐ 35
- Orgreave era ☐ 36
- TUC era ☐ 37
- NCB Xmas campaign ☐ 38
- fragmentation ☐ 39

How were food parcels made up and distributed ?

Were any other goods or commodities supplied ?
(what and by whom)

What were the most widespread forms of hardship experienced by members ? (card K rank)

- food ☐ 40
- clothing ☐ 41
- coal ☐ 42
- electric or gas ☐ 43
- housing ☐ 44
- financial ☐ 45
- other (specify) ☐ 46

What sort of people were <u>most affected</u> by hardship ?

- age groups (card 2) ☐ 47
- occupations (card 3) ☐ 48
- kinship (card 4) ☐ 49
- community (card 5) ☐ 50
- experience (card 6) ☐ 51
- politics (card 7) ☐ 52

How were hardship cases monitored ?

Were any special arrangements negotiated with local traders on behalf of strikers ? (specify)

When was a strike kitchen first organized in association with the pit, and by whom ?

Approximately how many meals per week were supplied during the different phases of the strike ? (card 1)

- mobilization ☐☐☐ 53-55
- Orgreave era ☐☐☐ 56-58
- TUC era ☐☐☐ 59-61
- NCB Xmas campaign ☐☐☐ 62-64
- fragmentation ☐☐☐ 65-67

Where did the kitchen receive most of its support from ? (card J rank):

- public donations ☐ 68
- street collections ☐ 69
- collections at meetings ☐ 70
- Labour Party ☐ 71
- trades councils ☐ 72
- other trade unions ☐ 73
- Miners' Support Group ☐ 74
- area/panel ☐ 75
- other (specify) ☐ 76

Did the branch, or active members, publish any leaflets, newsheets or papers during the strike ?

no (0) yes (1) DK (2) ☐ 77

(<u>if yes</u>, ask for copies)

7

STRIKE AFTERMATH

During the strike, did the NCB issue any "condition statements" about the pit ?

no (0) yes (1) DK (2) ☐ 1

(if yes, specify and ask for copies)

What was the overall condition of the colliery immediately after the strike ?

Were there any specific problems that prevented production being resumed ?

no (0) yes (1) DK (2) ☐ 2

(if yes, specify)

How many weeks before coal is/was produced after the strike ? ☐☐ 3 4

Since the strike, has the colliery been described as:

uneconomic	3
experiencing geological difficulties	2
nearing exhaustion	1 ☐ 5

Is the colliery now considered:

under immediate closure threat	3
a short-life colliery	2
a long-life colliery	1 ☐ 6

What was the composition of the branch committee, in terms of the men's ages (card 2) and occupations (card 3), during the strike ?

name/number ____________	occupation ☐ 7	age group ☐ 8
name/number ____________	occupation ☐ 9	age group ☐ 10
name/number ____________	occupation ☐ 11	age group ☐ 12
name/number ____________	occupation ☐ 13	age group ☐ 14
name/number ____________	occupation ☐ 15	age group ☐ 16
name/number ____________	occupation ☐ 17	age group ☐ 18
name/number ____________	occupation ☐ 19	age group ☐ 20
name/number ____________	occupation ☐ 21	age group ☐ 22
name/number ____________	occupation ☐ 23	age group ☐ 24
name/number ____________	occupation ☐ 25	age group ☐ 26
name/number ____________	occupation ☐ 27	age group ☐ 28
name/number ____________	occupation ☐ 29	age group ☐ 30
name/number ____________	occupation ☐ 31	age group ☐ 32
name/number ____________	occupation ☐ 33	age group ☐ 34
name/number ____________	occupation ☐ 35	age group ☐ 36
name/number ____________	occupation ☐ 37	age group ☐ 38

Are any of these no longer committee men ?

no (0) yes (1) DK (2) ☐ 40

(if yes, specify)

If yes, who has replaced them ?

If yes, why ?

Compared with before the strike, has attendance at branch meetings since the strike been:
higher (1) unchanged (2) lower (0) [ ] 41

Has the sort of people who attend branch meetings altered ?
no (0) yes (1) DK (2) [ ] 42

If yes, what sort of people now attend the branch more ?
- age groups (card 2) [ ] 43
- occupations (card 3) [ ] 44
- kinship (card 4) [ ] 45
- community (card 5) [ ] 46
- experience (card 6) [ ] 47
- politics (card 7) [ ] 48

Compared with before the strike, has the branch been:
more militant (1) unchanged (2) less militant (0) [ ] 49

In the future do you think your involvement in the union will:
increase(1) stay the same(2) decrease(0) ? [ ] 50

Since the strike do you expect your involvement in politics will:
increase(1) stay the same(2) decrease(0) ? [ ] 51

Do you think this is true of most people who were active in the strike ?
no (0) yes (1) DK (2) [ ] 52

How many men did the NCB threaten to sack at this pit ? [ ][ ][ ] 53 54 55

How many of these have been reinstated ? [ ][ ][ ] 56 57 58

Did any disputes arise, as a result of the overtime ban ?
no (0) yes (1) DK (2) [ ] 59

If so, were any unresolved at the time the strike began ?
no (0) yes (1) DK (2) [ ] 60

If yes, how were they resolved after the strike ?

In dealings with the pit management since the strike, is the branch:
stronger (1) unchanged (2) weaker (0) [ ] 61

Since the strike, does the manager consult the union:
more often (1) unchanged (2) less often (0) [ ] 62

Is there anything in particular about the strike at this colliery that you think we should be aware of ?

Is there anything about the strike in general that you think we ought to investigate ?

12

Appendix B

Questionnaire

Pages 318–21 are a facsimile reproduction of the questionnaire distributed to individual strikers.

MINERS' STRIKE QUESTIONNAIRE

This questionnaire is being conducted by Jonathan Winterton of Bradford University. On the advice of the Yorkshire Area Executive, its distribution was left until the strike ended. It is completely confidential. The results will be presented to the Area Executive and will be available for examination by any member.

Thank you for helping us in this way. Please try to answer all questions. The questionnaire contains a final section for completion by your wife or girlfriend. Please return the completed questionnaire to:

Age____years

Do you live:
alone/with parents/with wife/other ?

If you have any children, how many are:
- under school age ___
- at school ___
- unemployed ___
- work for NCB ___
- other paid employment ___

How far do you live from your place of work ? ___miles

Have any relatives of yours ever worked in coal mining ?
- grandfather/grandmother ? ___
- father/mother ? ___
- son/daughter ? ___
- brother/sister ? ___
- other relative ? ___

Do either of your next-door neighbours work for the NCB ? yes/no

Is the community where you live dependent upon your pit:
absolutely/partly/not at all ?

Is the house in which you live:
- rented from council/landlord ? ___
- rented/mortgaged from NCB ? ___
- mortgaged to a building society ? ___

Which colliery/workshop do you work at ? __________

What is your job ? ____________________ Wage class ? ____

How long have you been in this job ? ___years ___months

Have you ever been transferred from one pit to another ? Yes/No
- if yes, how many times ? ____
- when was the last time ? ____
- did any transfers involve moving from one NCB area to another ? yes/no

1

Do you regard your present job as being:
secure/insecure/don't know

Have you ever worked outside coal mining ? yes/no

Have you ever seriously considered leaving the industry ? yes/no

Were you offered redundancy before the strike ? yes/no

If you were offered redundancy now would you:
accept/refuse/consider ?

Do you hold a position on your local NUM branch ? yes/no

Do you serve on the area panel or executive ? yes/no

Before the strike, how often did you attend NUM branch meetings ?
regularly/never/occasionally

During the strike how many union meetings (including rallies) did you attend ? __ (approximately)

In the future do you think your involvement in the union will:
increase/decrease/stay the same

Which issues do you think were most important in bringing about the NCB's pit closure programme:
(Please rank: 1=most important, 2=second most important, etc.)

- new capacity like Selby __
- reorganization of exiting mines __
- new technology (MINOS) __
- lack of demand for coal __
- appointment of Ian MacGregor __
- defeat of Tory Government in 1974 __

Did you take an active part in the following strikes:

- 1969 unofficial Yorkshire strike ? __
- 1970 unofficial Yorkshire strike ? __
- 1972 national strike ? __
- 1974 national strike ? __

How often do you estimate you have taken local strike action since working for the NCB ? ___ times.

During the 1984-85 strike were you involved in the following ?
(Please rank: 1=most important, 2=second most important, etc.)

- picketing other areas ? __
- local picketing ? __
- driving pickets ? __
- manning a strike centre ? __
- organizing collections ? __
- publicity activities ? __

On average, how many days per week did you spend on the above activities ? ___ (days per week)

2

Did your involvement in the strike:
increase/decrease/stay the same

Were you arrested during the dispute ? yes/no
if so, what was the charge ?

How did you find out what was happening during the strike?
(Please rank: 1=most important, 2=second most important, etc.)
- television/radio ___
- newspapers ___
- Miner/Yorkshire Miner ___
- local strike bulletins ___
- union meetings ___
- strike centre ___
- friends or picket line ___

When did you first think that there would be a strike:
- before the overtime ban ? ___
- during the overtime ban ? ___
- not until the strike began ? ___

How long did you think it would last ? ____(months)

Where did you get income from during the strike?
(Please rank: 1=most important, 2=second most important, etc.)
- casual employment ___
- wife's earnings ___
- son's/daughter's earnings ___
- tax rebate ___
- supplementary benefit ___
- child allowance ___

Did you make any special financial arrangements:
- bank overdraft/loan ___
- building society (mortgage) ___
- hire purchase (suspension) ___
- rent or rate rebates ___
- special terms from local traders ___
- loans from relatives/friends ___

Did you, or your dependents, use the soup kitchens run at strike centres or by women's action groups ? yes/no

What did you hope the strike would achieve ?
(Please rank: 1=most important, 2=second most important, etc.)
- no pit closures ___
- no job losses ___
- 4 day week ___
- new technology agreement ___
- consolidation of bonus ___
- new plan for coal ___

Are you an active member of any political party ? yes/no

Since the strike do you expect your involvement in politics will·
increase/decrease/stay the same ?

3

WOMEN'S QUESTIONNAIRE

What is your relationship to the person who has completed the rest of this form: wife/living together/girl friend ?

Do you have a paid job ? yes/no
how many hours per week ? ___

As a result of the strike did you:
start work/increase hours/decrease hours ?

Are you a member of a trade union ? yes/no

Were you actively involved in:
picketing ? ___
strike centre ? ___
women against pit closures ? ___
women's support group ? ___
soup kitchen ? ___
rallies/marches ? ___
organizing collections ? ___

Did you receive any donations of food or money from outside supporters via the union ? yes/no

What possessions, if any, were you forced to sell or return (if rented or on hire purchase) in order to meet your basic needs?

During the strike did you feel:
miserable ? more/less/same
anxious ? more/less/same
irritated ? more/less/same
tired ? more/less/same

During the strike were there more arguments in the house over:
money ? yes/no
food ? yes/no
going out ? yes/no
politics ? yes/no
housework ? yes/no
children ? yes/no

Did the strike make your husband's/boyfriend's relationships:
with you better/worse/same
with friends better/worse/same
with children better/worse/same

Has the strike altered your political outlook ? yes/no
if so, how ?

4

List of interviews

anon. (Hatfield Main) Stainforth, 16 October 1985
Blake Barrett (Kellingley) Ferrybridge, 14 June 1985
Kevin Barron MP (Rother Valley) Thurcroft, 24 September 1985
George Bell (Shireoaks) Worksop, 22 July 1985
Rt Hon Tony Benn MP (Chesterfield) Westminster, 17 July 1985
Albert Bowns (Kiveton Park) Kiveton, 31 July 1985
John Brook (Stillingfleet) Thorpe Willoughby, 14 June 1985
Glen Burton (Fryston) Ferrybridge, 14 June 1985
Albert Butcher (Yorkshire Main) Edlington, 8 July 1985
Neil Butterfield (Royston Drift) Havercroft, 21 June 1985
Dave Buttle (Dodworth) Pogmore, 22 May 1985
Richard Caborn MP (Sheffield C.) Westminster, 16 July 1985
Steve Cadman (Savile) Woodlesford, 8 May 1985
Mick Carter (Cortonwood) Brampton, 26 July 1985
Frank Clarke (South Kirkby) South Kirkby, 17 May 1985
Peter Clarney (Barrow) Northern College, 14 October 1985
Jim Coulter (Woolley) Dodworth, 26 July 1985
Stan Crowther MP (Rotherham) Westminster, 16 July 1985
Karl Curry (Ledston Luck) Allerton Bywater, 18 June 1985
Bryan Davies (Sec., PLP) Westminster, 15 July 1985
Frank East (Fryston) Castleford, 26 April 1985
Neville Eaton (Treeton) Sheffield, 23 July 1985
Nigel Evans (Hickleton Main) Thurnscoe, 23 May 1985
Derek Fatchett MP (Leeds C.) Westminster, 16 July 1985
Martin Flannery MP (Hillsborough) Sheffield, 3 October 1985
Tony Fletcher (Dinnington) North Anston, 11 July 1985
Derek France (Silverwood) Rotherham, 27 August 1985
Alan Gosling (Manvers Main) Manvers Main, 28 August 1985
Richard Graystock (Emley Moor) Emley Moor, 4 July 1985
Alan Hale (Barnburgh) Goldthorpe, 27 August 1985
Walter Harrison MP (Wakefield) Westminster, 17 July 1985
John Hobson (Askern) Askern, 22 July 1985
Kevin Hughes (Brodsworth) Scawthorpe, 1 October 1985
Paul Ironmonger (Houghton Main) Little Houghton, 22 July 1985
Bernard Jackson (Wath) Wombwell, 1 October 1985
Phil James (Goldthorpe) Barnsley, 23 September 1985
Bob Johnston (Bullcliffe Wood) Wakefield, 24 July 1985

Rt Hon Gerald Kaufmann MP (Gorton) Westminster, 16 July 1985
Harry Leadbeater (Grimethorpe) Grimethorpe, 11 October 1985
Andy Leigh (Brookhouse) Hackenthorpe, 20 September 1985
Gary Long (Kellingley) Northern College, 21 October1985
Pete McCulloch (Rossington) Rossington, 12 July 1985
Allen McKay MP (Barnsley W. & Pen.) Westminster, 15 July 1985
Max Madden MP (Bradford W.) Westminster, 15 July 1985
Dave Marshall (Nostell) Crofton, 2 July 1985
John Mason (Wistow) Pontefract, 20 June 1985
William Michie MP (Heeley) Westminster, 17 July 1985
Jimmy Millar (Markham Main) Armthorpe, 16 October 1985
Ted Millward (Maltby) Maltby, 9 September 1985
Eric Mountain (Dearne Valley) Darfield, 14 October 1985
Gary Nicholson (Gascoigne Wood) Featherstone, 2 July 1985
Rt Hon Stan Orme MP (Salford E.) Westminster, 16 July 1985
Ian Oxley (Frickley) South Elmsall, 8 May 1985
Dave Parry (Thurcroft) Heeley, 11 July 1985
Neil Parry (Kinsley Drift) Barnsley, 9 May 1985
Nigel Pearce (Denby Grange) Middlestown, 29 July 1985
Jim Rae (Barrow) Worsborough, 30 July 1985
Martin Redmond MP (Don Valley) Westminster, 16 July 1985
Steve Roberts (Prince of Wales) Pontefract, 24 April 1985
Alf Robinson (Woolley) Wombwell, 17 May 1985
Bill Roby (Darfield Main) Wombwell, 25 July 1985
Roy Salter (Ferrymoor Riddings) Northern College, 2 May 1985
Steve Shaw-Wright (Ricall) Selby, 14 June 1985
Dennis Skinner MP (Bolsover) Westminster, 15 July 1985
Dave Southwell (Cadeby) Connisborough, 28 August 1985
Mick Spencer (Glasshoughton) Glasshoughton, 21 June 1985
Keith Storrs (North Gawber) Mapplewell, 24 July 1985
Colin Sweeting (Park Hill) Huddersfield, 13 September 1985
Clive Tenant (Ackton Hall) Featherstone, 24 April 1985
Inky Thompson (Highgate) Bolton-on Dearne, 19 September 1985
Frank Underwood (Manton) Manton, 31 July 1985
Steve Vickers (Sharlston) Featherstone, 17 April 1985
Alan Wassell (Wheldale) Normanton, 10 May 1985
Les Wigglesworth (Allerton Bywater) Allerton Bywater, 10 May 1985
Russ Williams (Bentley) Bentley, 22 May 1985
Maurice Wood (Kilnhurst) Swinton, 25 July 1985

Bibliography

In addition to items referred to in the Notes, the following bibliography is intended to be a comprehensive list of all publications to date relating to the 1984–85 miners' strike. Newspaper reports and features are omitted.

Aage, H. 'Economic arguments on the self-sufficiency of natural resources', *Cambridge Journal of Economics*, Vol. 8, 1984, pp. 105–13.

Abdel-Rahim, M. *Strike Breaking in Essex*, London, Canary Press, 1985.

Adam Smith Institute, *Energy Policy*, London, ASI, 1984.

Adeney, M. 'The miners' strike: all quiet on the Yorkshire front', *Listener*, 13 December 1984.

Adeney, M., and Lloyd, J. *The Miners' Strike 1984–5, Loss without Limit*, London, RKP, 1986.

Against All the Odds, Sheffield, NUM, 1984 (anthology of poems from the strike).

Ager, D. 'Geology and Mr Scargill', *New Scientist*, 17 May 1984, pp. 36–7.

Aikin, O. 'Legal aftermath of the miners' strike', *Personnel Management*, July 1985, pp. 24–7.

Algie, J., and Foster, W. 'Seams of agreement', *New Statesman*, 23 November 1984, pp. 14–16.

Allen, K. 'The coal industry', *The Nationalised Fuel Industries*, eds. G. L. Reid, K. Allen and D. J. Harris, London, Heinemann, 1973.

Allen, V. L. *The Militancy of British Miners*, Shipley, Moor Press, 1981.

Allen, V. L. 'Miners' man', *New Society*, 24 January 1985, pp. 136–7.

'Appomattox or civil war?', *Economist*, 27 May 1978, pp. 21–2.

Ashenfelter, O., and Johnson, G. E. 'Bargaining theory, trade unions and industrial strike activity', *American Economic Review*, Vol. 59, 1969, pp. 35–49.

Ashton, J., *et al. The Miners' Strike Assessed*, London, Communist Party, 1985.

Assault on the Unions, Report No. 34, London, Counter Information Services, 1984.

Bagwell, P. S. *End of the Line*, London, Verso, 1984.

Bailey, R. 'UK energy options', *National Westminster Quarterly Review*, August 1985, pp. 30–41.

Baldwin, G. B. *Beyond Nationalisation, the Labour Problems of British Coal*, Cambridge, Mass, Harvard UP, 1955.

Barker, C. 'Coal strike hits agencies', *Campaign*, 30 November 1984, p. 4.

Barnsley Women Against Pit Closures, *Women Against Pit Closures*, Barnsley, BWAPC, 1984.

Barnsley Women Against Pit Closures, *Women Against Pit Closures, Volume 2*, Barnsley, BWAPC, 1985.

Barnsley Metropolitan Borough Council, *Coal Mining and Barnsley, A Study of Employment Prospects*, Barnsley, BMBC, 1984.

Barratt Brown, M. *What Really Happened to the Coal Industry*? Nottingham, Institute for Workers' Control, 1972.

Barrett, M. *London in the Dark*? London, Greater London Council, 1984.

Beaton, L. *Shifting Horizons*, London, Canary, 1985.

Beharrell, P., and Philo, G., *Trade Unions and the Media*, London, Macmillan, 1977.

Benn, T. 'Don't let the Tories kill the coal industry', *Tribune*, 11 March 1983.

Bercusson, B. 'One hundred years of conspiracy and protection of property, time for a change', *Modern Law Review*, Vol. 40, 1977, p. 268.

Berkovitch, I. *Coal on the Switchback*, London, Allen and Unwin, 1977.

Berry, T., *et al.* 'NCB accounts – a mine of mis-information?', *Accountancy*, Vol. 96, No. 1097, January 1985, pp. 10–12.

Berry, T., *et al.* 'Management control in an area of the NCB, rationales of accounting practices in a public enterprise', *Accounting, Organisations and Society*, Vol. 10, No. 1, 1985 pp. 3–28.

Beynon, H. 'The miners' strike at Easington', *New Left Review*, No. 148, November/December 1984, pp. 104–15.

Beynon, H. (ed.) *Digging Deeper, Issues in the Miners' Strike*, London, Verso, 1985.

Beynon, H. 'Introduction', *Digging Deeper: Issues in the Miners' Strike*, ed. H. Beynon, London, Verso, 1985, pp. 1–26.

Beynon, H. 'Authority and change in the coalfields', *Journal of Law and Society*, Vol. 12, No. 3, 1985, pp. 395–403.

Beynon, H., Hudson, R. and Sadler, D. *(Mis)Managing Horden*, Work and Employment Research Unit, University of Durham, 1985.

Beynon, H., Hudson, R. and Sadler, D. 'Nationalised industry policies and the destruction of communities: some evidence from North East England', *Capital and Class*, No. 29, 1986, pp. 27–57.

Beynon, H., and McMylor P. 'Decisive power: the new Tory state against the miners', *Digging Deeper: Issues in the Miners' Strike*, ed. H. Beynon, London, Verso, 1985, pp. 29–45.

Bill, P., and Silke, E. 'Impact of the coal crisis', *Contract Journal*, 25 October 1984, pp. 12–13.

Blake, N. 'Picketing, justice and the law', *Policing the Miners' Strike*, eds. B. Fine and R. Millar, London, Lawrence and Wishart, 1985, pp. 103–19.

Blazyca, G., Harrison, J., and Morgan, B. 'Figures behind the strike, Nigel's sums', *New Statesman*, 10 August 1984, pp. 11–12.

Booth, A., and Pack, M. 'Baldwin, Thatcher and the aftermath of industrial disputes', *Political Quarterly*, Vol. 56, No. 3, 1985, pp. 271–8.

Booth, A., and Smith, R. 'The irony of the iron fist: social security and the coal dispute 1984–85', *Journal of Law and Society*, Vol. 12, No. 3, 1985, pp. 365–74.

Boyfield, K. *Putting pits back into profit*, London, Centre for Policy Studies, 1985.

Bradshaw, S. 'The impact of high technology on the politics of mining', *Listener*, 25 April 1985, pp. 2–4.

Brewer, D. 'The role of the accountant in the mining industry', *Mining Engineer*, August 1985, pp. 53–7.

BSSRS Technology of Political Control Group, *TechnoCop: New Police Technologies*, London, Free Association, 1985.

BSSRS Technology of Political Control Group, *TechnoCop: New Police Technologies*, London, Free Association, 1985.

British Coal, *Code of Conduct*, London, BC, 1987.

British Coal, *Report and Accounts 1986/7*, London, BC, 1987.

Britten, H. 'The miners' strike – what has it meant for women?', *Spare Rib*, No. 153, April 1985, pp. 9–10.

Brown, A. 'Hemsworth's pigs', *Spectator*, 21 July 1984, pp. 15–16.

Brown, A. 'Return to Hemsworth', *Spectator*, 9 February 1985, p. 14.

Brown, A. 'The miners' strike: back to what?', *Spectator*, 9 March 1985, pp. 11–12.

Bulmer, M. 'Sociological models of the mining community', *Sociological Review*, Vol. 23, No. 1, 1975, pp. 61–92.

Bulmer, M. (ed.) *Mining and Social Change*, London, Croom Helm, 1978.

Bunyan, T. *The Political Police in Britain*, London, Quartet, 1977.

Bunyan, T. 'From Saltley to Orgreave via Brixton', *Journal of Law and Society*, Vol. 12, No. 3, 1985, pp. 293–303.

Bunyard, P. *Nuclear Britain*, London, New English Library, 1981.

Burkham, C. 'Last stand at the Alamo', *New Statesman*, 27 April 1984, pp. 8–9.

Burns, A., *et al. An Interim Assessment of MINOS*, Report No. 4, Working Environment Research Group, University of Bradford, 1982.

Burns, A., *et al.* 'The miners and new technology', *Industrial Relations Journal*, Vol. 14, No. 4, 1983, pp. 7–20.

Burns, A., Newby, M., and Winterton, J. 'New technology and the restructuring of work in British coal mining', British Sociological Association Conference, University of Bradford, 2–5 April 1984.

Burns, A., Newby, M., and Winterton, J. *Second Report on MINOS*, Report No. 6, Working Environment Research Group, University of Bradford, 1984.

Burns, A., Newby, M., and Winterton, J. 'The restructuring of the British coal industry', *Cambridge Journal of Economics*, Vol. 9, No. 1, 1985 pp. 93–110.

Buxton, N. K. *The Economic Development of the British Coal Industry*, London, Batsford, 1978.

Callinicos, A., and Simons, M. *The Great Miners' Strike of 1984–5 and its Lessons*, London, SWP, 1985.

Campaign Group of Labour MPs, *Annual Report, June 1983–October 1984*, London, Campaign Group, 1984.

Campaign Group of Labour MPs, *Justice: The Miners' Strike 1984–85*, London, Verso, 1986.

Campbell, A., and Warner, M. 'Leadership in the miners' union: Arthur Scargill's rise to power', *Journal of General Management*, Vol. 10, No. 3, 1985, pp. 4–22.

Campbell, B. 'The other miners' strike', *New Statesman*, 27 July 1984, pp. 8–10.

Campbell, B. 'Politics old and new', *New Statesman*, 8 March 1985, pp. 22–5.

Campbell, M. 'Police strikeforce', *New Statesman*, 23 March 1984, p. 10.

'Can the TUC get Scargill off the hook?', *Economist*, 1 December 1984, pp. 23–4.

Canwell, N. 'How many pits really are uneconomic?', *Management Accounting*, Vol. 62, No. 7, 1984, pp. 36–7.

Carter, P. 'Striking the right note', *Marxism Today*, Vol. 29, No. 3, March 1985, pp. 28–31.

Central Electricity Generating Board, *Corporate Plan*, London, CEGB, 1978.

Central Electricity Generating Board, *Proof of Evidence P4: the need for Sizewell 'B'*, Evidence to Sizewell 'B' Enquiry, 1982.

Central Electricity Generating Board, *Annual Report and Accounts 1984–85*, London, CEGB, 1985.

'Challenge to deductions from strikers' benefit', *Legal Action*, January 1985.

Chandler, K. W. 'MINOS – a computer system for central control at collieries', Second International Conference on Centralised Control Systems, London, 20–3 March 1978.

Chapman, J., and Houghton, S. 'Coal in Crisis', Houghton Main NUM, 1985, (*mimeo*).

Cheshire, J. *The Future for Coal.* (Fabian Society briefing), London, Fabian Society, 1985.

Christian, G. S. 'The heating of public buildings: the case for coal', Working Paper No. 8 Vol. 2, Barnsley, Coalfield Communities Campaign, 1985, pp. 25–9.

Christian, L. 'Restriction without conviction: the role of the courts in legitimising police control in Nottinghamshire', *Policing the Miners' Strike*, eds. B. Fine and R. Millar, London, Lawrence and Wishart, 1985, pp. 120–36.

'City counts the cost of the miners' strike', *Investors' Chronicle*, 6 July 1984, p. 15.

Civil Liberties and the Miners' Dispute, London, National Council for Civil Liberties, 1984.

Clarney, P. 'Mineworkers' strike', *New Socialist*, No. 19, September 1984, pp. 26–7.

Clayton, R., and Tomlinson, H. 'Should the NUM be fined for Mr Scargill's contempt?' *Law Society's Gazette*, Vol. 81, No. 39, 31 October 1984, p. 2984.

Cleary, J. 'FIDO (Face Information Digested On-line) at Bold Colliery', *Mining Engineer*, November 1981, pp. 281–9.

Clegg, H. A. and Chester, D. N. *Future of Nationalisation*, Oxford, Blackwell, 1953.

Clegg, H. A. *The Changing System of Industrial Relations in Great Britain*, Oxford, Blackwell, 1979.

Clement, T. 'Long hot summer's day', *Police Review*, Vol. 93, No. 4818, 18 June 1985, pp. 1310–11.

Cliff, T. 'Patterns of mass strike', *International Socialism*, No. 29, 1985, pp. 3–61.

Clutterbuck, R. *Britain in Agony, the Growth of Political Violence*, London, Faber, 1978.

Clutterbuck, R. *Industrial Conflict and Democracy, The Last Chance*, London, Macmillan, 1984.

Coal, Church and Community, Easington, Theological Consultancy to the North-East Churches with Easington District Council, 1985.

Coal not Dole, London, Tommy Healey Fund, 1985.

'Coalfield women at the face: a roundtable discussion', *Marxism Today*, Vol. 28, No. 7, July 1984, pp. 28–30.

'Coal's strike-free future', *Economist*, 26 March 1985, pp. 13–14.

Coates, D. *The Labour Party and the Struggle for Socialism*, Cambridge, Cambridge UP, 1975.

Coates, K., and Topham, T. *Trade Unions and Politics*, Oxford, Blackwell, 1986.

Cockburn, C. 'Trade unions and the radicalising of socialist feminism', *Feminist Review*, No. 16, 1984, pp. 43–73.

Cohen, J. 'The divided village', *Listener*, 7 June 1984, p. 4.

Cole, J. 'The Scargill legacy: post-strike debts on the party balance sheets', *Listener*, 28 February 1985, pp. 4–5.

Commission of the European Communities, *The Effects of the Reduction of Manpower in the Mining Industry on Mining Social Security Systems and Pension Systems in Particular*, Social Policy Series, No. 23, Brussels, 1973.

Committee of Members of the Inns of Court Conservative and Unionist Society, *A Giant's Strength*, London, ICCUS, 1958.

Conservative Political Centre, *Fair Deal at Work*, London, Con. Party, 1968.

Conservative Political Centre, *Financing Strikes*, London, Con. Party, 1974.

Conservative Research Department, *Economic and Industrial Situation*, Notes on Current Politics No. 1, London, Con. Party, 21 January 1974.

Conservative Research Department, *The Coal Industry*, Politics Today, No. 19, 5 November 1984.

Cooper, C. C. 'Improving machine utilisation and reliability', COMMIT 82, Computer-based Mine Management Information Technology Exhibition and Symposium, Harrogate, 8–10 December 1982.

Cooper, D., and Hopper, T. (eds.) *Debating Coal Closures, Economic Calculation in the Coal Dispute 1984–5*, Cambridge, CUP, 1988.

Corrigan, S., Cunningham, C., and Thorburn, M. 'Fife women stand firm', *The Cutting Edge: Women and the Pit Strike*, ed. V. Seddon, London, Lawrence and Wishart, 1986, pp. 30–49.

'Cost of the dispute', *Labour Research*, Vol. 74, No. 3, March 1985, pp. 64–5.

'Cost of the strike: blank cheques can bounce', *Economist*, 13 October 1984, pp. 33–4.

Costello, M. 'Militant determination of the British miners', *World Marxist Review*, Vol. 27, No. 11, November 1984, pp. 104–8.

Coulter, J., Miller, S., and Walker, M. *State of Siege*, London, Canary, 1984.

'Counting the cost of the coal strike', *Investors' Chronicle*, 21 September 1984, pp. 8–9.

Cousin, G., Fine, Bob, and Millar, R. 'Conclusion: the politics of policing', *Policing the Miners' Strike*, eds. B. Fine and R. Millar, London, Lawrence and Wishart, 1985, pp. 227–36.

Coventry Miners' Wives Support Group, *Mummy . . . What did you do in the strike?* Coventry, CMWSG, 1986.

Crick, M. *Scargill and the Miners*, Harmondsworth, Penguin, 1985.

Critcher, C., *et al.* 'Report of a survey of attitudes to the miners' strike in Sheffield', Sheffield, Department of Communications, Sheffield City Polytechnic, 1984, (*mimeo*).

Curtis, R., Hogan, E., and Horowitz, S. *Nuclear Lessons*, Harrisburg, Pa., Stackpole Books, 1980.

Cutler, T., *et al. The Aberystwyth Report on Coal*, Economic History Department, University College of Wales, 1985.

Daly, L. *The Miners and the Nation*, Edinburgh, NUM (Scottish Area), 1968.

'Damaged mine costs climb in UK strike', *World Mining Equipment*, May 1985, pp. 40–1.

Danby, G. A. 'The use and abuse of rights', *Journal of Law and Society*, Vol. 12, No. 3, 1985, pp. 375–84.

Davies, B., *et al.* 'Taking stock of the year long strike', *Engineer*, 7 March 1985, p. 4.

Davies, G. 'Prospects for Sterling after the most expensive strike in British history', *Financial Weekly*, 1–7 February 1985, p. 6.

Davies, G. 'The real cost of the miners' strike', *Financial Weekly*, 8–14 March 1985, p. 4.

Davies, G., and Metcalf, D. *Pit Closures: Some Economics*, Weekend World, London Weekend Television, 1984.

Davies, J., and James, S. 'Women from the valleys turn activist', *The Cutting Edge: Women and the Pit Strike*, ed. V. Seddon, London, Lawrence and Wishart, 1986, pp. 16–29.

Deep Digs! Cartoons of the Miners' Strike, London, Pluto, 1985.

In Defence of the NUM, London, Socialist Action, 1985.

De Friend, R., and Rubin, G. R. 'Civil law and the 1984–85 coal dispute', *Journal of Law and Society*, Vol. 12, No. 3, 1985, pp. 321–32.

Dempsey, A., 'The role of the press in the formation of the Union of Democratic Mineworkers and the 1984/85 miners' strike', B.Sc. dissertation, University of Bradford, 1986.

Dennis, N., Henriques, F., and Slaughter, C. *Coal is Our Life*, London, Eyre and Spottiswoode, 1956.

Department of Economic Affairs, *National Plan*, Cmnd. 2764, London, HMSO, 1965.

Department of Employment, *Employment Code of Practice (Picketing) Order 1980*, SI 1980 No. 1757, London, HMSO, 1980.

Department of Employment, *Democracy in Trade Unions*, Green Paper, Cmnd. 8778, London, HMSO, 1983.

Department of Employment and Productivity, *In Place of Strife: A Policy for Industrial Relations*, White Paper Cmnd. 3888, London, HMSO 1969.

Department of Employment and Productivity, *Ryehope: A pit closes*, London, HMSO, 1970.

Department of Energy, *Coal Industry Examination: Interim Report June 1974*, London, HMSO, 1974.

Department of Energy, *Coal Industry Examination: Final Report 1974*, London, HMSO, 1974.

Department of Energy, *Coal for the Future: Progress with 'Plan for Coal' and Prospects to the Year 2000*, London, HMSO, 1977.

Department of Energy, *Energy Policy: A Consultative Document*, Green Paper, Cmnd. 7101, London, HMSO, 1978.

Department of Energy, *Proof of Evidence for Sizewell 'B' Enquiry*, London, HMSO, 1982.

Department of the Environment *et al. Coal and the Environment*, Commission on Energy and the Environment (Ch. Flowers), London, HMSO, 1981.

Dibblin, J., and Hollingsworth, M. 'What? No power cuts?', *New Statesman*, 1 February 1985, p. 14.

Dickinson, M. *To Break a Union*, Manchester, Booklist, 1984.

Diggins, L., Massey, D, and Wainwright, H. 'The comfort of strangers', *New Socialist*, No. 24, February 1985, pp. 22–3.

Dixon, S. C. 'A critical analysis of press coverage of the firemen's dispute', B.Sc. dissertation, University of Bradford, 1980.

Dolby, N. *Norma Dolby's Diary: An account of the Great Miners' Strike*, London, Verso, 1987.

Douglass, D., *Tell us Lies about the Miners: The Role of the Media in the Great Coal Strike of 1984/1985*, Doncaster, Direct Action Movement, 1985.

Douglass, D. *Come and Wet this Truncheon: The Role of the Police in the Coal Strike of 1984/1985*, Doncaster, Direct Action Movement, 1986.

Douglass, D. *A Year of Our Lives: A Colliery Community in the Great Coal Strike of 1984/85*, London, Hooligan Press, 1986.

Douglass, M. 'Mining for change', *Spare Rib*, No. 144, July 1984, p. 9.

Dunn, A. *The Strikers' Moneyguide*, London, Campaign for Alternative Industrial and Technical Systems, 1985.

Durcan, J. W., and McCarthy, W. E. J. 'The state subsidy theory of strikes: an examination of statistical data for the period 1956–1970', *British Journal of Industrial Relations*, Vol. 12, No. 1, 1974, pp. 26–47.

East, R., Power, H. and Thomas, P. A. 'The death of mass picketing', *Journal of Law and Society*, Vol. 12, No. 3, 1985, pp. 305–20.

East, R., and Thomas, P. A. 'Freedom of movement: Moss v McLachlan', *Journal of Law and Society*, Vol. 12, No. 1, 1985, pp. 77–86.

East, R., and Thomas, P. A. 'Road-blocks: the experience in Wales', *Policing the Miners' Strike*, eds. B. Fine and R. Millar, London, Lawrence and Wishart, 1985, pp. 137–44.

East, R. and Thomas, P. A. 'Stopping the miners', *New Law Journal*, Vol. 135, No. 6186, 18 January 1985, pp. 63–6.

Eden, R., *et al. Energy Economics: Growth, Resources and Policies*, Cambridge, Cambridge UP, 1981.

Electricity Council, *Medium term development plan 1985–92*, London, HMSO, 1985.

Employment Committee. Sixth Report, session 1984–85. *The Dismissal of National Coal Board Employees*, HC 416, London, HMSO, 19 June 1985.

Energy Committee, *Pit Closures*, Minutes of Evidence – National Union of Mineworkers, HC 80, London, HMSO, 25 November 1982.

Energy Committee, *Pit Closures*, Minutes of Evidence – National Coal Board, HC 80–i, London, HMSO, 1 December 1982.

Energy Committee, *Pit Closures*, Minutes of Evidence – Department of Energy, HC 80–ii, London HMSO, 7 December 1982.

Energy Committee, *Pit Closures*, Report, HC 135, London, HMSO, 21 December 1982.

Energy Committee, *The Coal Industry*, Vol. I, *Memoranda*, HC 165-I, London, HMSO, 19 January, 18 June, 5 November 1986.

Energy Committee, *The Coal Industry*, Vol. II, *Minutes of Evidence*, HC 165–II, London, HMSO, 5 November 1986.

Energy Committee, First Report, session 1986–87, *The Coal Industry*, HC 165 London, HMSO, 24 February 1987.

Evans, J., Hudson, C., and Smith, P. 'Women and the Strike: it's a whole way of life', *Policing the Miners' Strike*, eds. B. Fine and R. Millar, London, Lawrence and Wishart, 1985, pp. 188–203.

Ewing, K. D. 'The strike, the courts, and the rule books', *Industrial Law Journal*, Vol. 14, 1985, p. 160.

Ezra, D. *Prospects for Expansion in Coal*, Cadman Memorial Lecture, London, Royal Society 17 November 1976.

Ezra, D. 'A review of Plan for Coal', *Mining Engineer*, January 1981, pp. 479–87.

Feickert, D. 'Of men and MINOS', *Computing Europe*, November 1979, pp. 30–1.

Feickert, D. 'Towards a new future: campaigning for coal', *Digging Deeper: Issues in the Miners' Strike*, ed. H. Beynon, London, Verso, 1985, pp. 245–52.

Feickert, D. 'Turn on to coal', *New Socialist*, No. 25, March 1985, p. 15. (Edited version of above.)

Feickert, D. 'The midwife of mining', *Capital and Class*, No. 31, 1987, pp. 7–15.

Field, J. 'Union-free management: experiences and resources from North America', *Trade Union Studies Journal*, Summer 1983, pp. 14–15.

Field, J. 'Rule of the law', *New Socialist*, September 1984, pp. 28–9.

Field, J. 'Police monitoring: the Sheffield experience', *Policing the Miners' Strike*, eds. B. Fine and R. Millar, London, Lawrence and Wishart, 1985, pp. 204–16.

Field, J. 'Labour's Dunkirk', *New Socialist*, No. 26, April 1985, pp. 11–14.

Field, J. (ed.) *Coal Mining and the Community: Perspectives from the Warwickshire Coalfield Debate*, Open Studies Paper No. 1, Coventry, University of Warwick, 1986.

Fine, Ben, O'Donnell, K., and Prevezer, M. *Coal After Nationalisation*, Discussion Paper No. 138, Department of Economics, London, Birkbeck College, 1983.

Fine, Ben. 'The future of British coal', *Capital and Class*, No. 23, 1984, pp. 67–82.

Fine, Bob, and Millar, R. (eds.) *Policing the Miners' Strike*, London, Lawrence and Wishart, 1985.

Fitton, M. 'Union-free management: the importation of US union-busting to Britain', B.Sc. dissertation, University of Bradford, 1987.

Flanders, A. *Management and Unions*, London, Faber, 1970.

Fothergill, S. 'Nuclear power and the coal industry', Coalfield Communities Campaign Working Paper No. 10, Vol. 3, Barnsley, CCC, 1985, pp. 17–28.

Fothergill, S. Address to Coalfield Communities Campaign National Conference, Coalfield Communities Looking Forward, Wakefield, 19–21 February 1986, Barnsley, CCC, pp. 79–98.

Fothergill, S., and Gudgin, G. 'Job prospects in the coalfields', Coalfield Communities Campaign, Working Paper No. 2, Vol. 1, Barnsley, CCC, 1985, pp. 8–17.

Francis, H. 'Mining the popular front', *Marxism Today*, Vol. 29, No. 2, February 1985, pp. 12–15.

Francis, H. 'NUM united: a team in disarray', *Marxism Today*, Vol. 29. No. 4, April 1985, pp. 28–34.

Francis, H. 'Unfinished business: the breaking of the NUM?', *Marxism Today*, Vol. 29, No. 8, August 1985, pp. 22–4.

Francis, H. 'The law, oral tradition and the mining community', *Journal of Law and Society*, Vol. 12, No. 3, 1985, pp. 267–71.

Franks, S. 'The cost of the coal strike: who foots the bill?', *Listener*, 18 October 1984.

Freeman, M. *Our Day Will Come*, London, Junius, 1985.

Fryer, B. 'Trade unionism in crisis: the miners' strike and the challenge to union democracy', *Digging Deeper: Issues in the Miners' Strike*, ed. H. Beynon, London, Verso, 1985, pp. 69–85.

Fryer, B. 'Trade unionism, public policy and the law', *Freedom and Fairness: Empowering People at Work*, ed. K. Coates, Nottingham, Spokesman, 1986, pp. 63–95.

Fury, B. 'Women's work', *New Society*, 15 November 1984, p. 238.

Gamble, A. 'The strike the Tories wanted', *New Socialist*, No. 26, April, 1985, pp. 15–7.

Geary, R. *Policing Industrial Disputes, 1893–1985*, Cambridge, CUP, 1985.

Gerlis, A. 'Knuckles to briefcases', *New Statesman*, 5 November 1982, pp. 20–1.

Gennard, J. *Financing Strikers*, London, Macmillan, 1977.

Gennard, J., and Lasko, R. 'Supplementary benefit and strikers', *British Journal of Industrial Relations*, Vol. 12, No. 1, 1974, pp. 1–25.

Gennard, J., and Lasko, R. 'The individual and the strike', *British Journal of Industrial Relations*, Vol. 13, No. 3, 1975, pp. 346–70.

Gibbon, P. 'Analysing the British miners' strike of 1984–5', *Economy and Society*, Vol. 17, No. 2, 1988, pp. 139–94.

Gibbon, P., and Steyne, D. *Thurcroft: A Village and the Miners' Strike*, Nottingham, Spokesman, 1986.

Gibson, J. 'Coal-prospects and new industrial opportunities', Coalfield Communities Campaign, Working Paper No. 6, Vol. 2. Barnsley, CCC, 1985, pp. 11–16.

Gillman, P. 'Missing facts in Ian McGregor's history', *New Statesman*, 18 March 1983, p. 4.

Glasgow Media Group, *Bad News*, London, RKP, 1976.

Glasgow Media Group, *More Bad News*, London, RKP, 1980.

Glyn, A. *The Economic Case Against Pit Closures*, Sheffield, NUM, 1984.

Glyn, A. *One Million Jobs a Year*, London, Verso, 1985.

'Going back to what?', *Economist*, 9 March 1985, pp. 13–14.

Goodlad, A. (MP) 'The appropriate energy mix for Britain', Town and Country Planning Association Conference, A Responsible Energy Policy for Britain – The

Choices, Bristol, 15–16 October 1985.

Goodman, G. *The Miners' Strike*, London, Pluto, 1985.

Goodrich, C. *The Miners' Freedom*, Boston, Marshall Jones, 1925.

Gordon, P. 'If they come in the morning . . . the police, the miners and black peoples', *Policing the Miners' Strike*, eds. B. Fine and R. Millar, London, Lawrence and Wishart, 1985. pp. 161–77.

Gormley, J. *Battered Cherub*, London, Hamish Hamilton, 1982.

Grayson, J. 'Sexism struggle down the pits', *New Socialist*, No. 21, November 1984, p. 4.

Green, A. 'Research bibliography of published materials relating to the coal dispute 1984–85', *Journal of Law and Society*, Vol. 12, No. 3, 1985, pp. 405–14.

Green, D. 'The miners' strike: economics or politics?' *Economic Affairs*, April–June 1985, p. 44.

Gregson-Murray, S. 'Picket court justice', *Legal Action*, August 1984, p. 9.

Griffin, J. M., and Calarco, V. J. 'The global shift to lower rank, surface mineable coal', *The Coal Situation*, Chase Manhattan Bank, Vol. 5, No. 1, December 1984.

Groocock, V. 'If they need us to fight again . . . ', *New Statesman*, 7 June 1985, pp. 14–15.

Gudgin, G. 'Fuelling Britain: the future of coal', *Marxism Today*, Vol. 28, No. 6, June 1984, pp. 30–2.

Hagger, N. *Scargill the Stalinist? The Communist Role in the 1984 Miners' Strike*, London, Oak Tree Books, 1984.

Hain, P., and McCrindle, J. 'One and all: Labour's response to the miners', *New Socialist*, No. 20, October 1984, pp. 44–6.

Hall, T. *King Coal*, Harmondsworth, Penguin, 1981.

Hamilton, J. 'Give us a future, give us our jobs!', *The Cutting Edge: Women and the Pit Strike*, ed. V. Seddon, London, Lawrence and Wishart, 1986, pp. 211–21.

Hancock, H. 'Grimethorpe', *London Review of Books*, Vol. 6, No. 21, 15 November–6 December 1984, p. 5.

Hapgood, M. 'The Nottinghamshire miners: dispelling the myth of miners' solidarity', *Listener*, 24 May 1984, pp. 4–5, 19.

Harris, M. 'Waiting for General Winter', *New Society*, 30 August 1984, pp. 195–8.

Harrison, B. 'Pitfalls of academic accounting', *Accountancy*, Vol. 96, No. 1097, January 1985, p. 13. (Reply to Berry *et al.*, 'NCB accounts', *op. cit.*).

Harrison, R. 'The labour movement and the strike', Coal and Community Conference, University of Sheffield, 7–9 March 1986.

Harrison, T. "I've seen things I will never forget", *Listener*, 7 March 1985, pp. 6–7.

Hart, D. *The Soul Politic*, London, Adam Smith Institute, 1985.

Hartley, J., Kelly, J., and Nicholson, N. *Steel Strike*, London, Batsford, 1983.

Haynes, W. W. *Nationalisation in Practice: the British Coal Industry*, London, Bailey and Swinfen, 1953.

Head, J. 'From the heart', *New Statesman*, 15 June 1984, p. 13.

Heath, R. H. 'The National Power Loading Agreement in the coal industry and some aspects of workers' control', *Trade Union Register*, 1969, pp. 185–200.

Heathfield, P. 'What the miners are fighting for', *Tribune*, 20 April 1984.

Heathfield, P. 'The case for coal', *New Socialist*, No. 23, January 1985, pp. 14–17.

Henley Centre for Forecasting, *The Cost of the 1984 Coal Industry Dispute*, Newsnight, London, BBC TV, 17 September 1984.

Hepworth, R., *et al.* 'The effects of technological change in the Yorkshire coalfield, 1960–1965', *Economic Studies*, Vol. 4, 1969, pp. 221–37. (Republished as *Studies in the British*

Coal Industry, eds. D. M. Kelly and D. J. Forsyth, Oxford, Pergamon, 1969, pp. 221–37).

Hetherington, A. 'Is TV news impartial? Miners and the media'. *Listener*, 17 January 1985, pp. 5–6.

Hevener, J. W. *Which Side Are You On? The Harlan County Coal Miners, 1931–39*, Urbana, University of Illinois Press, 1978.

Hicks, I. 'Finance and investment policies of the National Coal Board', Economic Studies, Vol. 4, 1969, pp. 153–68. (Republished as *Studies in the British Coal Industry*, eds. D. M. Kelly and D. J. Forsyth, Oxford, Pergamon, 1969, pp. 153–68).

Hiller, E. T. *The Strike: A Study in Collective Action*, Chicago, Ill., Chicago University Press, 1928.

Hills, P., and Cope, D. 'Planning for an expanding coal industry', *The Planner*, Vol. 48, 1979, pp. 172–4.

Hillyard, P. 'Lessons from Ireland', *Policing the Miners' Strike*, eds. B. Fine and R. Millar, London, Lawrence and Wishart, 1985, pp. 177–87.

Holder, R. 'Better communications – thanks to coal strike', *Works Management*, April 1985, pp. 50–2.

Hollingsworth, M. 'Scargill's Sun support', *New Statesman*, 6 July 1984, p. 12.

Hollingsworth, M. 'A miner's bitter tale', *New Statesman*, 19 July 1984, pp. 13–14.

Hollingsworth, M. 'Using miners to bust the Union', *New Statesman*, 14 December 1984, pp. 8–10.

Hollingsworth, M. 'Who is paying the piper?', *New Statesman*, 21 December 1984, p. 14.

Hollingsworth, M. 'Blueing the black stuff', *New Statesman*, 1 March 1985, p. 12.

Hollingsworth, M. 'Press partisans', *New Socialist*, No. 26, April 1985, pp. 18–19.

Home Office, *Policing the miners' dispute; the National Reporting Centre*, London, Home Office, 1985.

Home Office, *Review of Public Order Law*, White Paper, Cmnd. 9510, London, HMSO, 1985.

Horton, E. 'Mining techniques in the 1980's, *Mining Engineer*, February 1983, pp. 451–5.

House, J. W., and Knight, E. M. *Pit Closure and Community*, Newcastle, Newcastle University, 1967.

Howell, D. '"Where's Ramsey MacKinnock?" Labour leadership and the miners', *Digging Deeper: Issues in the Miners' Strike*, ed. H. Beynon, London, Verso, 1985, pp. 181–98.

Howell, D. 'Labour and the miners', *New Socialist*, No. 25, March 1985, pp. 20–5. (Edited version of above).

Howell, D. 'Goodbye to all that? A review of literature on the 1984–85 miners' strike', *Work, Employment and Society*, Vol. 1, No. 3, 1987, pp. 388–404.

Howells, K. 'A stake in the heart of the NUM', *Radical Wales*, No. 3, 1984, pp. 20–1.

Howells, K. 'Stopping out: the birth of a new kind of politics', *Digging Deeper: Issues in the Miners' Strike*, ed. H. Beynon, London, Verso, 1985, pp. 139–47.

Howells, K. 'After the strike: Ned Thomas interviews Kim Howells', *Planet* No. 51, 1985, pp. 6–11.

Howells, K. 'British coal in the late 1980s', Study Conference on the British Coalfields, Northern College, 31 January – 2 February 1986.

Hoyles, M., and Hemmings, S. (eds.) *More Valuable than Gold*, London, Martin Hoyles, 1985, (By striking miners' children.)

Huddle, R. *et al. Blood, Sweat and Tears: photographs from the great miners' strike 1984–85*,

London, Artworker Books, 1985.

Hudson, R., and Sadler, D. 'Coal and dole: employment policies in the coalfields', *Digging Deeper: Issues in the Miners' Strike*, ed. H. Beynon, London, Verso, 1985, pp. 217–30.

Hudson, R., Peck, F., and Sadler, D. *Undermining Easington: Who'll pay the price of pit closures?*, Easington, Easington District Council, 1984.

Hudson, R., Sadler, D., and Wade, Eric 'Social costs of coal-mining decline: a position statement', Coalfield Communities Campaign Working Paper No. 4, Vol. 1, Barnsley, CCC, 1985, pp. 25–7.

Hughes, J., and Moore R. (eds.) *A Special Case?*, Harmondsworth, Penguin, 1972.

Hunt, D. (MP), 'The opportunities for coal', Coalfield Communities Campaign National Conference, Coalfield Communities Looking Forward, Wakefield, 19–21 February 1986, pp. 21–6.

Hutton, W. 'The Coal Board numbers game: coal-facts, figures and principles', *Listener*, 3 May 1984, pp. 4–7.

Hyman, R. *Strikes*, London, Fontana, 1972.

Hyman, R. 'Reflections on the miners' strike', *Socialist Register 1985/86*, eds. R. Miliband *et al.*, London, Merlin, 1986, pp. 330–54.

Ignatieff, M. 'Strangers and comrades', *New Statesman*, 14 December 1984, pp. 25–7.

Incomes Data Services, 'Miners' pay: Monopolies Commission Report reveals major changes', Report 405, July 1983, pp. 25–7.

Incomes Data Services, 'Miners' pay: the Coal Board replies', Report 408, September 1983, pp. 8–9.

Ingram, S. 'Strikes in the Yorkshire coalfield, 1958–1971: a multi-causal analysis', B.Sc. dissertation, University of Bradford, 1984.

'Investment in British mines', *Coal International*, July 1985, pp. 33–9.

Iron and Steel Trades Confederation, *New Deal for Steel*, London, ISTC, 1980.

Jackson, B., and Wardle, T. *The Battle for Orgreave*, London, Canary, 1986.

Jackson, S. 'Behind men? Not quite!' *Spare Rib*, No. 146, September 1984, p. 9.

James, P. *The Future of Coal*, London, Macmillan, 1982.

Jeans, J. S. 'The coal crisis and the paralysis of British industry', *Nineteenth Century*, Vol. 34, No. 201, November 1893, pp. 791–801.

Jeffrey, K., and Hennessy, P. *States of Emergency: British Governments and Strikebreaking since 1919*, London, RKP, 1983.

Jones, A., and Thompson, R. *After the Miners' Strike – What Next?*, London, International Marxist Group, 1972.

Jones, A. D. W., Quinn, J. J., and Woolley, K. M. *Evaluation of Coal-Winning Technology*, London, Technical Change Centre, 1984.

Jones, Ceri, 'New year strike by miners', *New Statesman*, 17/24 December 1982.

Jones, Ceri, 'Miners' strike called off', *New Statesman*, 14 January 1983.

Jones, Chris, 'Into year one: some personal observations on the miners' strike', *Bulletin on Social Policy*, No. 17, 1985, pp. 40–3.

Jones, Chris, and Novak, T. 'Welfare against the workers: benefits as a political weapon', *Digging Deeper: Issues in the Miners' Strike*, ed. H. Beynon, London, Verso, 1985, pp. 87–100.

Jones, D., *et al. Media Hits the Pits*, London, Campaign for Press and Broadcasting Freedom, 1985.

Jones, M. *Killed on the Picket Line 1984: The Story of David Gareth Jones*, London, New Park, 1985.

Jones, N. 'The miners' dispute: a media slanging match', *Listener*, 10 May 1984, pp. 4–5.

Jones, N. 'News management in dispute: NUM's relations with the media and NCB's objectivity: two casualties of the strike', *Listener*, 21 June 1984, pp. 4–5.

Jones, N. 'Miners and the media: Mrs Thatcher mounts a media cavalry charge', *Listener*, 26 July 1984, pp. 2–4.

Jones, N. 'The NCB's media management: the toughest contest is the media battle', *Listener*, 25 October 1984, pp. 7–8, 20.

Jones, N. 'The final stages of the miners' strike: the Government is winning the propaganda war', *Listener*, 24 January 1985, pp. 5–6.

Jones, N. 'Miners' strike: the importance of media presentation', *Listener*, 21 February 1985, pp. 5–6.

Jones, N. 'The miners' strike after a year: if anyone were going to take the blame then it would be Mr Scargill alone', *Listener*, 28 February 1985, pp. 2–4.

Jones, N. 'Special Branch surveillance', *Listener*, 7 March 1985, pp. 5–6.

Jones, N. *Strikes and the Media*, Oxford, Blackwell, 1986.

'Judges deny miners' rights', *Labour Research*, Vol. 73, No. 11, November 1984, p. 268.

Kahn, P. 'A new politics for coal', *New Socialist*, September 1985, pp. 26–30.

Kahn, P. 'Coal not dole: the British miners' strike of 1984–5', *Socialist Review*, Vol. 17, No. 3–4, 1987, pp. 57–88.

Kavanagh, S. 'Picketing miners: how the courts react', *Legal Action*, September 1984, pp. 4–5.

Kavanagh, S. 'The miners' welfare state?', *Legal Action*, October 1984, pp. 4–5.

Kay, J., and Millidene, K. 'Intruders, not peacekeepers', *The Cutting Edge: Women and the Pit Strike*, ed. V. Seddon, London, Lawrence and Wishart, 1986, pp. 176–202.

Kaye, T. 'A village at war: a report to the Yorkshire Area NUM concerning the policing of Fitzwilliam, West Yorkshire, on Monday 9th July 1984', University of Warwick, 1984 (*mimeo*).

Keenan, J. *Scargill's Strike*, Belfast, British and Irish Communist Organisation, 1984.

Kellner, P. 'What the public really thinks about the miners', *New Statesman*, 18 January 1985, p. 7.

Kellner, P. 'How Britain's tax payers helped beat the coal strike', *New Statesman*, 22 March 1985, p. 7.

Kelly, D. M., and Forsyth, D. J. (eds.) *Studies in the British Coal Industry*, Oxford, Pergamon, 1969. (Republication of *Economic Studies*, Vol. 4, 1969.)

Kelly, D. M. 'The process of mechanisation in British coal mining since 1945', *Economic Studies*, Vol. 4, 1969, pp. 135–51. (Republished as *Studies in the British Coal Industry*, eds. D. M. Kelly and D. J. Forsyth, Oxford, Pergamon, 1969, pp. 135–51.)

Kelly, D. M. 'Contraction of the coal industry: some aspects of the effects on manpower', *Economic Studies*, Vol. 4, 1969, pp. 199–220. (Republished as *Studies in the British Coal Industry*, eds. D. M. Kelly and D. J. Forsyth, Oxford, Pergamon, 1969, pp. 199–220.)

Kerevan, G., and Saville, R. *The case for deep-mined coal in Scotland*, Edinburgh, NUM (Scottish Area), 1984.

Kerevan, G., and Saville, R. 'A third world economy', *New Statesman*, 11 January 1985, p. 12.

Kerr, C., and Siegel, A. 'The interindustry propensity to strike: an international comparison', *Industrial Conflict*, eds. A. Kornhauser, R. Dubin and A. M. Ross, New York, McGraw-Hill, 1954, pp. 189–212.

Kettle, M. '1926 and now: if the miners lose again', *New Socialist*, No. 24, February 1985, pp. 24–7.

Kettle, M. 'The National Reporting Centre and the 1984 miners' strike', *Policing the Miners' Strike*, eds. B. Fine and R. Millar, London, Lawrence and Wishart, 1985, pp. 23–33.

Kettle, M., and Hodges, L. *Uprising! The police, the people and the riots in Britain's cities*, London, Pan, 1982.

Kidner, R. *Trade Union Law*, London, Stevens, 1979.

Kirk, G. *Schumacher on Energy*, London, Cape, 1982.

Kirsch, B., and Wolmar, C. 'Miners' pickets: the tally', *New Statesman*, 22 March 1985, p. 6.

Krieger, J. *Undermining Capitalism: State Ownership and the Dialectic of Control in the British Coal Industry*, Princeton, N.J., Princeton UP, 1983.

Labour Party, *Report of the Annual Conference of the Labour Party 1984*, London, LP, 1985.

Labour Party, 'Labour's Energy Policy', London, Labour Party, 1986 (*mimeo*).

Solidarity with the Miners, London, Labour Research Department, 1985.

Lambeth Women's Miners' Support Group, 'Striking new connections', *Spare Rib*, No. 153, April, 1985, pp. 32–4.

Lane, T., and Roberts, K. *Strike at Pilkingtons*, London, Fontana, 1971.

Laurence, J. 'Without the pits', *New Society*, 12 April 1984.

Law, C. 'Letter from a miners' wife', *New Socialist*, December 1984.

'The law on dismissal and the miners', *Labour Research*, Vol. 74, No. 5, May 1985.

'The lawyers go in', *Economist*, 19 May 1984, pp. 29–30.

Leonard, T. 'Policing the miners in Derbyshire', *Policing*, Vol. 1, No. 2, 1985, pp. 96–101.

Levy, C. *A Very Hard Year: the 1984–85 miners' strike in Mauchline*, Glasgow, Workers' Educational Association, 1985.

Livingstone, K. *The miners' strike: implications for Londoners – report by the Leader of the Council*, London, Greater London Council, 1984.

Loach, L. 'We'll be here right to the end . . . and after: women in the miners' strike', *Digging Deeper: Issues in the Miners' Strike*, ed. H. Beynon, London, Verso, 1985, pp. 169–80.

Loarridge, R. J. 'Contemporary technology in South African coal production and usage', B.Sc. dissertation, University of Bradford, 1986.

Lloyd, C. 'Policing street collections', *Marxism Today*, Vol. 28, No. 10, October 1984, pp. 5–6.

Lloyd, C. 'A national riot police: Britain's 'third force?' *Policing the Miners' Strike*, eds. B. Fine and R. Millar, London, Lawrence and Wishart, 1985, pp. 65–78.

Lloyd, C. 'Street collections and the coal dispute', *Journal of Law and Society*, Vol. 12, No. 3, 1985, pp. 285–92.

Lloyd, J. *Understanding the Miners' Strike*, London, Fabian Society, 1985.

Lloyd, J. 'The two faces of the miners' strike', *Personnel Management*, June 1985, pp. 32–6.

Lloyd, J. 'The long-term implications of the miners' strike', *Energy Focus*, July 1985, pp. 87–94.

Lustgarten, L. 'The long-distance policemen', *New Society*, 3 January 1985, pp. 24–5.

McCarthy, M. 'Buried treasure', *New Socialist*, No. 25, March 1985, pp. 11–13.

McCloskey, G. 'Coal: can UK capacity expand now the strike is over?', *Energy Economist*, April 1985, pp. 1–3.

McCormick, B. J. 'Strikes in the Yorkshire coalfield, 1947–1963', *Economic Studies*, Vol. 4, 1969, pp. 171–98. (Republished as *Studies in the British Coal Industry*, eds. D. M. Kelly and D. J. Forsyth, Oxford, Pergamon, 1969, pp. 171–98.)

McCormick, B. J. *Industrial Relations in the Coal Industry*, London, Macmillan, 1979.

McCormick, B. J. Address to the Coalfield Communities Campaign National Conference, Coalfield Communities Looking Forward, Wakefield, 19–21 February 1986, pp. 27–38.

McCrindle, J. 'More than just a memory' (interview with S. Rowbotham), *Feminist Review*, No. 23, 1986, pp. 109–24.

MacGregor, I. (with R. Tyler), *The Enemies Within: the Story of the Miners' Strike 1984–5*, London, Collins, 1986.

McHugh, P. *Unfinished Business: photo-text collections from the miners' strike 1984–5*, Accrington, Coppice, 1985.

McHugh, P., and Gwilt, R. *For the protection of miners: North-East Lancashire and the miners' strike 1984*, Preston, Lancashire Association of Trades Councils, 1985.

McIlroy, J. *Strike! How to Fight. How to Win*, London, Pluto, 1984.

McIlroy, J. 'On second fronts and allied commands', *New Statesman*, 24 August 1984.

McIlroy, J. 'The law struck dumb? Labour law and the miners' strike', *Policing the Miners' Strike*, Eds. B. Fine and R. Millar, London, Lawrence and Wishart, 1985, pp 79–102

McIlroy. J. 'Police and pickets: the law against the miners', *Digging Deeper: Issues in the Miners' Strike*, Ed. H. Beynon, London, Verso, 1985, pp 101–22.

McIlroy, J. 'The state and the miners' strike: contingency planning and state involvement in industrial disputes, Coal and Community Conference, University of Sheffield, 7–9 March 1986.

MacKey, K. 'Women against pit closures', *The Cutting Edge: Women and the Pit Strike*, Ed. V. Seddon, London, Lawrence and Wishart, 1986, pp. 50–62.

McMullen, J. *Employment Law Under the Tories*, London, Pluto, 1981.

McMullen, J. 'Legal strategy and the unions', *Policing the Miners' Strike*, eds. B. Fine and R. Miller, London, Lawrence and Wishart, 1985, pp. 217–26.

Manners, G. *Coal in Britain: An Uncertain Future*, London, Allen and Unwin, 1981.

Manners, G. 'The demand for British coal', Town and Country Planning Association Conference, Economic Prospects for the Coalfields: A Case Study of the Yorkshire Coalfield, Sheffield, 22 November 1985.

Marx, K. *Capital*, Vol. 1, London, Lawrence and Wishart, 1954.

'Martyrdom denied', *Economist*, 2 June 1984, pp. 19–20. (Orig. pub. 1887.)

Massey, D., and Wainwright, H. 'Beyond the coalfields: the work of the Miners' Support Groups', *Digging Deeper: Issues in the Miners' Strike*, ed. H. Beynon, London, Verso, 1985, pp. 149–68.

Massey, D., and Wainwright, H. 'Keep moving on . . . ' *New Socialist*, No. 26, April 1985, pp. 8–9. (Edited version of above.)

Maudling, R. *Memoirs*, London, Sidgwick and Jackson, 1978.

Merseyside Welfare Rights Resource Centre, *Aid for Strikers*, Liverpool, MWRRC, 1984.

Michels, R. *Political Parties*, Glencoe, Ill., Free Press, 1949. (Orig. pub. 1911.)

Miliband, R. *The State in Capitalist Society*, London, Weidenfeld and Nicolson, 1969.

Miller, Jean, 'Relations with the NUM: Barnsley', *The Cutting Edge: Women and the Pit Strike*, ed. V. Seddon, London, Lawrence and Wishart, 1986, pp. 227–40.

Miller, Jill, *You can't Kill the Spirit: women in a Welsh mining village*, London, Women's Press, 1986.

Miller, Sam, 'Grimethorpe now', *London Review of Books*, Vol. 7, No. 10, 6 June 1985, pp. 16–17.

Miller, Susan, 'War of attrition', *New Statesman*, 8 June 1984, pp. 10–11.

Miller, Susan, '"The best thing that ever happened to us": women's role in the coal dispute', *Journal of Law and Society*, Vol. 12, No. 3, 1985, pp. 355–64.

Miller, Susan, and Walker, M. *A State of Siege: policing the coalfields in the first six weeks of the miners' strike*, London, Yorkshire Area NUM and Greenwich NALGO, 1984.

The Miners' Case, London, Labour Research Department, 1984.

'Miners' case: a question of murder', *Labour Research*, Vol. 74, No. 8, August 1985, pp. 206–7.

'Miners: policing the dispute', *Labour Research*, Vol. 73, No. 9, September 1984, pp. 227–9.

'The miners: right on their side', *New Socialist*, September 1984.

'Miners' strike', *Labour Research*, Vol. 73, No. 6, June 1984, pp. 145–9.

'Miners' strike', *National Institute Economic Review*, August 1984, pp. 6–9.

'The miners' strike', *Poverty*, No. 58, August 1984, pp. 45–6.

'The miners' strike: a balance sheet – a roundtable discussion', *Marxism Today*, Vol. 29, No. 4, April, 1985, pp. 21–7.

'Miners' strike: economics of coal', *Labour Research*, Vol. 73, No. 12, December 1984, pp. 292–5.

'Miners' strike: picketing against the law', *Labour Research*, Vol. 73, No. 5, May 1984, pp. 116–17.

'Miners' strike: if the troops are called in', *Labour Research*, Vol. 73, No. 10, November 1984, pp. 272–3.

'Miners' strike: working miners', *Labour Research*, Vol. 74, No. 2, February 1985, pp. 40–1.

'Miners: up against the law', *Labour Research*, Vol. 74, No. 3, March 1985, pp. 62–4.

Minford, P. 'Labour market equilibration in an open economy', *Oxford Economic Papers*, November 1983, pp. 207–44. (Supplement: *The Causes of Unemployment.)*

Minford, P., and Kung, P. *The Costs and Benefits of Pit Closures*, Watford, Public Service Research Centre, 1984.

Minford, P., *et al.* 'The Liverpool macroeconomic model of the United Kingdom', *Economic Modelling*, January 1984, pp. 24–62.

Mining Department, *Production and Productivity Bulletin*, No. 16, March 1984, London, NCB.

Ministry of Fuel and Power, *Coal Mining. Report of the Technical Advisory Committee.* (Reid Report), Cmd. 6610, London, HMSO, 1945.

Ministry of Fuel and Power, *Fuel Policy*, Cmnd. 3438, London, HMSO, 1967.

Ministry of Power, *Fuel Policy*, White Paper, Cmnd. 2798, London, HMSO, 1965.

Monk, S. *Retraining Opportunities for Miners who have been made Redundant*, Coalfield Communities Campaign Special Report No. 1, Barnsley, CCC, 1986.

Monopolies and Mergers Commission, *National Coal Board: A Report on the efficiency and costs in the development, production and supply of coal by the NCB*, Cmnd. 8920, London, HMSO, 1983.

Moore, C. 'What about the workers?', *Spectator*, 9 March 1985, p. 4.

Moore, J. *Why Privatise?* London, Conservative Political Centre, 1983.

Morgan, K. O. 'A time for miners to forget history', *New Society*, 21 February 1985, pp. 283–5.

Morris, P. 'Benefits cut for miners with loans', *Community Care*, No. 510, 3 May 1984.

Moses, K. 'New mines – the way forward now', *Mining Engineer*, May 1982, pp. 641–5.

National Association of Citizens' Advice Bureaux, *A Report on the 1984/85 Coal Dispute and the CAB Service*, London, NACAB, 1985.

National Board for Prices and Incomes, *Payment by Results Systems*, Report No. 65, Cmnd. 3627, London, HMSO, 1968.

National Coal Board, 'NCB reply to the NUM claim for increased wages from November 1983', London, NCB (*mimeo*).

National Coal Board, 'New strategy for coal', Joint Policy Advisory Committee, Memorandum, 15 October 1985.

National Coal Board, *Plan for Coal*, London, NCB, 1974.

National Coal Board, *Plan 2000*, London, NCB, 1976.

National Coal Board, *Report and Accounts 1983/4*, London NCB, 1984.

National Coal Board, *Report and Accounts 1984/5*, London NCB, 1985.

National Coal Board, *Report and Accounts 1985/6*, London NCB, 1986.

National Union of Mineworkers, *Rules, Model Rules, Standing Orders*, London, NUM, 1978.

National Union of Mineworkers, *The Miners and the Battle for Britain*, London, NUM, 1980.

National Union of Mineworkers, *Campaign for Coal*, Sheffield, NUM, 1983.

National Union of Mineworkers, '1983 wage claim', Sheffield, NUM, 1983 (*mimeo*).

National Union of Mineworkers (Nottingham Area), *Rules and Standing Orders*, Nottingham, NUM (Nottingham Area), 1980.

National Union of Mineworkers (Nottinghamshire Area), *An Appraisal of the British Coalmining Industry and an Examination of the Possible Consequences following Pit Closures in the NUM (Nottinghamshire Area)*, Mansfield, NUM (Nottinghamshire Area), 1983.

National Union of Mineworkers (Research Department), 'Plan for Coal: Labour Party discussions', Sheffield, NUM, 1985 (*mimeo*).

National Union of Mineworkers (Research Department), 'Hours of work: 1908 Coal Mines Regulation Act', Sheffield, NUM, 1986 (*mimeo*).

National Union of Mineworkers (Research Department), 'The sale of the century: privatising power', Sheffield, NUM, 1987 (*mimeo*).

Neale, J. *Memoirs of a Callous Picket*, London, Pluto, 1983.

Newbold, A. L. 'Picketing miners and the courts', *Public Law*, Spring 1985, pp. 30–4.

Newsline, *The Miners' Strike 1984–85 in Pictures*, London, New Park, 1985.

Niclas Society, *Who Profits from Coal?*, Cardiff, Niclas, 1980.

North Yorkshire Women Against Pit Closures, *Strike 84–85*, Leeds, NYWAPC, 1985.

Northern Film and Television Archive, *A Catalogue of Films and Videotapes produced during the 1984/85 Miners' Strike*, Gateshead, the Archive, 1986.

Nottinghamshire County Council, Economic Development Committee, *Report on Employment Structure and Prospects in Nottinghamshire*, 24 July 1985.

National Working Miners' Committee, *The Miners' dispute: A Catalogue of Violence*, London, Nat WMC, 1984.

'The NUM dispute', *Colliery Guardian*, September 1985, pp. 416, 418.

O'Donnell, K. 'Brought to account: the NCB and the case for coal', *Capital and Class*, No. 26, 1985, pp. 105–23.

Okojie, P. 'Chief constables and political interference: the case of Anderton and Greater Manchester', *Policing the Miners' Strike*, eds. B. Fine and R. Millar, London, Lawrence and Wishart, 1985, pp. 54–64.

Orme, S. (Rt Hon), 'Energy and employment: the case for coal', Coalfield Communities Campaign National Conference, Coalfield Communities Looking Forward, Wakefield, 19–21 February 1986, pp. 65–77.

Ormerod, R. 'Corporate planning and its use of O. R. in the NCB: a personal view', *Journal of the Operational Research Society*, Vol. 34, No. 6, 1983, pp. 461–7.

'Other views on the coal strike', *IDS Focus*, December 1984, pp. 12–14 (London, Incomes Data Services).

Ottey, R. *The Strike: An Insider's Story*, London, Sidgwick and Jackson, 1985.

Page Arnot, R. *The Miners: One Union, One Industry*, London, Allen and Unwin, 1979.

Panitch, L. *Social Democracy and Industrial Militancy*, Cambridge, CUP, 1976.

Parker, T. *Red Hill: a mining community*, London, Heinemann, 1986.

Parkes, T. 'Does NUM plus NRC equal NPF?', *Police*, Vol. 17, No. 3, November 1984, pp. 14–16.

Parkinson, E. 'Coalfield communities and their environment', Coalfield Communities Campaign Working Paper No. 5, Vol. 2, Barnsley, CCC, 1985, pp. 1–10.

Paynter, W. *My Generation*, London, Allen and Unwin, 1972.

Pead, D. 'Diary of a dispute', *Police Review*, Vol. 93, No. 4815, 7 June 1985, pp. 1154–7.

Pead, D. "Where can we get 2,000 officers by 4 p.m.?", *Police Review*, Vol. 93, No. 4816, 14 June, pp. 1212–13.

Peck, F. 'Horden colliery: the costs of closure', *Northern Economic Review*, No. 11, 1985, pp. 42–50.

Percy-Smith, J., and Hillyard, P. 'Miners in the arms of the law: a statistical analysis', *Journal of Law and Society*, Vol. 12, No. 3, 1985, pp. 345–54.

Phillips, D. 'Warm welcome, sharp realities: Gwent', *The Cutting Edge: Women and the Pit Strike*, ed. V. Seddon, London, Lawrence and Wishart, 1986, pp. 241–8.

'From picket to riot', *Economist*, 23 June 1984, p. 25.

The Pit Strike in Perspective, London, Ernest Bevin Society, 1984.

Pitt, M. *The World on Our Backs*, London, Lawrence and Wishart, 1979.

Platt, J. *British Coal: A Review of the Industry, its Organisation and Management*, London, Lyon Grant and Green, 1968.

Policing of the demonstration in support of the miners, London, Greater London Council, 1985.

'Policing the pits: dangers for us all', *Policing London*, No. 13, July/August 1984, pp. 77–82.

Political and Economic Planning, *The Market for Coal*, London, PEP, 1960.

Political and Economic Planning, *Questions of Fuel Policy*, London, PEP, 1965.

Political and Economic Planning, *A Fuel Policy for Britain*, London, PEP, 1966.

Pollard, M. *The Hardest Work under Heaven: The Life and Death of the British Miner*, London Hutchinson, 1984.

Pond, W. (P. Wintour), 'Why Arthur Scargill wanted a confrontation', *New Statesman*, 11 March 1983, p. 4.

Pond, W. (P. Wintour), 'Last stand or orderly retreat', *New Statesman*, 23 March 1984, pp. 8–10.

Pond, W. (P. Wintour), 'Shaking up the Union', *New Statesman*, 20 January 1984, pp. 8–9.

Posner, M. V. *Fuel Policy: A Study in Applied Economics*, London, Macmillan, 1973.

Presthus, R. *The Nationalisation of the British Coal Industry*, Chicago, University of Chicago, 1948, (microfilm).

The Proposed Closure of St John's Colliery, Maesteg: Report of An Independent Public Inquiry, Maesteg, St John's NUM and Communities Action Committee, 1985.

Radice, H. 'The national economy: A Keynesian myth?', *Capital and Class*, No. 22, 1984, pp. 111–40.

Railworkers and Miners, London, Socialist Action, 1985.

Reed, D, and Adamson, O. *Miners' Strike 1984–1985: People versus State*, London, Larkin, 1985.

Rees, G. 'Regional restructuring, class change, and political action: preliminary comments on the 1984–85 miners' strike in South Wales', *Environment and Planning D: Society and Space*, Vol. 3, 1985, pp. 389–406.

Reichert, K. 'Coal in the Community', Coalfield Communities Campaign National Conference, Coalfield Communities Looking Forward, Wakefield, 19–21 February 1986, pp. 9–20.

Reid, G. L., Allen, K. and Harris, D. J. (eds.), *The Nationalised Fuel Industries*, London, Heinemann, 1973.

Reid, J. 'What Scargill means', *New Society*, 17 January 1985, pp. 91–3.

Reiner, R. 'Is Britain turning into a police state?', *New Society*, 2 August 1984, pp. 51–6.

Reiner, R. 'Policing strikes: an historical 'U' turn', *Policing*, Vol. 1, No. 3, 1985, pp. 138–48.

Report of the Public Inquiry into the Proposed Closure of Polkemmet Colliery, Whitburn, Edinburgh, NUM (Scottish Area), 1985.

Richards, F. *The Miners' Next Step*, London, Junius, 1984.

Rigg, P. 'Miners and militancy: a study of branch union leadership', *Industrial Relations Journal*, Vol. 18, No. 3, 1987, pp. 189–200.

Rimlinger, G. 'International differences in the strike propensity of coal miners: experience in four countries', *Industrial and Labour Relations Review*, Vol. 12, 1959, pp. 389–405.

'The riot charge?', *New Law Journal*, Vol. 135, No. 6215, 9 August 1985, pp. 789–90.

Robens, A. (Lord), *Ten Year Stint*, London, Cassell, 1972.

Roberts, P. W. 'Regional and local planning implications of energy policy: lessons from the past and issues for the future', Coalfield Communities Campaign Working Paper No. 9, Vol. 3, Barnsley, CCC, 1986, pp. 1–16.

Robinson, B. 'Economic background to the coal dispute', *Economic Outlook*, Vol. 9, No. 1, October 1984, pp. 23–32.

Robinson, B. 'How large a coal industry?', *Economic Outlook*, Vol. 9, No. 3, December 1984, pp. 1–4.

Robinson, B. 'The coal industry after Scargill and after OPEC III', Coal and Community Conference, University of Sheffield, 7–9 March 1986.

Robinson, B. 'The coal industry after OPEC III', *Economic Outlook*, Vol. 10, No. 7, April 1986, pp. 1–4.

Robinson, C. *The Energy 'Crisis' and British Coal*, Hobart Paper 59, London, Institute of Economic Affairs, 1974.

Robinson, C. 'Coal policy in Britain', *Economic Review*, Vol. 2, No. 4, March 1985, pp. 2–6.

Robinson, C., and Marshall, E. *What Future for British Coal?* Hobart Paper 89, London, Institute of Economic Affairs, 1981.

Robinson, C., and Marshall, E. 'The coal industry and coal policy in Britain', University of Surrey, 1983 (*mimeo*).

Robinson, C., and Marshall, E. *Can Coal be Saved?*, Hobart Paper 105, London, Institute of Economic Affairs, 1985.

Robinson C. and Sykes, A. *Privatise Coal*, Policy Study No. 85, London, Centre for Policy Studies, 1987.

Rose, H. 'The miners' wives of Upton', *New Society*, 29 November 1984, pp. 326–9.

Rose, H. 'Securing social citizenship', *New Socialist*, No. 25, March 1985, pp. 16–18.

Rose, J. 'Intimidating behaviour', *New Statesman*, 17 August 1984.

Rose, J. 'Miners and magistrates', *New Statesman*, 31 August 1984, pp. 15–16.

Rose, J. 'Take no prisoners', *New Statesman*, 21 June 1985, pp. 13–14.

Ross, A. M. *Trade Union Wage Policy*, Berkeley, Cal., University of California Press, 1948.

Rost, P., and Pargeter, M. 'Pits, privatisation and politics', *Economic Affairs*, Vol. 15, No. 4, July–September 1985, pp. 39–41.

Rouffiniac, J. *Harringay Supporting the Miners 1984–1985*, London, Harringay Trades Union Council Support Unit, 1985.

Royal Commission on Trade Unions and Employers' Associations 1965–1968, *Report* (Ch. Donovan), Cmnd. 3623, London, HMSO, 1968.

Rozenberg, J. 'The legacy of legal changes', *Listener*, 7 March 1985, pp. 5–6.

Rutledge, I. 'Changes in the mode of production and the growth of 'mass militancy' in the British mining industry 1954–1974', *Science and Society*, Vol. 41, 1977, pp. 410–29.

Rutledge, I. 'How the Coal Board is subordinating our industry to private capital', *Tribune*, 10 January 1986.

Salt, C., and Layzell, J. *Here We Go: Women's Memories of the 1984/85 Miners' Strike*, London, London Political Committee, Co-operative Retail Services, 1985.

Samuel, R. 'Friends and outsiders', *New Statesman*, 11 January 1985, pp. 14–15.

Samuel, R. 'Ethics and the strike', *New Society*, 28 February 1985, pp. 329–30.

Samuel, R. 'A plan for disaster', *New Society*, 7 March 1985, pp. 367–8.

Samuel, R. 'A managerial power cut', *New Socialist*, No. 26, April 1985, pp. 4–7.

Samuel, R., Bloomfield, B. and Boanas, G. *The Enemy Within: Pit Villages and the Miners' Strike of 1984–5*, London, RKP, 1986.

Saville, J. 'An open conspiracy – Conservative politics and the miners' strike 1984–85', *Socialist Register 1985/86*, eds. R. Miliband *et al.*, London, Merlin, 1986, pp. 295–329.

Saxena, S. K. *Nationalisation and Conflict; Example of British Coal Mining*, The Hague, Nijhoff, 1955.

Scargill, A. *Miners in the Eighties*, Barnsley, NUM (Yorkshire Area), 1981.

Scargill, A., *et al.*, 'Scargill takes on the telly men', *New Socialist*, October, 1984.

Schumacher, E. F. 'Coal – the next fifty years', NUM Study Conference, Britain's Coal. 25–6 March 1960, London, NUM pp. 6–17.

Schwarz, B. 'Let them eat coal: the Conservative Party and the strike', *Digging Deeper: Issues in the Miners' Strike*, ed. H. Beynon, London, Verso, 1985, pp. 47–67.

Scott, W. H. *et al. Coal and Conflict*, Liverpool, Liverpool University Press, 1963.

Scraton, P. 'From Saltley Gates to Orgreave: a history of the policing of recent industrial disputes', *Policing the Miners' Strike*, eds. B. Fine and R. Millar, London, Lawrence and Wishart, 1985, pp. 145–60.

Scraton, P. 'The State v. The People: an introduction', *Journal of Law and Society*, Vol. 12, No. 3, 1985, pp. 251–66.

Scraton, P. '"If you want a riot, change the law": the implications of the 1985 White Paper on Public Order', *Journal of Law and Society*, Vol. 12, No. 3, 1985, pp. 385–93.

Scraton, P. *The State of the Police*, London, Pluto, 1985.

Scraton, P., and Thomas, P. (eds.), *The State v. The People: Lessons from the Coal Dispute*, Special Issue of *Journal of Law and Society*, Vol. 12, No. 3, 1985.

Searle-Barnes, R. G. *Pay and Productivity Bargaining*, Manchester, MUP, 1969.

Seddon, V. (ed.), *The Cutting Edge: Women and the Pit Strike*, London, Lawrence and Wishart, 1986.

Selbourne, D. 'In the trenches of South Yorkshire', *New Society*, 10 January 1985, pp. 48–51.

Sheffield Policewatch, *Taking Liberties: Policing during the Miners' Strike, April–Oct 1984*, Sheffield, Sheffield Policewatch, 1984.

Shepherd, C. 'The reconstruction of the Barnsley Area of the NCB', Midlands Institute of Mining Engineers, January 1984.

Shinwell, E. *Conflict without Malice*, London, Odhams, 1955.

Simmonds, D., and Franey, R. 'Keeping the home fires burning', *New Socialist*, No. 18, July/August 1984, pp. 18–19.

Simpson, R. 'Benefits: the row over the miners', *New Society*, 29 November 1984, pp. 348–9.

Slaughter, C. 'The strike of Yorkshire mineworkers in May 1955', *Sociological Review*, Vol. 6, 1956, pp. 241–59.

Smith, C. T. B., *et al. Strikes in Britain*, Department of Employment Manpower Paper No. 15, London, HMSO, 1978.

Smith, D. 'The longest strike: "in 1926 it was the employers; now it's the state" ', *Listener*, 15 November 1984, pp. 11–12.

Smith, N. 'The duties of management', *Solid Fuel*, March 1985, p. 29.

Smith, N. 'Sand in the negotiating machinery?', *Personnel Management*, April 1985, p. 3.

Socialist Organiser, *Magnificent Miners*, London, SO, 1985.

Solidarity with the Miners, London, Labour Research Department, 1985.

South Yorkshire Police Committee, *Policing Policy during the Strike of the National Union of Mineworkers (Yorkshire Area), 1984*, Report of an inquiry by the Special Sub-Committee of the South Yorkshire Police Committee, 4 February 1985.

Spencer, S. *Police Authorities during the Miners' Strike*, London, Cobden Trust, 1985.

Spencer, S. 'The eclipse of the police authority', *Policing the Miners' Strike*, eds B. Fine and R. Millar, London, Lawrence and Wishart, 1985, pp. 34–53.

Stead, J. *Never the same again: women and the miners' strike*, London, Women's Press, 1987.

Stephenson, R. 'The UDM and the miners' strike: democratic breakaway or union busting?', B.A. dissertation, University of Bradford, 1986.

Steuben, J. *Strike Strategy*, New York, Gaer, 1950.

Stewart, M. 'The miners' strike', *London Review of Books*, Vol. 6, No. 16, 6–19 September 1984, pp. 3–4.

'Strike costs', *Economist*, 9 March 1985, p. 30.

'A strike to the finish: a roundtable discussion', *Marxism Today*, Vol. 28, No. 9, September 1984, pp. 9–15.

Strikers' Handbook, London, Federation of Claimants' Unions, 1984.

Striking Back, Cardiff: Welsh Campaign for Civil and Political Liberties and NUM (South Wales Area), 1985.

Sutcliffe, L., and Hill, B. *Let Them Eat Coal: the Political use of Social Security during the Miners' Strike*, London, Canary, 1985.

Sutherland, I. 'The sacked miners of Scotland', *New Society*, 5 July 1985, pp. 7-9.

Sweet, C. *The Price of Nuclear Power*, London, Heinemann, 1983.

Sweet, C. 'Wanted: a new Plan for Coal', *Tribune*, 23 March 1984.

Sweet, C. 'What they really plan for coal', *New Statesman*, 9 November 1984, pp. 16–18.

Sweet, C. 'Why coal is under attack: nuclear powers in the energy establishment',

Digging Deeper: Issues in the Miners' Strike, ed. H. Beynon, London, Verso, 1985, pp. 201–16.

Sweet, C., and Coote, A. 'Frank Chapple's fantasy', *New Statesman*, 20 November 1981, pp. 6–8.

Taylor, A. *The Politics of the Yorkshire Miners*, London, Croom Helm, 1984.

Taylor, S. 'Grub up for the miners!', *The Cutting Edge: Women and the Pit Strike*, ed. V. Seddon, London, Lawrence and Wishart, 1986, pp. 79–96.

Temple, D. *British Miners and the Capitalist Crisis*, London, New Park, 1983.

Thinking about the Miners' Strike, London, Ernest Bevin Society, 1984.

Thomas, D. 'The unexpected miners', *New Society*, 22 March 1984.

Thomas, D. 'The Mansfield march', *New Society*, 17 May 1984, pp. 253–4.

Thomas, D. 'The secret war against coal', *New Socialist*, No. 19, September 1984, pp. 31–2.

Thomas, D. 'How many pits should go?', *New Society*, 31 January 1985, pp. 185–6.

Thomas, D. 'Paying the new price of coal', *New Society*, 12 July 1985, pp. 49–51.

Thomas, P. A., and Todd, P. N. 'Mass picketing: RIP', *New Law Journal*, Vol. 135, No. 6199, 19 April 1985, pp. 379–81.

Thornett, A. (ed.) *The Miners' Strike in Oxford*, Oxford, Oxford and District Trades Union Council, 1985.

Towers, B. 'Posing larger questions: the British miners' strike of 1984–85', *Industrial Relations Journal*, Vol. 16, No. 2, 1985, pp. 8–25.

Townsend, J. *Innovation in Coal Mining Machinery: the Anderton Shearer Loader*, Occasional Paper No. 3, Science Policy Research Unit, University of Sussex, 1976.

Trade Union and Community Resource and Information Centre, *Benefits in Dispute*, Leeds, TUCRIC, 1982.

Trades Union Congress, *Guide on the Conduct of Disputes*, London, TUC, 1979.

Trades Union Congress, *Review of Energy Policy*, London, TUC, 1981.

Trades Union Congress, *Report to General Council on Fuel Stocks at Power Stations*, London, TUC Industrial Relations Department, 1984 (*mimeo*).

Trades Union Congress, *Report of the 116th Annual Trades Union Congress*, London, TUC, 1984.

'Training for the long fight', *Economist*, 16 June 1984, pp. 19–20.

Triple Alliance, *What is the Future?*, London, Triple Alliance, 1980.

Triple Alliance, *Coal, Rail and Steel Fighting for the Future*, London, Triple Alliance, 1983.

Turner, H. A., Clack, G. and Roberts, G. *Labour Relations in the Motor Industry*, London, Allen and Unwin, 1967.

Turner, L. *Coal's Contribution to UK Self-Sufficiency*, London, Heinemann, 1984.

Turner, M. 'The future of the coal industry and its labour force', B.Sc. dissertation, University of Bradford, 1983.

Turner, R. L. 'Post-war pit closures: the politics of de-industrialisation', *Political Quarterly*, Vol. 56, No. 2, April–June 1985, pp. 167–74.

Twenty-seven miners at work tell their story, London, Aims of Industry, 1985.

Upham, M. 'British Steel: retrospect and prospect', *Industrial Relations Journal*, Vol. 11, No. 2, 1980, pp. 5–21.

'After victory, a new start for coal', *Economist*, 9 March 1985, pp. 29–33.

Waddington, D. 'Maltby: a study in community resistance', Coal and Community Conference, University of Sheffield, 7–9 March 1986.

Waddington, P. *The Effects of Manpower Depletion during the NUM Strike, 1984–85*, London, Police Foundation, 1985.

Waddington, P. 'The effects of manpower depletion during the NUM dispute', *Policing*, Vol. 1, No. 3, 1985, pp. 149–60 (edited version of above).

Wade, Elaine, 'The miners and the media: themes of newspaper reporting', *Journal of Law and Society*, Vol. 12, No. 3, 1985, pp. 273–84.

Wade, Eric, *Coal Mining and Employment: A study of Blyth Valley*, Borough of Blyth Valley Council, 1985.

Walker, M. *A Turn of the Screw*, London, Canary, 1985.

Walker, M. 'Miners in prison: workers in prison: political prisoners', *Journal of Law and Society*, Vol.12, No 3, 1985, pp. 333–43.

Walker, M., and Miller, Susan, *The Iron Fist: A State of Siege Vol. 2: policing the coalfields in the second six weeks of the miners' strike*, London, Yorkshire Area NUM and Greenwich NALGO.

Waller, R. J. *The Dukeries Transformed*, Oxford, Clarendon, 1983.

Warren, D. *The Key to my Cell*, London, New Park, 1982.

Watson, D. 'The miners' strike: life on the picket line settles into a routine', *Listener*, 19 April 1984, pp. 4–5.

Watson, M., Head, J. and Webb, T. *Striking Figures: the story of Normanton and Altofts Miners Support Group 1984–1985*, Huddersfield, Artivan and Striking Figures, 1986.

Wedderburn, K. W. (Lord), *The Worker and the Law*, Harmondsworth, Penguin, 1986 (third edition).

Weekes, B. *et al. Industrial Relations and the Limits of Law*, Oxford, Blackwell, 1975.

West Yorkshire Metropolitan County Council, *Prospects for the West Yorkshire Coalfield*, Wakefield, WYMCC, 1984.

Whitfield, D. 'Coal: a privatisation postponed?', *Capital and Class*, No. 20, Spring 1985, pp. 5–14.

Williams, E. 'Emlyn Williams, President of the South Wales Area of the National Union of Mineworkers, talks to Gwyn Williams', *Radical Wales*, No. 4, Autumn 1984, pp. 4–7.

Williams, R. 'Mining the meaning: key words in the miners' strike', *New Socialist*, No. 25, March 1985, pp. 6–9.

Wilsher, P., Macintyre, D., and Jones, M. *Strike: Thatcher, Scargill and the Miners*, London, Coronet, 1985.

Winterton, J. 'The trend of strikes in British coal mining', *Industrial Relations Journal*, Vol. 12, No. 6, 1981, pp. 10–19.

Winterton, J. 'Cutting coal by computer', *Tribune*, 11 March 1983.

Winterton, J. 'How the bonus scheme has cheated the miners', *Tribune*, 6 April 1984.

Winterton, J. 'Security blankets', *New Statesman*, 6 April 1984, pp. 12–13.

Winterton, J. 'Police activities', *New Statesman*, 27 April 1984, p. 8.

Winterton, J. 'Computerised coal: new technology in the mines', *Digging Deeper: Issues in the Miners' Strike*, ed. H. Beynon, London, Verso, 1985, pp. 231–43.

Winterton, J. 'The source of the crisis', Coalfield Communities Campaign, Working Paper No. 1, Vol. 1, Barnsley, CCC, 1985, pp. 1–7.

Winterton, J. 'The crisis in British coal mining', *Insurgent Sociologist*, Vol. 13, No. 1–2, 1985, pp. 53–62.

Winterton, J. 'Investment and jobs in coal: the implications of NCB restructuring policies for Yorkshire', Town and Country Planning Association Conference, Economic Prospects for the Coalfields: A Case Study of the Yorkshire Coalfield, Sheffield, 22 November 1985.

Winterton, J. 'The miners' strike: lessons from the literature', *Industrial Tutor*, Vol. 4,

Nos. 4–5, 1987, pp. 92–106.

Winterton, J. 'The politics of new technology in British coal mining', Conference of Socialist Economists, Sheffield, 10–12 July 1987.

Winterton, J. 'Private power and public relations: the effects of privatisation upon industrial relations in British Coal', Conference on New Forms of Ownership, Cardiff Business School, 14–15 September 1988.

Winterton, J. 'The size and structure of the British coal industry in the 1990s: a projection', International Symposium on Modern Mining Technology, Taiwan (People's Republic of China), 11–15 October 1988.

Winterton, J. and Rentoule, J. 'Confronting the new technology', *New Statesman*, 20 January 1984, pp. 9–10.

Winterton, J., and Winterton, R. *Implications of NCB Retructuring Policies for West Yorkshire*, Wakefield, West Yorkshire Metropolitan County Council, 1985.

Wintour, P. 'If at first you don't succeed . . . ', *New Statesman*, 5 November 1982, pp. 10–11.

Witham, J. *Hearts and Minds: the story of the women of Nottinghamshire in the miners' strike, 1984–1985*, London, Canary, 1986.

'Women fight for pits', *Community Action*, No. 68, December 1984, pp. 22–30.

'Women in the miners' strike: towards year one', *Spare Rib*, No. 151, February 1985, pp. 6–8.

Workers; Revolutionary Party, *The miners and the Case for A General Strike*, London, New Park, 1984.

Workers' Revolutionary Party, *Struggle for power: lessons of the miners' strike*, London, New Park, 1985.

Worsborough Community Group, *The Heart and Soul of it*, Barnsley, Bannerworks, 1985.

Wright, Peter *Policing the Coal Industry Dispute in South Yorkshire*, Report to South Yorkshire Police Committee for the Chief Constable, Sheffield, South Yorkshire Police, 1985.

Wright, Peter *Enquiry into Incidents at Stainforth and Armthorpe on 21 August 1984*, Report to South Yorkshire Police Committee from the Chief Constable, Sheffield, South Yorkshire Police, 25 September 1985.

Wright, Peter *Charges of Riot and Unlawful Assembly*, Report to South Yorkshire Police Committee from the Chief Constable, Sheffield, South Yorkshire Police, 25 September 1985.

Wright, Phil 'Bottomless pits?', *New Statesman*, 30 March 1984, pp. 12–13.

Wright, Phil 'Coal stocks: glaring contradictions', *New Statesman*, 27 July 1984.

Wright, Phil 'Interim projections of coal stocks at power stations in the UK', University of Sheffield Division of Continuing Education, 1984 (*mimeo*).

Wright, Phil 'Coal communities and energy policy', Coalfield Communities Campaign, Working Paper No. 3, Vol. 1, Barnsley, CCC, 1985, pp. 19–24.

Wright, Phil, and Rutledge, I. D. 'Steaming ahead', *New Statesman*, 1 June 1984, pp. 13–15.

Wright, Phil, and Rutledge, I. D. 'Coal worldwide: the international context of the British miners' strike', *Cambridge Journal of Economics*, Vol. 9, No. 4, 1985, pp. 303–26.

Wright, Phil, and Rutledge, I. D. 'The oil companies world coal and the British miners' strike', *International Labour Reports*, No. 7, January–February 1985, pp. 7–8.

Young, S. 'The scope for locally-based social and economic initiatives in the coalfield communities', Coalfield Communities Campaign Working Paper No. 7, Vol. 2, Barnsley, CCC, 1985, pp. 17–23.

Index

(Names of collieries are in bold type)